Ecology and Agriculture in the Himalayan Region

About the Author

Dr. Yasir Saeed Hanafi having acquired Ph.D. degree in Geography from Aligarh Muslim University, Aligarh is presently associated with the Department of Geography of the same University. He has credit of editing several books and publishing a number of research papers in journals of repute apart from attending seminars and conferences of nationai and international level.

Ecology and Agriculture in the Himalayan Region

Problems and Prospects of Agricultural Development in North-Western Himalaya

Yasir Saeed Hanafi

CONCEPT PUBLISHING COMPANY PVT. LTD.
NEW DELHI-110059

ISBN-13 : 978-81-8069-788-3

First Published 2012

Published and Printed by

Concept
Since 1974

Concept Publishing Company Pvt. Ltd.
Regd. Office:
A/15-16, Commercial Block, Mohan Garden
New Delhi-110059 (India)
Phones : 25351460, 25351794, *Fax* : 091-11-25357109
Email : publishing@conceptpub.com
Website: www.conceptpub.com

Editorial Office:
H-13, Bali Nagar, New Delhi-110 015, India

Cataloging in Publication Data--*Courtesy:* D.K. Agencies (P) Ltd. <docinfo@dkagencies.com>

Hanafi, Yasir Saeed.
Ecology and agriculture in the Himalayan region : problems and prospects of agricultural development in north-western Himalaya / Yasir Saeed Hanafi.
p. cm.
Includes bibliographical references (p.) and index.
ISBN 9788180697883

1. Agricultural ecology--India. 2. Agricultural ecology--Himalaya Mountains Region. 3. Sustainable agriculture--India. 4. Sustainable agriculture--Himalaya Mountains Region. 5. Natural resources--India. 6. Natural resources--Himalaya Mountains Region. 7. Agricultural diversification--India. 8. Agricultural diversification--Himalaya Mountains Region. 9. Hill farming--India. 10. Hill farming--Himalaya Mountains Region. I. Title.

DDC 630.954 22

Dedicated
To

MY PARENTS

Late Saeedul Haque Jami

and

Late Shakila Afzal Jami

Foreword

Area Study has assumed considerable significance since the Publication of the monumental reports of the National Commission on Indian Agriculture in 1976. The Commission has drawn special attention to the development of Hill Region.

Dr. Yasir Saeed Hanafi has made a commendable study of the Agricultural Development in North-Western Himalaya in respect of regional inequalities as influenced by agro-ecology. The study becomes very significant owing to the strategic location of the region.

Dr. Hanafi examines in depth the Physical landscape of the area, including the changing land use pattern and forest cover over the last three decades and has clearly brought out that man has interfered in the natural ecological balance. His major contribution which would interest the planners and all those interested in the development of the Himalaya, is his dedicated study of the inequalities in agricultural development and agricultural potential in the region and diversification. In fact there are very few such exhaustive and in-depth study of the Himalaya as Dr. Hanafi has done. Indian and foreigners interested in the Himalaya, I am sure, would like to have a copy of this valuable book.

(Prof. M. Shafi)
Professor Emeritus

Preface

At present, the agro-ecology of the Himalayan region is threatened by two-fold human interactions. First, the quest to increase the area under cultivation and second to get maximum possible agricultural return from every unit of land. This in turn has given rise to problems of agro-ecological nature in this region.

For agricultural operation, the clearance of forest areas is badly affecting agro-ecological balance. Consequently, the cultivation is now pushed to forest areas, marginal lands and up-slopes without taking into account the suitability of these lands for agriculture and vegetal cover is being lopped and cleared for fuel and fodder and finally the patches are being brought under grazing. This unscientific and irrational land transformation process have brought about very serious environmental crises in terms of erosion of top-soils and reduction in groundwater recharge in this ecologically sensitive region.

To get maximum return from agricultural land, heavy doses of fertilizers, insecticides, pesticides and irrigation are being applied without any proper knowledge of actual requirement of the agricultural land. This has resulted in the problems like development of toxic soils, elimination of favourable pests and insects, toxicities in water flow in the out drainage river channels multiplication of pesticide-resistant pests, concentration of harmful salts in the agricultural fields etc.

The ecological balance of flora, fauna and forest is being drastically disturbed by the rapid increase in human population. The urbanization and industrialization at the cost of natural habitats, further results in loss of biodiversity which once lost cannot be recovered. Forest play important role in maintaining ecological balance but ever increasing demand for fuelwood, fodder and timber necessitated by rapid increase in human as well as livestock population, diversion of forest land non-forest uses without ensuring compensatory afforestation and essential environmental safe-guards exerting an ever increasing influence on the environment and ecological balance of the region.

This book on 'Ecology and Agriculture of Himalayan Region: Problems and Prospects of Agricultural Development in North Western Himalaya' tries to present the concept of Agro-ecology and its prospects especially in the Himalayan region. It also tries to present a brief account of constraints of Agro-climatic diversity and man's interventions in this mountainous region. It also presents the possibilities of agriculture diversification and measures of eco-conservation in hilly areas. It is not framed only for the students and scholars of Geography but also helpful for the students of Agriculture, Botany, Biogeography and Environmental studies.

Yasir Saeed Hanafi
Department of Geography
Aligarh Muslim University,
Aligarh (U.P.) India.

Acknowledgements

I have immense pleasure in expressing gratitude from the depth of my heart to a number of Individuals and institutions who helped me to complete this piece of work and rendered invaluable assistance, guidance, and encouragement in so many ways.

First of all, with utmost humility and humbleness I thank God for the opportunity. He provided me to contribute, in my own little way, to the field of knowledge which is one of His qualities.

I take this opportunity to record my deep sense of gratitude to Prof. Ali Mohammad (Supervisor) with whose inspiration, able guidance, invaluable assistance, constructive criticism at all the stages of work and his familiarity with the region also helped me a lot to make this task possible.

I am equally thankful to Prof. Asma Aziz and Prof. Azim Uddin Qureshi, Chairman, Department of Geography, A.M.U., Aligarh, for their immense encouragement to complete this piece of work on top priority for which they gave me all possible help in a very short period of time.

I also wish to record my sincere gratitude to all my respected teachers in the Department of Geography, A.M.U., Aligarh, for their time to time cooperation.

My special thanks goes to the departmental staff in particular to Mr. Khwaja Mubeen Ahmad. Mr. Mazharul Haque, Mr. Asad Tariq, Mr. Naseemul Islam Ansari and Mr. Munne Khan for their constant help and support during the entire period.

I am also thankful to my close friends Mr. Mansoor Ali and Mr. Ahmad Mujtaba Siddiqui for their all kind of support, help and personal encouragement they provided me throughout the period of this work.

My special thanks also goes to my colleague Mr. Zulfiqar Ali Shah and his family for their hospitality and sweet gesture they provide me on every stage of my field surveys and gave me company at every step in the militancy stricken areas of Jammu and Kashmir, otherwise it could not be possible for me to have such a close look of the region, understanding of the various agro-ecological phenomena and observe the actual situation on the spot.

I must not forget the invaluable help of Mr. Rajaneesh Kumar who devoted his precious time in drawing the maps using GIS resource techniques which help me to plot the various features of the region on paper very efficiently.

I am also thankful to a number of individuals and their institutions for their help like Prof. Muzaffar A. Siddiqui (Provost, Sulaiman Hall, A. M.U., Aligarh), Prof. S.M. Uzair (D.S.W., A.M.U., Aligarh). Mr. Darshan Lal Sharma (Superintendent, N.S.S., Poonch), Mr. Asadullah Asad (Statistical Assistant, Finance Commissioner Office, Srinagar), Mr. Mohd Yaseen (Dy. Director, Economics and Statistics with FCs, Srinagar), Mr. Mohd Yusuf Roshi (Junior Statistical Assistant, Directorate of Economics and Statistics, Srinagar) and Mr. Zahoor Iqbal Rather (Scientist, Pollution Control Board, Srinagar), Mr. Syed Husnain (Teacher, Rajouri), Dr. P.C. Tiwari (Lecturer) and Ms. Bhagwati Joshi (Kumaun University, Nainital).

I am also thankful to all the persons who helped me in many ways during this entire piece of work and I might not mention them in the above lines.

Last but not the least, I must mention my wife, *Tabassum*, who also shared the pains in the process of preparation of this work, and remained a source of encouragement and inspiration throughout the period and of course to my son, *Arham*, who watched me with patience and astonishment while writing these words.

Yasir Saeed Hanafi

Contents

List of Tables

List of Figures

List of Coloured Plates

(between Page No. 234-235)

1. Farm yard manure is the main source of soil fertility in this mountainous region. Which is the demand of eco-friendly approach of agricultural development.
2. Livestock are important sources for household income and fertility of soil as well in the region.
3. Urbanization enveloped the hill ranges of Shimla. In ancient time no doubt the forest covered these ranges in the same manner. Can we call them urban hills or urban havoc?
4. Unwanted Hydrophytes is a big threat for Shikara sailing in Dal lake besides they consume water and causes shrinkage of it.
5. Because of human ignorance towards ecological balance uncontrolled forest cutting in the region results in degraded forest (a case of Rajouri district)
6. A view of scattered population settlement with their ricultural fields to look upon. (Salwa and Banola village of Mendhar, Poonch)
7. Households on agriculture fields is a common practice in the region. (Shahadra Sharief, Rajouri)
8. Terraced farming is the ultimate way of agriculture practices in mountainous region. Sowing of paddy is in progress.
9. Destruction of forest to fulfill the fuel needs of the household reached to the young trees and young ones are engaged in this un-ecological act of deforestation
10. Conventional method of fodder reserve. Fodder for off-seasen is generally kept on trees. In this way it is not damaged by soil moisture and canopy of the tree also protect it from snow.
11. Edible crops gaining importance in the region because of nutrition value in balanced diet. Mustard field at Barkot (Uttarkashi)
12. Population pressure forced encroachment of fertile and imited agricultural land for urbanization (a case of Rajouri district)

13. Urban encroachment of agricultural land results in barren lands and degraded forest.
14. Urban development is a major threat for the ecology of most of the tourist centers in hilly regions. Nainital (lake) is rinking and landslides are common phenomena in the region.
15. Wide scope for horticultural development with a variety of fruits and vegetables in the region. A variety of large size lemon in Karn Prayag (Chamoli)
16. New Tehri is developed as a compensation for the resident of Old Tehri which will submerge when the Tehri dam start its functioning.

1
Introduction

The Problem

In the beginning, man was shaped by the environment which controlled and shaped its evolution and man remained in equilibrium with the environment. But in the quest of its development, man broke this equilibrium and since then he is exerting an ever increasing influence on his environment. Due to growing urbanization and industrialization, man distorted its environment.

All developmental activities, irrespective of their scale nature and magnitude affect the environment. The impact of their developmental activities may have short-term as well as long-term implications. The short-term implications may be negligible as compared to the benefits of the developmental programmes but the long-term implications may further create new ecological and environmental problems, the solutions for which may not be easy to find. Some of these implications may be summarized in terms of vast areas of deforested lands, denudation of slopes, soil erosion and silting of rivers, regular occurrence of floods and droughts, impoverishment of important flora and fauna, changed climate, polluted water unfit to drink and contaminated air unfit to breath.

The ecological balance of flora, fauna and forests is being drastically disturbed by the rapid increase in human population. The urbanization and industrialization costs the decrease of natural habitats, which further-results in loss of biological diversity. Biodiversity, once lost cannot be recovered. Forests play an important ecological role, the serious depletion of which attributed to a number of factors. These include the ever increasing demand for fuel-wood, fodder and timber necessitated by rapid increase in human as well as livestock population, diversion of forest land to non-forest uses without ensuring compensatory afforestation and essential environmental safeguards.

Besides these, land degradation is a major problem which results in loss of its intrinsic qualities or a decline in its potential utilities and capabilities. Land degradation may be caused due to natural or human factor or due to the combination of both. Soil erosion by rain and river that takes place in hilly areas causes landslides and floods, while cutting trees for firewood, agricultural implements and timber grazing by large number of livestock over and above the carrying capacity of grasslands, traditional agricultural policies. Soil erosion results in huge loss of nutrients in suspension or solution, which we removed from one place to another, causing depletion or enrichment of nutrients. Besides the loss of nutrients from topsoil, there is also degradation through the creation of gullies and ravines which makes the land unsuitable for agricultural production.

The process of agricultural resource of development has a large and long-term impact on the natural component of a region through the destruction of original ecosystem. The unscientific and irrational agricultural transformation process now operating in the ecologically fragile Himalayan region, in terms of the extension of cultivation to up slope, marginal and sub-marginal lands, clearing of forest areas under the increasing pressure of population and changing pattern of agricultural landuse and cropping has introduced a variety of new conditions to the traditional agro-ecology of the region.

Increasing population concentration in this mountainous tract deserves specific and adequate attention for protecting the environment and sustainable development of natural resources because the fragile ecosystems of these mountainous region are vulnerable to all kinds of anthropogenic interferences.

During recent years, the growth of population and technological development tends to affect adversely the ecology both by utilizing the environmental resources beyond the carrying capacity of the nature and by altering the resource use pattern. The scarce agricultural resources of the region are now under great biotic stress. More than 70 per cent of the population of the region is solely dependent on agriculture which is being practiced in most compelling circumstances in almost complete absence of other means of livelihood in the majority of the areas. The land use pattern of the region has been undergoing a process of rapid change and transformation with more and more areas come under wasteland, pastures and cultivation. This process of irrational utilization of natural resources has not only depleted and degraded the critical environmental components in this ecologically sensitive and economically underdeveloped mountainous tract but also threatened the agro-ecology of the region itself.

Presently, the agro-ecology of this region is threatened by two-fold human interactions. First, the quest to increase the area under cultivation and second to get maximum possible agricultural return from a unit area. This in turn has given rise to three main problems of agro-ecological nature in the region. First, for agricultural operation, the clearance of forest areas is badly affecting agro-ecological balance. Consequently, the cultivation is now pushed to forest areas, marginal lands and up slopes without making into account the suitability of these lands for agriculture and vegetal cover is being lopped and cleared for fuel and fodder and finally the patches are being brought under grazing. This unscientific and irrational land transformation process has brought about very serious environmental crises in terms of erosion of top soils and reduction in groundwater recharge in this ecologically sensitive region. The process and magnitude of soil erosion is a function of a number of factors operating in combination such as intensity of rainfall, amount of surface run-off and infiltration, chemical and physical properties of rocks, gradient of slope, drainage density and of course the process of land transformation followed by man.

In the Himalayan agro-ecological region, erodibility of land by overland flow depends on the degree of slope and its length. There will be no erosion until the depth of water and slope acquire an erosion force. The process of soil erosion is intensified by the volume of water accumulating on a long slope and the consequent increased velocity of run-off (Bennet, 1955). It has been proved experimentally that if the degree of slopes is doubled the erosion per unit area increases two and a half times and on an average soil erosion increases one and a half time per unit area when the length of the slope doubled (Khonke and Bertrand, 1959). Surface run-off in another factor of soil erosion, lesser will be the rate of infiltration and permeability, the more will be the rate of overland flow and greater will be the volume of surface run-off and ultimately the higher will be the magnitude of soil erosion.

Second, to get maximum return from agricultural land, heavy doses of fertilizers, insecticides, pesticides and irrigation are being applied without any proper knowledge of actual requirement of the agricultural land. This has resulted in the problems like development of toxic soils, elimination of favourable pest and insects, toxicated water flow in the out drainage river channels, multiplication of pesticide-resistant pests, concentration of harmful salts in the agricultural fields, waterlogging, salinity and alkalinity etc.

Thirdly, good quality agricultural land is being brought under non-agricultural uses, like roads, buildings and water channels, etc. because of the increase of the urbanization.

In the Himalayan mountain geosystem about 60 per cent of the total geographical area should be under permanent vegetal cover keeping in view the smooth functioning of geo-eco-cycle. Against this ecological standard, forest cover in north-western Himalaya as a whole confined to only about 44.67 per cent during 1992-95. Large scale transformation of forested sites was carried out to meet the ever increasing demand for agricultural land and pastures due to rapidly increasing human and cattle population. This phenomenon pushed the uneconomic agriculture (high input and low output) in marginal lands, steep slopes, forested sites and even on other categories of land which is not suitable for this purpose. This human transformation process has resulted into cleared forests and terraced slopes, damaging the entire forest ecology. More than two-third of the total population of the region depends on agriculture for its livelihood and in this agricultural economy cattle are very closely interrelated with crop farming. Cattle are the only medium of energy transfer and energy input (in terms of manuring soils and labour) to the Himalayan agro-ecology.

Hence, it is required to maintain relatively higher cattle population to meet the increasing demand of manure and labour for agricultural expansion. In the recent years, the pressure of grazing animals has increased far beyond the carrying capacity of pastures and forests in the region. It has been estimated that only 0.13 to 0.30 hectares per cattle grazing land is available against the ecologically recommended standard of minimum 6 hectares per cattle in the Himalayan ecological region. (Ashish, 1983).

Concept and Definition

Nature in its diverse ecological conditions offers its diverse earth units with completely different possibilities of utilization.[1] It not only imposes limits but also indicates the direction of specialization. Thus efforts are needed to investigate the agro-ecology of a region to meet the ever growing human need in harmony with the ecological balance. Agro-ecology is a twin word 'Agro' means related to agriculture and agriculture is come from the Latin term 'agercultura' which consists of two words, namely, 'ager' meaning field and 'cultura' meaning to cultivate. Agriculture is defined by different scientists in different words such as "use of human efforts with the object of inquiring product of plants or animal origin" (Whittlesey, 1936), "Man's husbandry of land" (Symon, 1967), "the practice of cultivation of soil in order to produce crops is called agriculture" (Grigg, 1969), "a way of living as well as an occupation" (Gregor, 1970). In *Encyclopaedia Britannica*, agriculture is the science or art of cultivating the soil, growing and harvesting the crops and livestock.

The term 'ecology' seems to have almost as many meanings as there are ecologist. Ecology is one of the three major divisions of biology, the other two being morphology and physiology. The emphasis in morphology is on how organisms are made?, in physiology how they function? and in ecology how they live? Ecology, is a very simple term, a science that studies the interdependent, mutually reactive and interconnected relationships between the organism and environment on the one hand and among the organisms on the other hand. In 1959, Geoffrey Saint Hillaire used the term 'ethology' to refer to the study of relationships between the organism and the environment, but the term was never generally accepted by the earlier ecologists. Later ethology has been considered as an important part of ecology dealing with the field of animal behaviour. Nine year later, Reiter introduced the 'oekologie' derived from two Greek words 'Oikos' meaning ' home or house or household or something like that' and Logos meaning a 'discussion or study.' Ernst Haeckel had falsely credited with the derivation of the term because he defined it in the literature a year later. The term has since been anglicized to ecology (C.B. Knight, 1965, p. 8). Since then its published definitions now range from traditional study of the relationships between organism and their environment (Hughes and Walker, 1965), to the embracing structure and function of nature (Odum, 1963), which virtually identifies it with the whole of natural science.

Ecology, to great extent concerned with the habitat, and it, cosequently, applied habitat biology. Elton (1949), succinctly defined the habitat as an area possessing uniformity of physiography, vegetation, climate or any other quality the investigator assumes is important. Habitat, though they are restricted in an area will exhibit considerable variability in size from a coniferous forest or oceanic area measured in terms of hundred of square miles to an exposed surface of rock or the lower surface of a twig a few square centimeters or a fraction of square millimeter in area. The later two examples are called micro-habitat by most ecologists (Knight, 1965, Basic concept of Ecology, p. 3).

Habitat refers to the sum total of all external conditions affecting the development, special response and growth of plants (organisms). Since the term habitat was first used by botanists, and especially by ecologists, it is best to apply it to the description of the physiological conditions influencing the distribution and growth of plants as contrasted with social environment which deals with the influence of a variety of factors other than those concerned with the direct growth requirements determining the distribution of crop plants[2] (Klages, 1942). In the words of Paul, A. Colinvaux (1973), ecology is invented to convey the idea of the study of animals and plants in relation to their habits and habitats.[3] Stability and unstability of an agro-ecological region can be understood from the following concepts of ecology. According to Jordan (1975), an adequate definition of ecology must specify the unit of study that is unique or basic to ecology and such a basic unit obviously is the *'ecosystem.'* This term was first proposed by British ecologist A.G. Tansley in 1935. Thus, ecology may be defined as "the study of ecosystem" (Kumar, 1975). An ecosystem in turn is an integrated unit, consisting of interacting plants and animals whose survival depends upon the maintenance of abiotic as well as biotic structures and functions. The basic and most important concept of an ecosystem is that everything is somehow or the other related to another thing else, and such relationships include interlocking functioning of organisms among themselves as well as with their environment. Smith (1966), has emphasized the following general characteristics of most ecosystems:

- The ecosystem is a major structural and functional unit of ecology.
- The structure of an ecosystem is related to its species diversity, the more complex ecosystem have high species diversity.
- The function of the ecosystem is related to energy flow and material cycling through and within the system.
- The relative amount of energy needed to maintain an ecosystem depends on its structure. The more complex the structure is, the lesser the energy it needs to maintain itself.
- Ecosystem matures by passing from less complex to more complex states. Early stages of such succession have an excess of potential energy and a relatively high energy flow per unit biomass. Later (mature) stages have less energy accumulation and its flow through more diverse components.
- Both the environment and the energy fixation in any given ecosystem are limited and cannot be exceeded without causing serious undesirable effects.
- Alteration in the environments represent selective pressures upon the population to which it must

> adjust. Organisms which are unable to adjust themselves under changed environmental conditions does not survive and becomes extinct or vanishes from the Environment.

Ecology is one of the most popular branch of biology and differs from most other branches of biology is that it can be properly appreciated or studied only through a multi-disciplinary approach involving close cooperation from experts in several disciplines e.g., physics, geography, engineering, mathematics, statistics, zoology, botany, microbiology, limnology and chemistry. In fact most of our present-day environmental problems have been caused by the lack of appreciation of the multidisciplinary perspective and through a narrow perception of the world (Kumar, 1977, p. 1). The term ecology is now widely used in the field of social, as well as biological science, whilst the subject of ecology covering as its does the relationship of the organism to its environment, has many facets (Southwood, 1978).

The definition of ecology stated above is, however, not accepted universelly. In certain countries, e.g., the U.S.A., Canada and England, 'ecology' includes within its scope both the description of communities and the more experimental ecological investigations, e.g., energetics. On the other hand in European Continent, the term 'ecology' (also sometimes called eco-physiology) is restricted to the experimental part only whereas the descriptive part is styled as sociology e.g., plant sociology or animal sociology. The chief objective of an experimental ecologist or eco-physiologist is to understand the mass and energy turnover within different ecosystems and to estimate quantitatively the primary and secondary production. (Kumar, 1977, p. 3).

Barret (1978), advocates restriction of the term ecology to define that discipline which attempts to understand the structure and function of ecosystems, whereas 'environmental sciences' be defined as an interdisciplinary science which attempts to study the impact of man on the structure and function of social and ecological systems as well as management of these systems for human survival and benefits. One should have to know the following three concepts of ecosystems to understand the agro-ecology of a region in a sound manner:

(a) *The Concept of Climax*

The climax concept has long been and indeed still is, central to many ecological principles. The climax is the final or it is sometimes described the nature stage or end point of ecosystem biomass possible, under prevailing environmental conditions, has been attained. The ecosystem reached a steady state which implies a relative balance or equilibrium in respect of its biomass, species populations, inputs of energy and nutrients and biological productivity.

(b) *The Concept of Stability*

One of the most important attributes of the climax or mature ecosystem, is that of stability. This is the ability of a system to maintain a relatively constant condition in terms of its species composition, biomass and productivity, with minor fluctuations around a mean value (the equilibrium point) and to return to this steady condition fairly rapidly after interval or external disturbances.

(c) *The Concept of Fragility*

Another important characteristic of ecosystem related to stability is that of fragility, this is an ease with which an ecosystem can be disrupted. It will depend, first on the relative resilience of the system and second on the type of disturbance to which it is subjected.

Resilience is the term used to describe the ability of an ecosystem to persist and to maintain itself in the face of disturbances due to weather or climate, chemical factors, organisms or human. The relative resilience of an ecosystem depends on:

(a) Its amplitude or how far it can deviate from the equilibrium or steady state (before complete disruption and replacement by another system occurs).
(b) Its elasticity or the rate with which it can recover from disturbances. Fragility will obviously vary from one ecosystem to another.

Fragility also depends upon the type of disturbance, in terms of intensity and duration of which the ecosystem is subjected.

It is evident that the concept of agriculture and ecology, should be clear from the foregoing discussion. Ecology with some prefixes modifies its definition e.g., Human ecology is the study of man's relation to environment, especially man's effects on the biosphere and implications of these effects for man. Population ecology is the study of the progress that affect the distribution and abundance of animal and plant populations. The concept of agro-ecology began in 1930, when Bensin proposed the term agro-ecology to apply for 'the detailed studies of commercially important crop plants by the use of ecological methods.' He proposed a systematic collection of data so that the main agricultural regions (agrochoras) of the world and the characteristics of local cultivated varieties of important crops (chorotypes) may be described and recorded by the employment of standardized methods and by a prescribed and uniform terminology.[4] The agro-ecology is defined as 'the study of the physical characteristics of environment, climate and soil, in relation to the development of agricultural plants, and to the yield of such plants from the quantitative (amount of the product), qualitative (quality of the product), and generative (characters of the seeds) point of view' (Azzi, 1956)[5]. Recent concept of agro-ecology is 'the study of optimization of agricultural production with least deteriotation of natural resources and environment.' Crops like other plants have distinct ecological preferences of altitude, climate, soil, photo-period at different stages of life cycle, but they can usually be grown over a wide range of environments. Man can often modify the environment by cultural and other methods to provide better conditions for growth and productivity.

Significance of the Study

The north-western Himalaya because of the increasing population concentration deserves specific and adequate attention for protecting the environment and sustainable development of natural resources. The fragile ecosystems of this mountainous agro-ecological region are vulnerable to all kinds of anthropogenic interference. The extensive soil erosion, frequent landslides because of deforestation, extension of cultivation, overgrazing and construction activities.

The study of this region has become more significant in recent years with massive growth in population and the scarce agricultural resources of the region are now under great biotic stress. More than 70 per cent of the population of the region solely depends on agriculture which is being practiced in most compelling circumstances in the almost complete absence of other means of livelihood in the majority of the area. The land use pattern of the region has been undergoing a process of rapid change and transformation with more and more area under wasteland, pastures and cultivation. This present process of irrational utilization of natural resources has not only depleted and degraded the critical environ-mental components in this ecologically sensitive and economically viable mountainous tract but also threatened the agro-ecology of the region in a worst manner.

This study also has its significance for the agro-ecological development of the region because it assess and correlate various parameters like land capability, temperature, rainfall, relief, geohydrological balance, rate of soil erosion, extent of deforestation, magnitude of surface and subsurface water pollution because of suggested biochemical techniques, population structure and existing pattern of agricultural resources utilization.

Significance of the present study also become evident as agro-ecology studies the agro-ecosystem in the biosphere. Agro-ecosystems[6], studies mainly under 'production and conservation ecology.' In production ecology such parameters are studied that increase the agricultural production, while in conservation ecology, such parameters are studied that explain the basic prerequisite components of agriculture. Agro-ecological study of the region become inevitable as crops like other organisms have distinct ecological preferences of altitude, climate, soil, photo-period and micro-nutriets at different stages of their life cycles, a few examples of such preferences are cited here for deficient, optimum and toxic responses of crops.

A very clear relationship of temperature with yield of spring wheat were plotted on a graph (Fig. 1.1) bearing the yield expressed in quintal per hectares and temperature in °C, taken from the work of Girolamo Azzi. The curve which geometrically co-ordinates all the points on the graph are the true representation of yield's course in respect of temperature. This curve may be divided into three parts, a central and practically horizontal part comprising wheat we may call the favourable situation or optimum response and two laterals in rapid decline corresponding to unfavourable situations through the excess and deficiency.

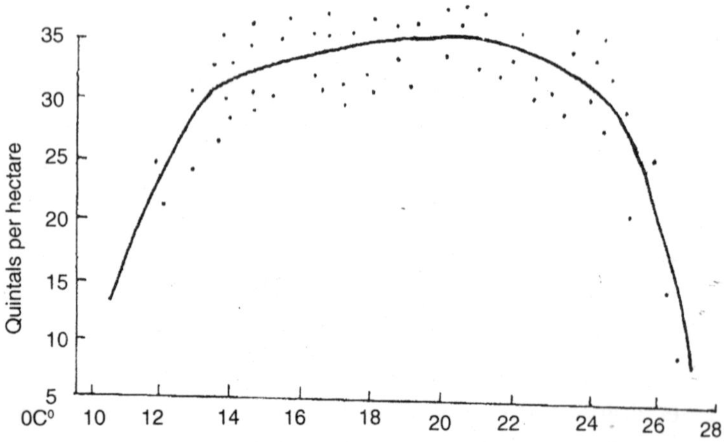

Fig. 1.1 : Yield of spring wheat in relation to temperature

Another example which highlights the response of duration of temperature on development of root drawn by Talma and cited by Klages. The effect of the length of the exposure to given temperatures on the development of root in *Lepidium sativum* can be seen very clear in Fig. 1.2. The temperature for optimum rate of development in the root system of *Lepidium sativum* are different for different period of exposures. A short exposure of three and half hours, showed optium development at 30°C, with doubling of time period i.e., 7 hours, the optimum temperature for development in root system is found at 29°C, and when the period of exposure lengthened to 14 hours, the highest rate of activity was at 27.2°C. It is noted that, the increase in temperature, especially with the longer periods of exposure, exerts influence depressing to growth. The growth depressing to factors become more and more active with the approach of the optimum and beyond it account for the rapid downward trend of the growth curves.

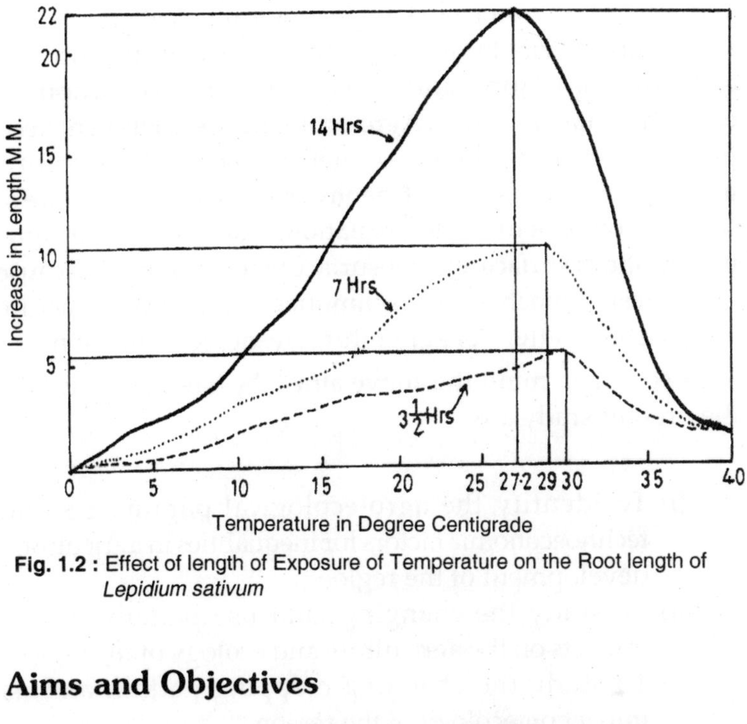

Fig. 1.2 : Effect of length of Exposure of Temperature on the Root length of *Lepidium sativum*

Aims and Objectives

In order to check the agro-ecological threats from becoming unmanageable problems in the region, the eco-conservation and sustainable eco-friendly economic activities becomes inevitable. As a result of urgent and dual problem of conservation and increase in production, agricultural research in the last decade has tended to move from 'maximization of production' towards 'optimal use of agro-eco-system.'[7] This approach has been more beneficial in the developing countries, where it has ensured enhancement of productivity without disbalancing much of the ecological setup for agricultural activities.[8]

It is, therefore, imperative to integrate agricultural development policies with environmental management and planning in ecologically sensitive Himalayan tract. The agro-ecological planning in this context is referred to as scientific and integrated framework for ecologically sustainable and economically viable development of agriculture through ensuring optimum and most rational utilization of environmental resources. Agro-ecological development need information on various parameters like land capability, temperature, rainfall, relief, geohydrological balance, rate of soil erosion, extent of deforestation, extension of farming, magnitude of surface and subsurface water pollution because of suggested biochemical techniques, population structure and existing pattern of agricultural resource utilization.

Keeping in mind the above aims, the basic objectives of the present study are:

(i) To identify the agro-ecological parameters and technoeconomic factors for inequalities in agricultural development of the region.
(ii) To study the changing land use pattern and its impacts on the agriculture and ecology of the region.
(iii) To study the changing cropping pattern and its impact on ecology of the region.
(iv) To identify the driving forces for deteriorating agro-ecological scenario of the region.
(v) To identify the suitable ecological region of specific crops.
(vi) To identify the potential agricultural region for planning purposes on priority basis.

Database

The problems facing in the data collection during the entire work piece of work are of different nature, such as to take the complete geomorphic unit of north western Himalaya as the study region become too large and come under the administrative limits of three different state which further sets problems like the point or period which I choose for data analysis, not necessarily tally in the records of all three states because they record and publish data according to their convenience. The frontier of this region is surrounded by Pakistan, Afghanistan, China and Nepal. So a large number of information and materials of the region are either restricted or censored for political reasons, besides it a large portion of the region is a matter of conflict with the neighbouring countries and to get the exact information and real situation of such areas is a difficult task. Militancy in the region further sets hurdles for the research work.

Considering the above problems of the area, the data has been collected from the following secondary sources:

(a) The data of area production and yield of principle crops has been collected from the Director of Economics and Statistics, New Delhi; Director of Economics and Statistics, Srinagar; Financial Commissioner's office, Tankipora, Srinagar; Directorate of Economics and Statistics, Shimla and Lucknow.

(b) Meteorological data has been collected from the Indian Meteorological Department, Lodhi Road, New Delhi. Also from the statistical abstracts of Himachal Pradesh, Jammu and Kashmir and Uttar Pradesh.

(c) Livestock and agricultural machinery data has been collected from Livestock Censuses, Directorate of Economics and Statistics, New Delhi.

(d) Data of soils has been collected from various reports of Director, Soil and Landuse Survey, IARI, New Delhi.

(e) Water resource data has been collected from Water Commission of India, New Delhi.

(f) Data of Landuse Statistics has also been collected from the Directorate of Economic and Statistics, New Delhi, Lucknow, Shimla and Srinagar.

(g) Water Resource Development Atlas of India and National Atlas of India have also been used in final plotting and verifications of various data.

Methodology

The study has been conducted on the basis of secondary sources of data and their verification by personal surveys at micro-level. Since, the region is too large the *district has been taken as the unit of study*. There are 34 districts in the region. 14 in Jammu and Kashmir, 12 in Himachal Pradesh and 8 mountainous districts of Uttar Pradesh Himalaya now known as Uttaranchal.

During the period of 1950s till now, there were formed several new districts from the existed ones. As a result, the up to date data of all the districts is not possible, in such cases generally the data of newly formed districts has clubbed for convenience but here only those districts are taken into consideration which are given in 1991 Census of India. Out of these 34 districts, only those districts are taken for data analysis which have been existing since 1950s and their data is also available for all point of time.

Decadal growth in agricultural development of various parameters has been calculated by simple growth rate formula:

$$R = \{(b/a) \times 100\} - 100$$

where, b = any value of later date

a = any value of previous date

The trend of agricultural development has been calculated by the logarithmic trend line equation which is given below:

$$\text{Log } Yt = \text{Log } a + \times \text{ Log } b$$

$$\text{where log } a = \sum (\log Y)/n$$

$$\text{Log } b = \sum (x \log Y)/\sum x^2$$

The trend line values have been plotted on the smooth line curve to show the actual situation of the trend. To determine the rate of change following equation has been used:

$$r = R - 1$$

where, r is the rate of change

R is the antilog of log b

Finally, the collected data of various agricultural parameters has been standarized by using standard score (Z-score) method. On that basis composite development index has been developed to find out the inequalities in agricultural development of the region.

Organization of the Study

The present thesis has been divided into seven chapters. The *first chapter* is the introductory chapter highlights the statement of the problem, the concept of Agro-ecology and significance of the study besides the traditional background items. *Second chapter* deals with the agro-ecological components of the region, significance of prevailing climatic constraints in agricultural practices. Also agro-ecological consequences of the changing landuse pattern and livestock pressure analysed in this chapter. The emphasis of the *Third chapter* is on the understanding of natural ecological balance, ecological parameters in the light of ecological laws, ecological requirement of different crops and how the agricultural practices responded to different ecological parameters. *Fourth chapter* analyzes the trend of agricultural development in terms of agricultural inputs and agricultural development in the region and also to record the possibilities and potentials of agricultural development in the region. *Sixth chapter* focuses on conservation of ecology and sustainable agricultural development as well as integrated approaches of sustainable agricultural development. The *seventh chapter* conclude the investigations and findings of the thesis.

Location and Extent of the Region

North-western Himalaya, as delineated in this study covers an area of about 3,29,032 sq. kms, lying within the Himalayan system west of river Kali that separate Pithoragarh district of Uttaranchal (India) from the Kingdom of Nepal. This region is a complete geomorphic unit and well distinct agro-ecological or agro-climatic region different in all its physio-cultural and socio-economic conditions. This region extending over about eight and half degrees of longitude, the western most point of this region lies just west of Chunar Khan pass on 72°40′ of east longitude while its eastern most point lies on the north-eastern tip of Pithoragarh district which is about 50 km. West of Mangsha Dhura pass on 81° 02′22″ of East longitude, occupying almost one-third of the total East-West extent of the Himalaya. The northern and southern limits of the region stretching over eight degrees of latitude. The northern most point of the region is lying between Kilik Dawan and Khunjerab pass on 37°05′ of north latitude while southern most point lies about 50 km. South of Tanakpur on south-eastern corner of Nainital on 28° 43′ 24″ of north latitude.

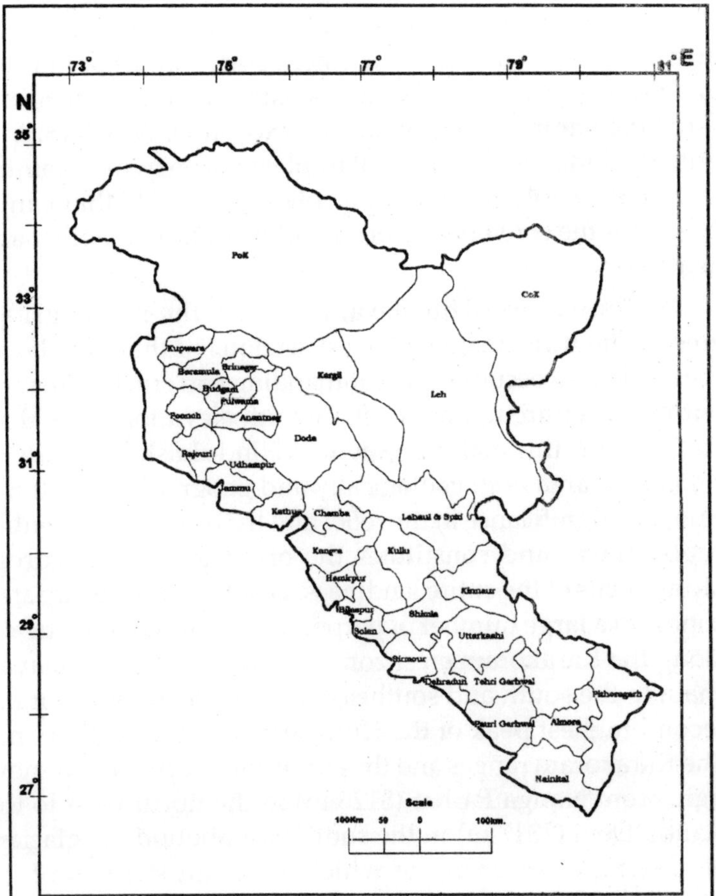

Fig. 1.3 : North-western Himalaya : Administrative Setup

North-western Himalaya is a single geomorphic region but this region comes under the administrative limits of three northern states of India. A major western portion lies within Jammu and Kashmir, occupying an area of 2, 22, 236 sq. kms. or 67.5 per cent of the total area of this geomorphic region, another 17 per cent or 55,673 sq. kms. of area of mountains comes under the administrative powers of Himachal

Pradesh, while the remaining 15.5 per cent constituting an area of 51,123 sq. kms. is the northern portion of Uttar Pradesh and now come under the administrative limits of newly formed Uttaranchal state with two exceptions as Saharanpur is not included in it because of its plain nature while Nainital included in totality including plain area of Udhamsingh Nagar for the convenience of secondary data availability and analysis.

North-western Himalaya, is by far, the most curious among the three main east-west sections of the Himalayan chain and also occupying a commanding and crucial situation within the entire mountain chain with elevation more than 7000 meters, this majestic crown of Hindukush Himalaya is a complex area both geologically and geographically. It has a unique significance in the relief and hydrology of the entire subcontinent and constitutes the orographic and hydrographic focus of the entire landmass. Besides the vast glaciated zone and a large number of perpetually snow covered peaks along the Karakoram crest zone and the greater Himalayan chain to the south and southeast, the region account for the second highest peak of the Himalayan i.e., mt. K_2 (8611 m). The Karakoram ranges and the entire zone of perpetual snow right from Nanga Parbat (8126 m) in the north-west to the Nanda Devi (7817 m) in the south-east abounds in glaciers of varying sizes, many of which are giant streams of ice exceeding 30 km in length. Some longest glaciers of the world are also found here like Siachen with 72 km. and Biafo with 62 km of lengths. According to an estimate, about 16,000 sq. kms. area of Karakoram and 9500 sq. kms. area in Greater Himalayan ranges in the north-western Himalaya is under glaciated ice.

The region as a whole dominated by several giant and smaller parallel mountain ranges running north-west to south-east like Siwalik ranges in the extreme south with elevation ranges between 450 to 950 m, parallel to it is Dholadhar range with elevation more than 2000 m. Above this Dholadhar range which is significant only in Himachal Pradesh east of Ravi, the Pir Punjal ranges starting from the north of Muzaffarabad in the north-west up to Manali in Kullu district of H.P., a continuous chain with an elevation ranges between 4000 m and 5000 m. The range parallel to it in the north is the Great Himalayan ranges running to the full length of north-west Himalaya right from Indus near Chilas in the extreme north-west up to the Pithoragarh near Nanda Devi to a length of about 1000 kms. and its elevation is generally found more than 5000 m above mean sea level. The Zaskar range north of it runs parallel from Wakha sector of Kargil district in north-west up to the northern boundary of Pithoragarh with China in the south-east with an elevation of about 6000 m. and the Ladakh range found north-east of it with an average elevation more than 6000 m. In the extreme north Great Karakoram ranges with number of snow clad peaks and Glaciers like Hispar, Biafo, Balono and Siachen Glaciers with few peaks of more than 7500 m occurr in these ranges.

Between these parallel ranges there are considerable but widely distributed stretches of wide and fertile valleys throughout the length and breadth of the region like Kashmir valley, Kullu valley, Dun Valley, etc. Their evolution being associated with a variety of causes e.g. some of these valleys are the result of fluvial action i.e., vertical and lateral erosion by rivers, it is evident that these have been vast lacustrine areas in the past.

The most important among them, is the Kashmir valley, which is an oval 'dun' of techtonic origin. The valley is about 135 kms. long and 14 kms. wide, thus encompassing a total area of 15,853 sq. kms. It is 200 kms. long and 125 kms. wide when aerial distances are measured from crest to crest. The mean elevation of the valley of Kashmir is 1830 m with heavy concentration of human activities between 1510 m and 1800 m. The valley is surrounded by the Great Himalaya (Himadri) and the Pir Punjal range with 12 passes. Banihal pass (2832 m), Zoji la (3529 m), Deosai pass (3765 m), Burzil pass (4199 m), Kamri pass, Babusar pass (4173 m), Janod, Kepan, Tithwal, Hajipur, Baramula and Pir Punjal Pass. Dehradun is the biggest and most developed dune of the Uttaranchal. It is about 35 km long and 25 km. wide. It rises from 360 m along the Yamuna and Ganga to 660 m in the middle and 900 m near the foothills.

The north-western Himalayan region occupies an strategic position in South Asia with its border touching Pakistan in the west, China in the north, Tibet in the east and Nepal in the south-east. The valley of Ab-i-Panja in Afghanistan separates India from Tajikistan.

It is over the majestic heights of its snow-capped mountains that several passes providing route link with adjacent countries. Notable among them are Babusar pass in Chilas, Burzil pass in Gilgit, Karakoram pass (5575 m) in the north Ladakh while Lanak La pass (5486 m) in eastern Ladakh are of utmost importance.

The north-western portion of Kashmir has more massive mountains and within a radius of 112 km of Gilgit town, there are 21 peaks over 6000 m with Nanga Parbat at the Highest with 8126 m. The north-eastern portion of Ladakh has an average elevation over 6000 m, where Mt. K_2 or Mt. Godwin Austin (8611 m) on Karakoram range is the second highest peak of the world.

In north-western Himalaya, the region lies south of Kashmir is Himachal Pradesh with intricate mosaic of mountain ranges, hills and valleys. The Dhauladhar range looks in supreme majesty over the Kangra Valley, while Pir Punjal, Great Himalaya and Zaskar ranges looks over Chamba, Lahaul Spiti, Kullu and Kinnaur. There is general increase in elevation from west to east and south to north. The Great Himalayan range (5000 m-6000 m) runs along the eastern boundary with certain passes, such as Kangla (5248 m), Bara Lacha la (4512 m), Kanzam (4551 m), Parang (5548 m) and Pin Parbati (4802 m) while Zaskar range is the eastern most range with peaks over 6500 m, like Riwo Phargyul (6791 m) and Shilla (7026 m).

To the south-east of it, lies the U.P. Hill ranges or Uttarakhand region, consists of snow capped glacier peaks like Bandar Punch (6315 m), Gangotri (6614 m), Kedarnath (6940 m), Chaukhamba (7138 m), Kamet (7756 m), Nanda Devi (7817 m), Dunagiri (7066 m), Trisul (7120 m) and Nandakot (6861 m).

Review of the Work done So Far

Agro-ecology, 'the study of physical charateristics of environment, climate and soil in relation to the optimum agricultural production with least deterioration' is a discipline which is not recognized all of a sudden but its roots goes to the writings of early Greek philosophers. Aristotle classified animals according to their habits and habitats, but he didn't consider the interrelations of plants, animals and their environment.[9] It was however, Theophrastus, an associate and student of Aristotle who in 4th century B.C. first described the interrelations between organisms and their environment. He is therefore referred as first ecologist.[10] The population regulation of insects-pests has long been under study e.g. the introduction of Mynah bird from India to the Islands of Mauritius in 1962 reduced the red locust threat to a negligible problem by 1770.[11] Similarly, the introduction of predatory ants from nearby mountains into date palm orchards to control other species of ants in south-western Arabia has been mentioned in the writings of Forskal in 1775.[12] The 18th century botanist Richard Bradley understood the basic similarities between animals and plants production, and proposed the methods of obtaining maximum agricultural yields for trees, vineyards, rabbits, poultry and fish.[13]

A botanist Justin Liebig, in 1840, stated that any

environmental factor that most closely approached the critical minimum for a species will tend to be a limiting factor for that species.[14] Liebig's idea has since been incorporated into the ecological literature and called the 'Law of Minimum.'

Zones of the flanks of high mountains became of particular interest to ecology as a result of the work in Arizona by one man C. Hart Merriam. He was charged with a biological survey of part in Arizona in 1889, in the pioneer days when the region was still biologically unknown. He was particularly impressed with the vegetation belts he encountered on the ascent of the 13,000 feet San Francisco Mountain.[15]

Investigation during the beginning of 20th century have set ecology and ecological relationships more and more on a specific and on experimental basis. In 1903, Davenport[16], published a paper on the animal ecology of Coldspring Harbor while in 1907, Forbes, wrote about the distribution of Illinois fishes.[17] E. Warming (1909), a Danish plant ecologist, recognized the interdependence and close relationships that often exists between plant and animal communities and cautiously recommended dealing with plant and animal communities an entity rather than separately.[18]

On the question of limiting factors, a very considerable literature has developed, and numerous theories have been advanced, since the appearance of Blackman's initial paper in 1905. Blackman's axiom of limiting factor is in reality an elaboration of Liebig's law of minimum which in its essence may be as "the yield of any crop always depends on that nutrient constituent which is present in the minimum amount."[19] V.L. Shelford in 1913, expanded Liebig's observations to include the maximum limits of an environmental factor. Thus 'Shelford law of tolerance' states that "a value below a critical minimum or a quantity or factor in excess of the critical maximum would exclude certain organisms from environmental areas."[20]

The branch of science known as Agricultural Ecology, first mention in a report presented to the "Accademia dei Lincei" by a commission of inquiry composed of Messrs. L. Luzzatti, L. De Marchi and R. Pirotta; on the scientific activities of Prof. Girolamo Azzi in 1920. The commission has reported that, "this branch of science (Ecology), has a purely scientific character and scope, that it consider the plant as an organism thriving in its environment, and it can have great practical significance when dealing with plants cultivated by man for his own requirements, it then becomes agricultural ecology."[21] Following this report in 1921, the Ministry of Agriculture has created a chair in agricultural ecology, the first in the world at the Institute of Superiore Agrario, Perugia.

The study of agricultural ecology has been started from 1929 when Girolamo Azzi went to establish for the first time a course in this subject at the faculty of agriculture, University of Buenos Aires, Argentina.[22]

Tansley 1926, used the term in its widest meaning as the study of organisms as they exist in their natural homes, or as the economy, household affairs of organisms.[23] Weavers and Clements in 1929, stated that "every plant is a product of the conditions under which it grows." It indicates in general and often in a specific manner what other species would do if grown in the same place.

The term 'Agro-ecology' for the first time proposed by B.M. Bensin in 1930, to apply for the detailed studies of commercially important crop plants by the use of ecological methods.[24] He proposed a systematic collection of data so that the main agricultural regions (agrochoras) of the world and the characteristics of local cultivated varieties of important crops (chorotypes) may be described and recorded by the employment of standardized methods and by a prescribed and uniform terminology. It is observed that Bensin deal only commercially important crops so a large number of other crops excluded from the field of agro-ecology.

One of the classic works 'Principles of Animal Ecology' by Alee, *et al.* (1949), deals with all the major aspects of ecological work.[25] 'Natural Communities' by Dice (1952), properly deals with the community in terms of physical factors territoriality, food, succession and evaluation.[26] A most scholarly work by Andrewartha and Birch, 'The Distribution and Abundance of Animals' (1954), covers a number of phases of ecological thought never before presented in a single volume.[27] C.B. Knight (1965), in his book "Basic Concept of Ecology" designed introductory ecology also useful for the person who belongs to other disciplines like wildlife management, forestry and agriculture. He explained the very basic concepts of ecology and defined various ecological terms which should be known to every young ecologist.[28] Peter Farb in 1968, gave an ecological account for the distribution of plants and animals on the globe in these words "all animals and plants live where they do for a very good reason. If they are absent from an area that support similar organisms, there are good reasons for that too. They may be prevented from getting there by such visible but equally insurmountable barriers of insufficient water, or the wrong kind of food, or subtle conditions of climate or soil. Sometimes, they are inhibited by the presence of man, more often by the presence of another competing type. Sometimes, an animal or plant appears entirely suited to an environment and is still absent from it. This may be because of the absence of another on which it depends. The distribution of the econite plant is a good example, it is wholly depended on the bumble-bee for cross-pollination, and as a result is never found beyond the range of the bumble-bee itself."[29] Cox, C.B. *etal.* (1973), deal very well the organism's ecological relationships with soil, temperature etc. and causes of their distribution why these different species found in different areas of the globe.[30] Colinvaux[31] in the same year tried to solve the problems like "why did the vegetation of different parts of the world look different." In 1974, Robert W. Poole gave a brief account of quantitative methods to determine the causes of some ecological phenomena.[32] Mann (1982)[33] and Seed (1986)[34] gave a detailed account of coastal ecology in their books 'Ecology of Coastal Waters' and 'The Ecology of Rocky Coasts' respectively and covered almost all the issues of problems and prospects of coastal biomass and ways for their conservation to maintain ecological balance. Sir Edmond Hillary in 1984, covered various issues of changing face of the earth like expanding deserts, extinction of flora and fauna, water and atmospheric pollution, climatic change, politics of ecology and their consequences in future.[35] The Great U-Turn by Goldsmith (1992), compiled a series of essays written during the last sixteen years with radical approach, and published in 'The Ecologist'. The general themes of the essays are ecology of unemployment, ecology of health, ecology of wars and social and ecological interpretation to the fall of Roman Empire.[36] Hayward (1994), offers a new and accessible introduction to the implications of ecology for social and political thought. A range of debates about values in nature, the meaning of sustainable development and such questions as whether human rights are compatible with ecological responsibilities are discussed in a very judicious manner.[37]

A Brief Account of the Work done in the Region

Agro-ecological studies in Indian context and of Himalayan region in particular are done by non-Indian up to the independence specially in the writings of great hunters and forest officers which was later on took over by the Indians. A brief compilation of such studies is as follows:

Semple (1911)[38], while describing the stock grazings on the slopes of Kashmir valley pointed out towards an ecologically beneficial change in the landuse pattern. In the same year Semple, in a paper "Influences of Mountainous Environment" had tried to stress the need of integrated approach of mountain slope and valley agriculture planning.

Mason (1929)[39], while describing the routes of western Himalaya had pointed out that the deterioration of living environment in the caravan nodal centres would become a stressing problem if allowed to sprawl on the fringes of virgin forests. This shows that in the early decades of this century, geographers and environmentalists have been consciously aware of the after-effects of reckless handling of ecological issues.

The study of Agricultural landuse of Kashmir valley with ecological methods was carried out by Hoon (1955)[40], when he described the agricultural landuse on the basis of soil characteristics. But he limited the study to hilly areas only where slope, as seemed to him was the main restricting factor of soil formation. Later, Koul (1956)[41], presented altitude based soil zones for agricultural landuse. Datta and others (1957)[42], studied the agro-ecological conditions of the valley for agricultural landuse when they analyzed the Karewa soils and rainfall characteristics for saffron cultivation. Singh (1958)[43], while referring towards the depletion of forests in the valley made an ecological analysis of agricultural landuse and soil erosion. Stamp (1960)[44], described the importance of Kashmir valley ecology in the whole region of western Himalaya and possibilities of crop diversification in the Kashmir valley. Raina (1962)[45], while discussing the scope for rural landscape in the Kashmir valley has analysed the role of rainfall and temperature in shaping the cropping pattern in different relief divisions of the Kashmir valley. Jalal (1977)[46], has taken into account productivity per unit (PPU) for land capability classification in Pithoragarh district of U.P. Hudson (1979), classified the land of U.P. hills into six capability classes based on soil depth and slope. Singh (1980), Singh and Kaur (1983) and Chadha (1989) gave accounts to conserve the Himalayan ecology in more or less similar ways. The poor environmental management in creating shortage of food, fuel and fodder, out migration and poverty seems to conspired with nature catastrophe, influencing the worst injuries to the highlands. The rapid deforestation is causing soil erosion and ubranization along with unplanned tourism development in ecologically fragile Himalayan region. The solutions of the said problems are suggested in eco-development i.e., development without destruction, the environmental consciousness and to know how much exploitation of natural resource and any economic activity is too much? Shah (1982a, 1982b, 1983, 1991), emphasizing the need of a pragmatic approach based on environmental and economic considerations has made remarkable contribution towards land capability classification in Himalaya. Sen (1986)[47], indicates that the existing landuse pattern in the hill is deteriorating the ecosystem and suggested that, the land unsuitable for cultivation of foodgrains should not be used for regular cultivation but to be put under permanent vegetation like Orchards and other perennial crops in the hills. Raza (1987), made an attempt to develop an understanding between diversities and disparities of the nature.[48]

Mukherjee (1987)[49], identifies the major causes of imbalances in the ecosystem and suggests remedial measure for its conservation and restoration. It is pointed out that the human intervention into the natural realm for the exploitation of resource base in terms of mining, deforestation and agricultural landuse has resulted in serious ecological disorders. The work of Dhaliwal and Kler (1995)[50], "Principle of Agricultural Ecology" is really a masterpiece for the researchers not only in agro-ecology but also in other disciplines i.e., life sciences and environmental sciences. It deals the very basic concepts of ecology, agro-ecosystem, current status of various natural resources and their management to feed the growing population of the world. Misra (1998), said about the three dimensions of environmental sustainability in biodiversity which has to be seen from three perspectives i.e., genetic diversity, species diversity and ecosystem diversity.[51] Singh (1999)[52], compiled the work of Canadian International Development Agency (CIDA) and Shastri Indo-Canadian Institute (SICI) on Sustainable Development of Mountain Environment in India and Canada. The focussed regions for this study are Himachal Pradesh in India while Arrow Lake region of British Columbia in Canada. It covers various issues of natural resource mismanagement and their judicious conservation. Rawat, Kumar and Singh (1999)[53], edited a book on the theme "Environmental Management and Regional Development in the Himalaya" with special reference to U.P. Himalaya, in which they discussed a large number of burning issues related to the sustainable development of the region and their cure with the application of most advanced techniques of Remote Sensing and GIS at micro-level. Mohan (2000)[54], highlights the multidimensional nature of ecological issues that have been emerged as a consequence of rapid economic growth with special reference to Indo-Gangetic Divide and adjoining region.

NOTES

1. S.S. Visher, *Recent Trends in Geography*, Scientific monthly, Vol. 35, No. 5, July 1972, p. 439.
2. K.H.W. Klage, (1942), *Ecological Crop Geography*, Macmillan Company, New York, p. 73.
3. Colinvaux, P.A. (1973), *Introduction to Ecology*, John Wiley & Sons, New York.
4. Klage, K.H.W. (1954), *Ecological Crop Geography*, The Macmillan Company, New York.
5. Girolamo Azzi, (1956), *Agricultural Ecology*, Constable and Company, Ltd. London, p. xiii.
6. Risser, R.P.G. *et al.* (1984), *Landscape Ecology*, The U.S.A. Natural History Special Pub., Illinois, vol. 2, pp. 18-21.
7. Gulinck, H. (1986), "Landscape Ecological Aspects of Agro-Eco-system", Agriculture, Eco-system, Environment, Elsevier Science Pub., Amsterdam, Vol. 16, pp. 79-86.

8. *Ibid.*, p. 83
9. Southwick, C.H. (1976), *Ecology and the Quality of Our Environment*, D. Van Nostrand Company, New York.
10. Moutia, L.A. and Mamet, R. (1946), *The Growth of a Science*, University of Colorado, Stud. 26 : 3-14.
11. Ramaley, F. (1940), A Review of 25 years of Economic Entomology in the Island of Mauritius, *Bull. Ent. Res.* 36 : 439-472.
12. Dhaliwal, G.S. and D.S. Kler (1995), *Principles of Agricultural Ecology*, Himalayan Publishing House, New Delhi, p. 6.
13. *Ibid.*, p. 7.
14. Liebig, J. (1840), *Chemistry in its Application to Agriculture and Physiology*, Taylor and Walton, London.
15. Merriam, C.H. (1890), *Results of Biological Survey of the San Francisco Mountain Region and Desert of Little Colorado*, Arizona, North America Fauna, 1-136.
16. Davenport, C.B. (1903), *The Animal Ecology of the Cold Spring Harbor Sand Split with Remark on the Theory of Adaption*, Decennial Publ. Univ. Chicago, 10 : 157-176.
17. Forbes, S.A. (1907), On the local distribution of certain Illinois fishes : An essay in Statistical Ecology, *Bull. Ill. Lab. Nat. Hist. Survey*, 15 : 537-550.
18. Warming, J.E.B. (1909), *Ecology of Plants*, Oxford Press, London, pp. i-xi.
19. Klages, K.H.W. (1942), *Ecological Crop Geography*, Macmillan Company, New York, pp. 105-6.
20. Shelford, V.E. (1913), *Animal Communities in temperate America*, University of Chicago Press, Chicago.
21. Report in the Official Gazette of the Accademia dei Lincei, vol. XXIX, 5th Series, Part II, Ist Quarter, Deliberation of 4th June, 1920.
22. Azzi, G. (1956), *Agricultural Ecology*, Constable and Company Ltd. London, p. xii.
23. Tansley, A.G. (1926), *Practical Plant Ecology*, Allen and Unwin, London.
24. Bensin, B.M. (1930) Possibilities for International Cooperation in Agro-ecological investigations, *Internatl. Agr. Mo. Bull. Agr. Sci. and Pract.* (Rome), 21 : 277-284.
25. Allee, W.C., A.E. Emerson, O. Park and K.P. Schinidt, (1949), *Principles of Animal Ecology*, W.B. Saunders, Philadelphia, pp. i-xii, 1-873.

26. Dice, L.R. (1952), *Natural Communities*, University of Michigan Press, Michigan, pp. i-x, 1-547.
27. Andrewartha, H.G. and L.C. Birch (1954), *The Distribution and Abundance of Animals*, University of Chicago Press, Chicago, pp. i-xv, 1-782.
28. Knight, C.B. (1965), *Basic Concept of Ecology*, The Macmillan Company, New York.
29. Peter Farb (1968), *Ecology, Time-Life International*, p. 11.
30. Cox, C.B. and P.D. Moore (1973), *Biogeography : An ecological and evolutionary approach*, Blackwell Scientific Publications, Oxford.
31. Colinvaux, P.A. (1973)., *Introduction of Ecology*, John Wiley & Sons, Inc. New York.
32. Poole, W.R. (1974), *An Introduction to Qualitative Ecology*, McGraw Hill, Kogakusha, Ltd. Tokyo.
33. Mann, K.H. (1982), *Ecology of Coastal Waters*, Blackwell Scientific Publication, England.
34. Moore, P.G. and R. Seed (1986), *The Ecology of Rocky Coasts*, Columbia University Press, New York.
35. Sir E. Hillary (1984), *Ecology 2000*, Multimedia Publications, London.
36. Goldsmith, E. (1992), *The Great U-Turn : Essays on Ecology*, Ashish Publishing House, New Delhi.
37. Hayward, T. (1994), *Ecological thought*, Polity Press, U.K.
38. Semple, E.C. (1911), *Influences of Geographical Environment*, Russel and Russel Pub., London.
39. Mason, K. (1929), *Routes in the Western Himalaya*, Surveyor General of India Publications, Calcutta.
40. Hoon, R.C. (1955), A study of the Soils in the Hill Areas of Kashmir, *Forest Record Journal*, Jammu and Kashmir State, Sericulture Deptt., Srinagar, vol. 3.
41. Koul, M.L. (1956), Study of Soils at Different Altitudes in Kashmir State, *Journal of Indian Agriculture Research Institute*, Vol. xxi.
42. Datta, N.R. *et al.* (1957), Soil Conditions for the Growth of Saffron of Pampore-Kashmir, *Indian Journal of Agricultural Sciences*, Vol. xxv.
43. Singh, H. (1958), Land Erosion in Kashmir Valley, *Indian Forester*, Vol. xxxiv.

44. Stamp, L.D. (1960), *India, Pakistan, Ceylon and Burma,* Methuen and Co. London.
45. Raina, A.N. (1962), Scope of Rural Landuse in Kashmir Valley, *Indian Council of Agricultural Report,* Vol. xxxiv.
46. Jalal, D.S. (1977), Classification and Potential Productivity of Arable Land in Kumaon Himalaya with special reference to District Pithoragarh, Uttarakhand, *Bharti,* Kumaon University, vol. 1 (3), pp. 65-75.
47. Sen, C. (1986), Landuse Pattern in Hills of Uttar Pradesh: An Ecological Abuse, *The National Geographical Journal of India,* Vol. XXXII, No. 2, pp. 111-117.
48. Raza, M. (1987), India : Diversities and Disparities, *Contemporary Affairs,* vol. 1, No. 1, pp. 36-41.
49. Mukherjee, S. (1987), Landuse Maps for conservation of ecosystems, *Geographical Review of India,* Vol. xlix, No. 3, pp. 23-28.
50. Dhaliwal, G.S. and D.S. Kler (1995), *Principles of Agricultural Ecology,* Himalayan Publishing House, New Delhi.
51. Misra, R.P. (1998), Sustainable Development: The Ecological Perspectives, in eds. H.S. Sharma and S. Chattopadhayay's *Sustainable Development : Issues and case studies,* Concept Publishing Co. Pvt. Ltd. New Delhi.
52. Singh, R.B. (1999), *Sustainable Development of Mountain Environment in India and Canada,* Oxford and IBH Publishing Co. Pvt. Ltd. New Delhi.
53. Rawat, M.S.S., Kumar, K. and R.N. Singh (1999), *Environmental Management and Regional Development in the Himalaya,* Department of Geography, HNB Garhwal University, Srinagar, Garhwal.
54. Mohan, M. (2000), *Ecology and Development,* Rawat Publications, New Delhi.

2

Agro-ecology and its Constraints of the Region

AGRO-CLIMATIC DIVERSITY

The location of north-western Himalaya at the high altitude in north-western nook of the subcontinent is enclosed within mountain range gives it a distinctive character with its own climatic peculiarities. A moderate climate is found only in central longitudinal valleys at 1500 m to 2100 m., the rest of the places suffer either from 'hot climate' in Siwalik or Kandi belt or the extreme cold in the north.

The weather or climate of any place is the sum of its atmospheric conditions e.g. temperature, pressure, wind, precipitation and humidity, etc. Therefore, the elements of weather and climate are the same, but weather is a state of atmospheric conditions at a given time and space, whereas climate is a composite picture or generalization of the variety of day-to-day weather condition. It is therefore, obvious that weather varies from day-to-day and climate differs from place to place depending upon the variations in amount, intensity and areal distribution of their elements. Climate is a key factor in understanding many processes taking place on the surface of the earth. Diversified climatic conditions existing in this agro-ecological region, practically regulate the natural processes and human activities in the entire region. Climatic complexities at micro-level are mainly generated by altitude and aspects of slope. Temperature and precipitation are the two main atmospheric variables to form the climate of any region.

Rainfall is recorded by the Indian Meteorological Department at 556 surface and 624 Hydro-met observatories. In addition there are about 5737 rainfall recording stations maintained by the various state governments who publish the data emanating therefrom. Daily rainfall is recorded at 08:30 hours every day throughout the country. This amount is the total rainfall for the preceding 24 hours ending at 08:30 hours of the date of measurement. The 'actual monthly' rainfall of the subdivision for a year is the mean of the monthly rainfall of the available stations. The sum of these for 12 months is the 'actual annual' rainfall of the subdivision for that year. The 'monthly normal' rainfall of the subdivision is the mean of the monthly normal of the corresponding stations. The 'annual normal' for the subdivision is the sum of these 12 monthly normal. Normal is an average of all data since 1901 or from the year of commencement of the stations whichever is later.

The climatic changes with the elevation cause striking meteorological contrasts. In north-western Himalaya, variations in the exposure to sunlight and rain bearing winds result in intricate local variations in climate. However, there are also broad areas in which the climate varies only with elevation. The high altitude climate differs from middle altitude climate. As the elevation increases, the seasonal variation in temperature practically disappear. The foothills of Siwalik, like the adjoining Kandi belt of rolling plain have subtropical monsoon climate. The monsoonal rhythm is expressed as seasonal rainfall, combined with southerly winds from summer monsoon and practically little rainfall experiences in winters of the mountain valleys.

The western part of the north-western Himalaya which

happens to be the farthest from the sea is also latitudinally and altitudinally higher than the rest and thus shows a climatic pattern quite dissimilar to the areas lying eastward where conditions are governed by the monsoon system. The climate of north-western Himalaya is more temperate, often referred to as extra-tropical mountain climate. Likewise, the continental effects reflected in the large annual and di-annual and di-temperature are more prominent towards west than the central or eastern part of the Himalaya. There are striking contrast between the pattern of annual precipitation in the western and eastern parts e.g. north-western Himalaya experiences greater amount of precipitation during winter months from the western disturbances while central and north-eastern Himalaya makes the larger proportion of annual rainfall by the monsoon winds. Srinagar in NW Himalaya receives 331 mm out of 661 mm of annual rainfall during winter months from January to April. On the other hand, Darjeeling in north-eastern Himalaya receives 2582 mm out of 3100 mm of annual rainfall during monsoon months from July to September.

Variations of Annual Rainfall

The distribution of rainfall varies rapidly with altitude and its amount to a great extent is modified by the location of the place to the windward or leeward sides of the high ridges. It is generally observed that the amount of rainfall increases with the height up to 2500 m but a decrease in rainfall is recorded on ascending above this elevation due to the exhaustion of water vapours in the air because every high ridge or peak cuts-off the supply of vapours from lower ground to the upper reaches. The data regarding rainfall is not available for full range of heights in the region but the rainfall data from the stations of all districts are trying to analyze and a correlation can be observed with the agricultural productivity.

It is obvious that physiographic diversities and altitudinal variations play an important role in determining the quantity of rainfall at different locations. The higher amount of rainfall is recorded along the south facing slopes which come first to the contact of monsoon winds.

To analyse the distribution of rainfall in north-western Himalaya, the amount of rainfall in millimeters for the year 1996 is divided into five categories viz. less than 800 mm, 800 – 1200 mm, 1201 – 1600 mm, 1601 – 2000 mm and more than 2000 mm. Out of all the 34 districts, data of 7 districts namely, Kupwara, Rajouri, Jammu, Udhampur, Kargil, Leh and Bilaspur is not available from any station. In rest of the districts, Pulwama, Solan, Lahaul and Spiti and Tehri Garhwal observed very low rainfall i.e., less than 800 mm while 800 – 1200 mm of rainfall is recorded in 10 districts which are prominent on the map in the three pockets i.e., Baramula, Srinagar, Budgam and Anantnag in Kashmir valley while Kullu, Shimla and Kinnaur in Himachal Pradesh and Chamoli and Almora are prominent in Kumaon and Garhwal Himalaya (Fig. 2.1). There are 7 districts which fall in the medium rainfall of 1201 – 1600 mm namely Poonch of Jammu region. Chamba and Hamirpur of Himachal Pradesh while Uttarkashi, Dehradun, Garhwal and Nainital of Uttaranchal region. Four districts recorded high rainfall during this period namely, Pithoragarh, Sirmour, Kangra and Kathua. Only two districts viz. Doda and Mandi recorded very high rainfall i.e., more than 2000 mm.

As a whole, it is observed that the rainfall is unevenly distributed over the whole region in different seasons of the year. Relief also plays an important role in such variations of rainfall distribution.

Variations of Seasonal Rainfall

The climate of north-western Himalaya is more temperate and often referred to as extra-tropical mountain climate.

There is striking contrast between the pattern of annual precipitation, as some seasons found dry with scanty rainfall while in some other seasons there is plenty of rainfall. To analyze the seasonal variation in rainfall, the whole year is divided into four seasons viz. January-February, March-May, June-September and October-December based on IMD 1996. Categories of rainfall are formed accordings to the range of precipitation in a particular season.

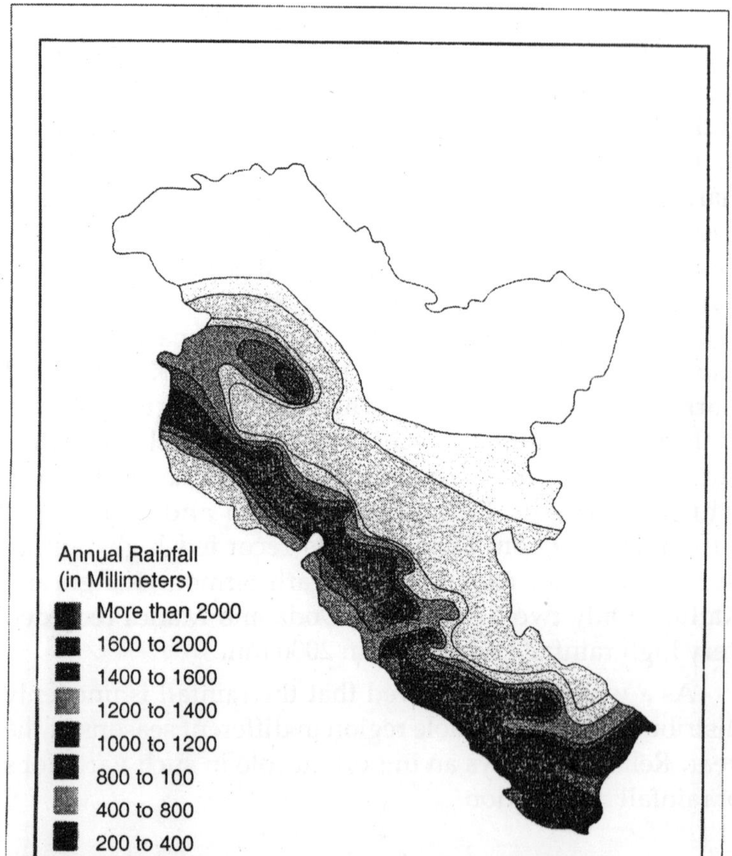

Fig. 2.1 : North-western Himalaya : Average Annual Rainfall

Rainfall during January-February

In the period of January-February the rainfall cannot exceed 300 mm and so there are five categories of rainfall formed for analysis viz. below 25 mm, 25-50 mm, 50-100 mm, 100-200 mm and more than 200 mm. It is revealed from (Fig 2.2) that the area under first three categories of below 100 mm are mostly found in the north of Kargil and Leh region, while the belt of more than 200 mm rainfall found amidst the 100 – 200 rainfall zones. Poonch, Rajouri, Budgam,

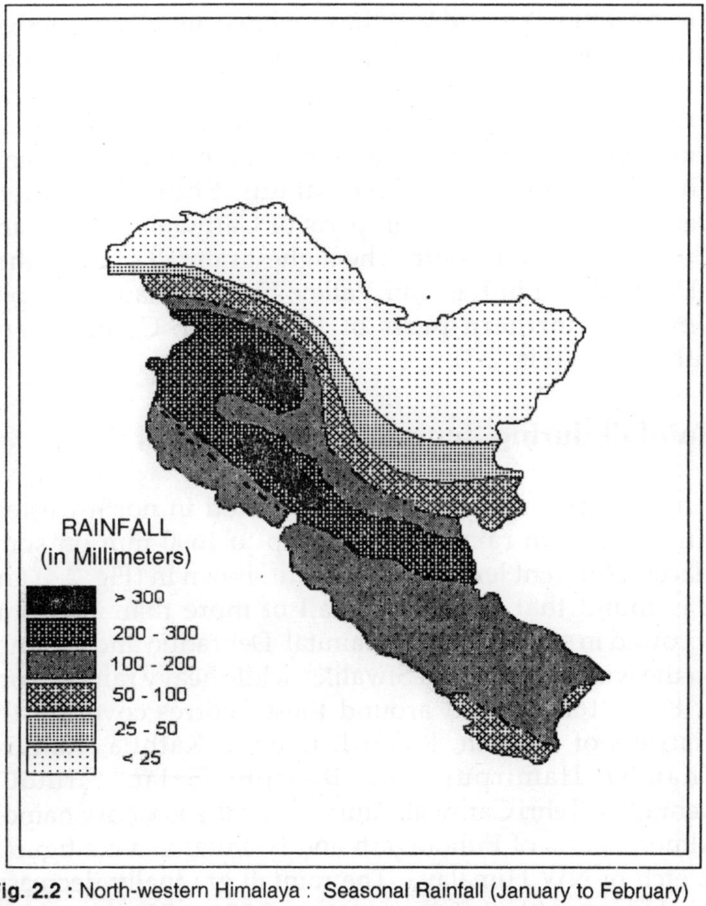

Fig. 2.2 : North-western Himalaya : Seasonal Rainfall (January to February)

Pulwama, Udhampur, parts of Doda, Chamba and Lahaul and Spiti recorded more than 200 mm of rain while districts south of Chamba and Lahaul and Spiti records 100-200 mm of rain and a narrow belt covering northern Doda, southern Kargil and Leh also observed 100–200 mm of rain during January–February.

Rainfall during March to May

This period of March to May again records rainfall below 300 mm but it is found that the previous rainfall zone moves upward the zone of 50 – 100 mm of rainfall become prominent on both north and south of 100 – 200 mm of rainfall zone. 100 – 200 mm rainfall is recorded in southern parts of Leh and Kargil in the north, while Jammu, Kathua, south-western Chamba, parts of Kangra, Mandi, Shimla, northern Dehradun, Tehri Garhwal, parts of Garhwal, Almora and Pithoragarh in the South. The highest rainfall of more than 200 mm found in Kashmir Valley, Doda, Lahaul and Spiti, Kinnaur, northern parts of Uttarkashi, Chamoli and Pithoragarh (Fig. 2.3).

Rainfall during June to September

June to September is the rainiest period in north-western Himalaya with rainfall exceeds up to 1600 mm on some places. Different level of rainfall are shown in (Fig. 2.4) and it is found that highest rainfall of more than 1600 mm occurred in some places of Nainital, Dehradun and Chamba on the windward side of Siwaliks, while heavy rainfall zone of 800 – 1600 spread around these centres covering the districts of Poonch, Rajouri, Jammu, Kathua, Kangra, Chamba, Hamirpur, Una, Bilaspur, Solan, Sirmour, Dehradun, Tehri Garhwal, Pauri Garhwal, southern Chamoli, Almora, parts of Pithoragarh and Nainital in a contiguous stretch of NW Himalaya. The rainfall gradually decreases northward with different zones of 200 – 800 mm, 100 –

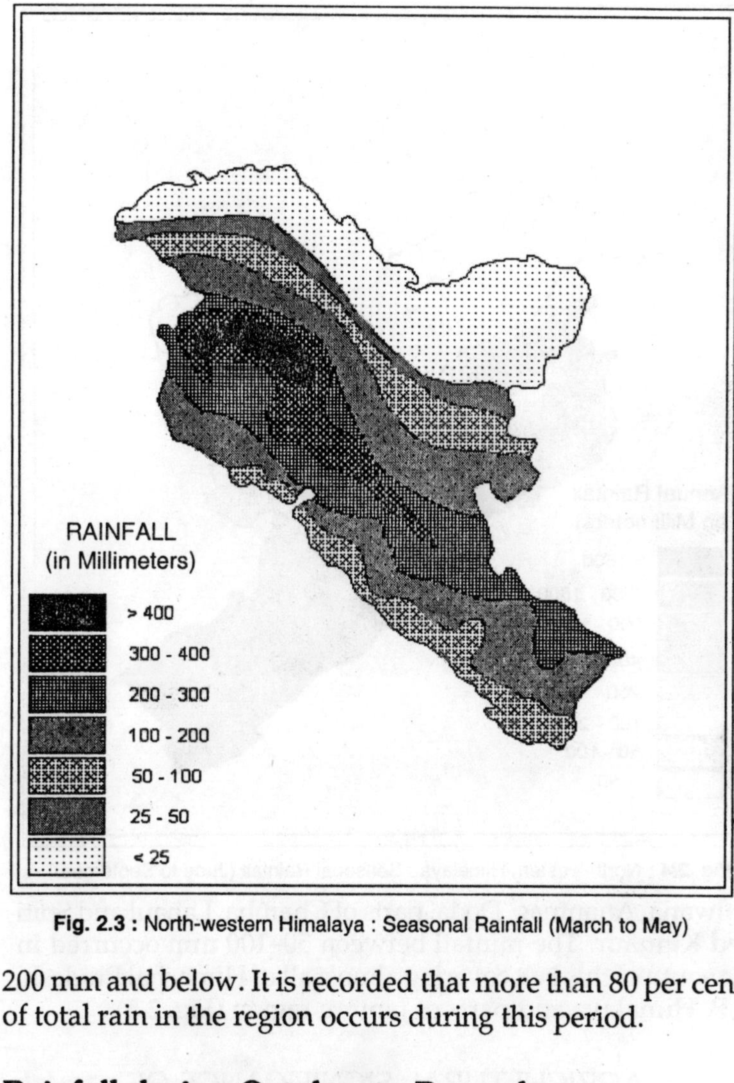

Fig. 2.3 : North-western Himalaya : Seasonal Rainfall (March to May)

200 mm and below. It is recorded that more than 80 per cent of total rain in the region occurs during this period.

Rainfall during October to December

In this period, the rainfall hardly exceeds 200 mm in this region which is observed in northern Doda and Kargil while 100 – 200 mm of rainfall occurred in the districts of Budgam,

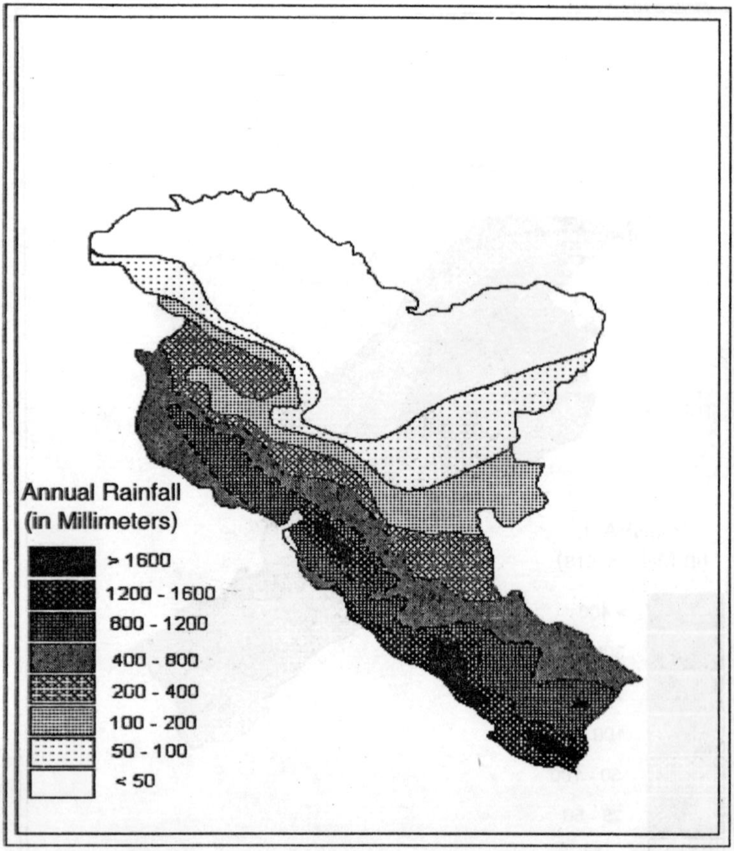

Fig. 2.4 : North-western Himalaya : Seasonal Rainfall (June to September)

Pulwana, Anantnag, Doda, parts of Chamba, Lahaul and Spiti and Kinnaur. The rainfall between 50–100 mm occurred in the south of this belt covering almost all of Himachal Pradesh, U.P. Himalaya and parts of Jammu region (Fig. 2.5).

AGRICULTURAL SIGNIFICANCE OF TEMPERATURE DIVERSITY

A new impetus in climatology was advocated by Thornwaite (1953), who stressed the horizontal rather than the vertical gradient in atmosphere and proposed the term *Topoclimatology*.

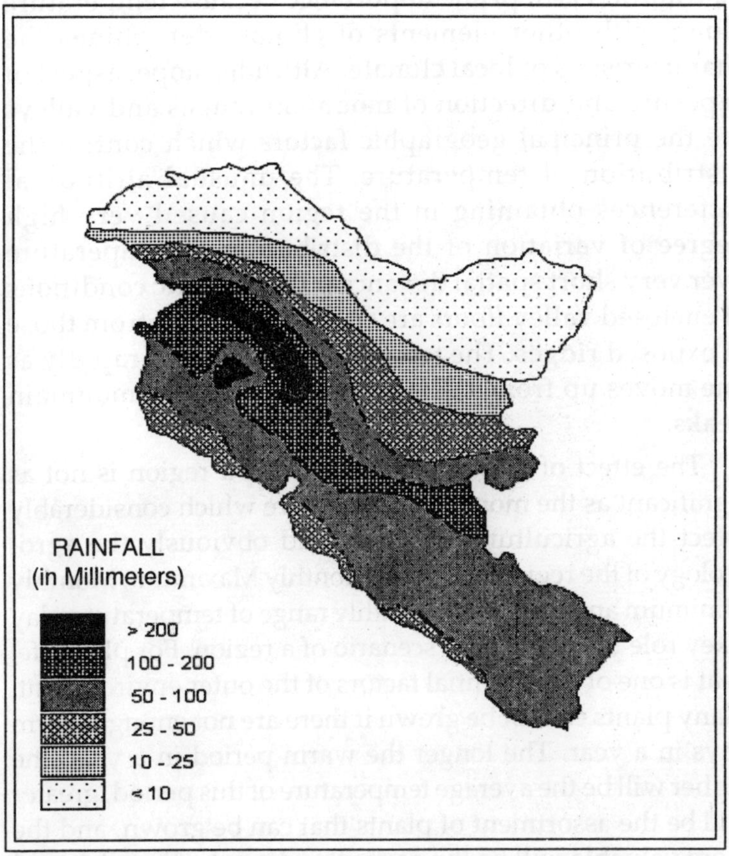

Fig. 2.5 : North-western Himalaya : Seasonal Rainfall (October to December)

This suggested a more geographical approach with an appreciation of the influence of relief, aspect, vegetation and landuse contrasts.

Temperature is an independent variable among the climatic elements. It causes change in the pressure distribution and consequently in wind direction and its velocity which controls atmospheric humidity, condensation, formation of cloud and their drafting in the sky, precipitation and storms. So a detailed study and analysis of temperature conditions over the region and its areal and seasonal distribution is of prime importance because temperature along with other elements of climate determines the characteristics of local climate. Altitude, slope, aspect of exposure and direction of mountain ranges and valleys are the principal geographic factors which control the distribution of temperature. The unusual altitudinal differences obtaining in the region caused very high degree of variation of the distribution of temperature over very short spatial distances. The climatic conditions of enclosed valley floors are entirely different from those of exposed ridges. The temperature decreases rapidly as one moves up from valley bottom to elevated mountain peaks.

The effect of annual temperature of a region is not as significant as the monthly temperature which considerably affect the agricultural practices and obviously the agro-ecology of the region. Limits of Monthly Maximum, Monthly Minimum and Daily and Monthly range of temperature play a key role in agricultural scenario of a region. For plant life, heat is one of the essential factors of the outer environment. Many plants cannot be grown if there are not enough warm days in a year. The longer the warm period in a year, the higher will be the average temperature of this period, greater will be the assortment of plants that can be grown, and the more valuable will be the agricultural production obtained from the given area.

The intensity of radiation pressure or the temperature of the air determines the rate at which the life processes takes place in the plant organism. Each species and variety of plant and each age group has its own lower and upper limits of temperature. Within a certain range, the plant will be partly or entirely destroyed. Within another range, the plant will cease to grow although it remains alive. Finally, there are limits between which the growth of a plant and its development is optimum. The influence of a given temperature on a plant depends on the intensity of the radiation pressure, on the duration of its action and also on the nature and condition of the plant itself.

Variations of Seasonal Temperature and its Agro-Ecological Significance

To look at the pattern of seasonal temperature in north-western Himalayan region, the data of monthly maximum, monthly minimum, monthly average and their monthly range of temperatures have been taken into account from the Climatological Tables published by IMD for an average of 1950-1980, and assesses their relationship to the agro-ecology of the regions. The data of monthly temperatures from 18 stations have been taken into account which are at different heights. The smooth line curves drawn for monthly maximum, monthly minimum and average monthly temperature shows two different forms of types :

1. Symmetrical Hill Type
2. Unsymmetrical Hill Type

Table 2.1 : Variations of Temperature and Rainfall at some Meteorological Stations

Jammu (1951-78) Height 367 M				
Months	*Min.*	*Max.*	*Avg.*	*Rainfall*
J	7.4	18.8	13.1	50.0
F	9.6	21.9	15.8	46.4
M	13.6	26.6	20.1	53.2
A	19.0	32.9	26.0	26.3
M	24.4	38.6	31.4	16.0
J	26.8	40.6	33.7	51.8
J	24.5	35.5	30.0	283.4
A	24.0	33.7	28.9	344.5
S	23.0	33.6	28.3	123.9
O	18.4	31.7	25.1	38.1
N	12.6	26.8	19.7	11.9
D	8.5	21.1	14.8	42.2

Table 2.1. (Contd...)

Dehradun (1951-80) Height 682 M				
Months	*Min.*	*Max.*	*Avg.*	*Rainfall*
J	5.8	19.4	12.6	51.6
F	7.9	21.9	14.9	48.1
M	12.2	26.7	19.5	51.3
A	16.9	32.3	24.6	17.3
M	20.5	35.8	28.2	36.2
J	23.0	34.9	29.0	225.0
J	22.5	30.5	26.5	718.6
A	22.0	29.5	25.8	734.0
S	20.6	29.7	25.2	321.7
O	15.9	28.5	22.2	47.5
N	10.4	25.1	17.8	11.3
D	6.8	21.3	14.1	23.2
Mandi (1954-80) Height 761 M				
Months	*Min.*	*Max.*	*Avg.*	*Rainfall*
J	2.8	18.5	10.7	74.3
F	4.2	21.1	12.7	60.2
M	9.0	25.8	17.4	81.8
A	13.5	30.8	22.2	47.6
M	17.3	34.7	26.0	51.1
J	20.1	36.0	28.1	130.0
J	21.1	31.8	26.5	500.0
A	20.5	31.0	25.8	427.6
S	18.4	30.7	24.6	186.8
O	12.3	28.8	20.6	45.8
N	6.8	24.9	15.9	13.8
D	3.1	20.6	11.9	23.2
Tehri (1956-80) Height 770 M				
Months	*Min.*	*Max.*	*Avg.*	*Rainfall*
J	4.6	19.6	12.1	119.8
F	7.1	22.3	14.7	53.7
M	11.0	27.6	19.3	54.4
A	15.6	33.0	24.3	29.3
M	19.0	36.3	27.7	43.1
J	22.8	36.5	29.7	79.7
J	23.4	33.6	28.5	220.8
A	23.2	32.5	27.9	230.4
S	20.9	32.2	26.6	126.6
O	15.4	30.5	23.0	36.1
N	9.3	26.0	17.7	9.2
D	4.8	21.6	13.2	28.9

Table 2.1. (Contd...)

Bhuntar (1960-80) Height 1096 M				
Months	*Min.*	*Max.*	*Avg.*	*Rainfall*
J	1.3	14.9	8.1	92.0
F	3.4	17.1	10.25	92.7
M	6.6	21.8	14.2	108.9
A	10.0	27.3	18.65	63.9
M	12.5	30.9	21.7	59.4
J	17.1	32.9	25.0	57.3
J	19.6	30.9	25.25	146.5
A	19.4	30.1	24.75	146.5
S	16.2	30.0	23.1	47.1
O	9.9	27.5	18.7	20.1
N	4.3	22.8	13.55	21.6
D	1.4	17.3	9.35	29.6
Dharamshala (1951-80) Height 1211 M				
Months	*Min.*	*Max.*	*Avg.*	*Rainfall*
J	5.9	14.5	10.2	114.5
F	7.7	16.6	12.15	100.5
M	11.8	21.1	16.45	98.8
A	16.3	26.2	21.25	48.6
M	20.1	30.5	25.3	59.1
J	21.8	31.4	26.6	202.7
J	20.7	27.2	23.95	959.7
A	20.2	26.3	23.25	909.2
S	18.7	26.3	22.5	404.8
O	15.3	24.8	20.05	66.3
N	10.7	20.7	15.7	16.7
D	7.4	16.7	12.05	54.0

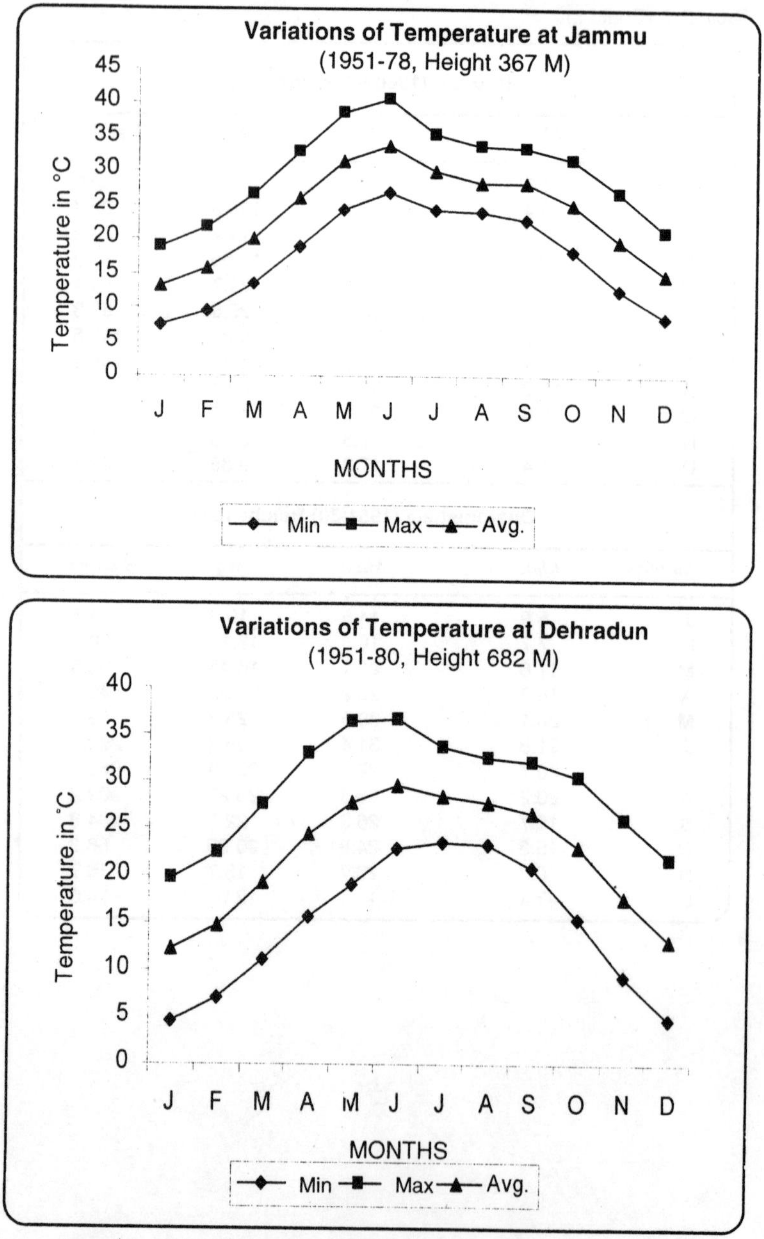

Fig. 2.6 : Variation of Temperature at some Meteorological Stations

(Fig. 2.6 : Contd...)

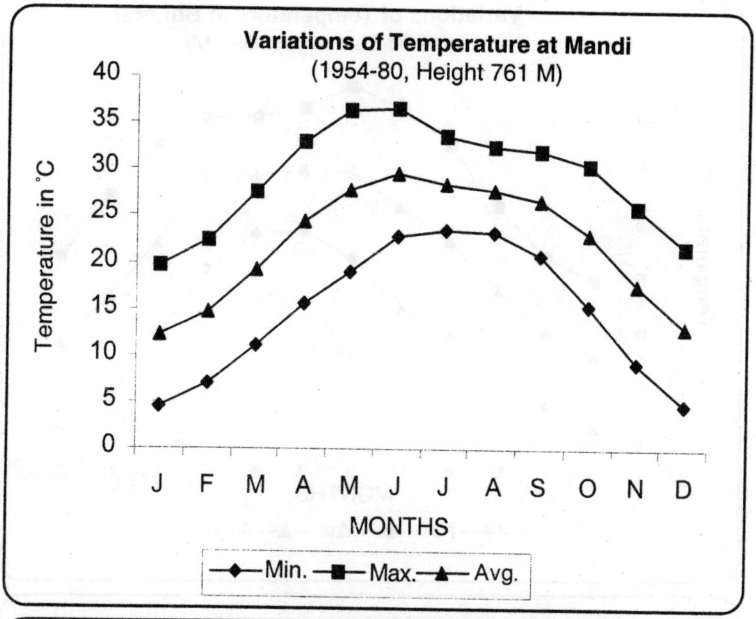
Variations of Temperature at Mandi
(1954-80, Height 761 M)
40
35
30
25
20
15
10
5
0
Temperature in °C
J F M A M J J A S O N D
MONTHS
Min.
Max.
Avg.

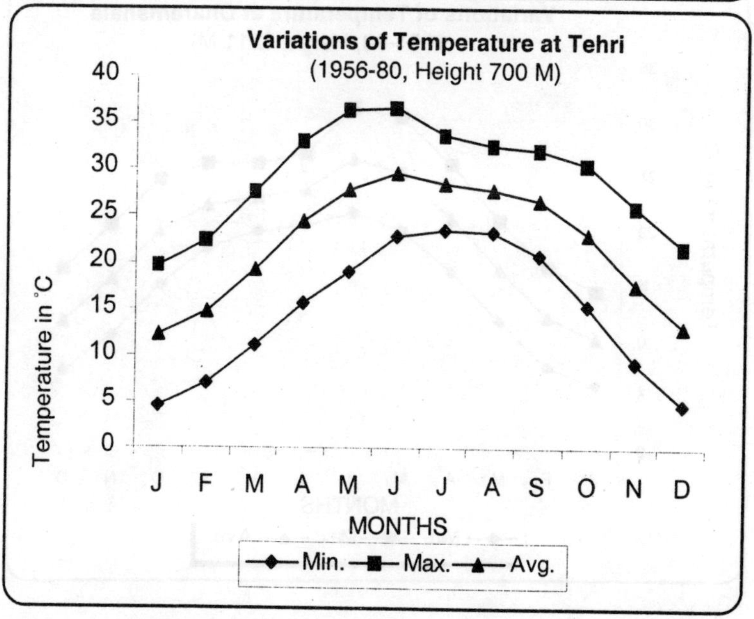
Variations of Temperature at Tehri
(1956-80, Height 700 M)
40
35
30
25
20
15
10
5
0
Temperature in °C
J F M A M J J A S O N D
MONTHS
Min.
Max.
Avg.

(Fig. 2.6 : Contd...)

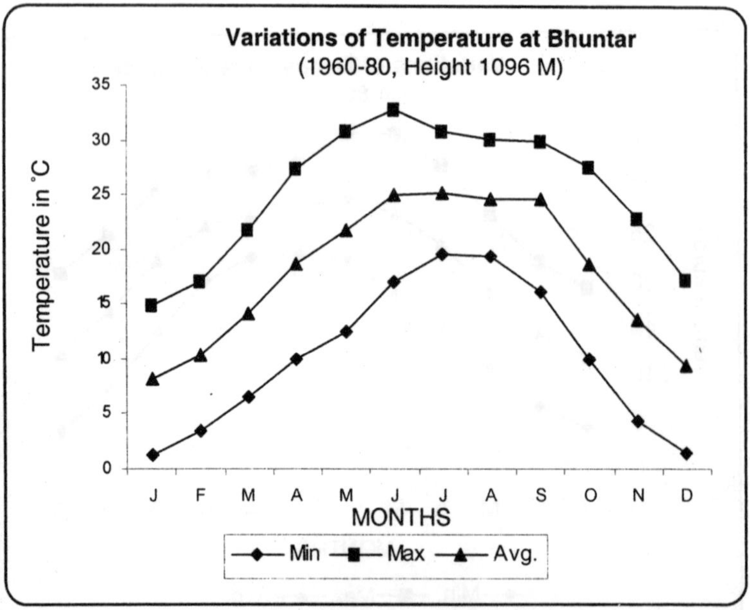
Variations of Temperature at Bhuntar
(1960-80, Height 1096 M)
Temperature in °C
35
30
25
20
15
10
5
0
J F M A M J J A S O N D
MONTHS
Min
Max
Avg.

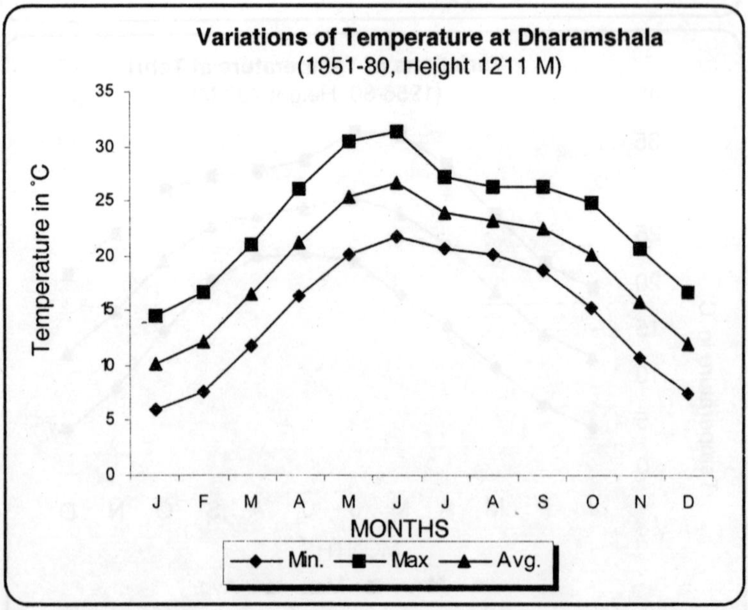
Variations of Temperature at Dharamshala
(1951-80, Height 1211 M)
Temperature in °C
35
30
25
20
15
10
5
0
J F M A M J J A S O N D
MONTHS
Min.
Max
Avg.

The general pattern of Symmetrical Hill type is such that the temperature increases smoothly up to June-July and then declines, more or less with the same magnitude.

On the basis of agricultural (suitability) practices there are three different seasons observed in India and their duration is as follows:

- ☞ Kharif Season : July to October
- ☞ Rabi Season : October to March
- ☞ Zaid Season : March to July

It is observed in general that during Zaid season the temperature increases to reach the maximum while during Kharif season it shows stagnation after a little decline then in Rabi season it continuously decline to reach the minimum. As a result the curve looks like swimmer in diving position.

Variation of Temperature in Kharif Season

Kharif, is the main growing season in India. It is the monsoon or rainy season on which most of Indian farmers depend to fulfil their irrigation needs for the success of agricultural practices. It is evident from the data that at a large number of stations experiences monthly maximum temperature ranges between 18°C to 28°C namely, Joshimath, Quazigund, Dharamshala, Manali, Mukteshwar, Mukhim, Mussourie, Nainital, Dalhousie and Shimla but the temperature also go beyond 28°C at some stations, viz., Srinagar, Jammu, Bhuntar, Tehri, Dehradun and Mandi, while at Leh and Gulmarg maximum temperature could not reach even 18°C at any time during this season. (Table 2.2) and (Fig. 2.7)

Table 2.2 : Variations of Temperature and Rainfall at some Meteorological Stations of North-western Himalaya

Srinagar (1951-80) Height 1587 M				
Months	*Min.*	*Max.*	*Avg.*	*Rainfall*
J	(2.3)	4.7	1.2	56.5
F	(0.6)	7.8	3.6	64.9
M	3.8	13.6	8.7	98.5
A	7.7	19.4	13.6	87.5
M	10.7	23.8	17.3	71.9
J	14.7	29.2	22.0	37.2
J	18.2	30.0	24.1	48.7
A	17.5	29.7	23.6	69.7
S	12.9	27.8	20.4	33.3
O	6.1	21.9	14.0	36.4
N	0.9	14.7	7.8	27.0
D	(1.6)	8.2	3.3	43.3
Quazigund (1962-80) Height 1739 M				
Months	*Min.*	*Max.*	*Avg.*	*Rainfall*
J	(4.0)	4.4	0.2	146.4
F	(1.8)	7.5	2.9	180.4
M	2.8	13.6	8.2	202.2
A	7.1	19.0	13.1	142.3
M	10.0	22.8	16.4	113.0
J	14.3	27.7	21.0	64.4
J	17.2	28.0	22.6	95.8
A	16.4	27.5	22.0	74.3
S	11.3	25.8	18.6	61.9
O	6.0	21.3	13.7	41.3
N	1.1	14.9	8.0	49.8
D	(2.3)	7.6	2.7	96.1
Joshimath (1958-80) Height 1875 M				
Months	*Min.*	*Max.*	*Avg.*	*Rainfall*
J	2.0	11.7	6.9	77.8
F	3.3	13.0	8.2	107.6
M	6.4	17.9	12.2	113.0
A	10.9	22.5	16.7	59.1
M	14.0	25.4	19.7	52.9
J	16.6	26.4	21.5	111.5
J	16.9	24.5	20.7	267.1
A	16.8	23.8	20.3	232.7
S	14.8	23.2	19.0	124.3
O	10.4	21.1	15.8	40.4
N	6.3	17.0	11.7	14.9
D	3.9	13.8	8.9	24.3

Table 2.2 : (Contd...)

Nainital (1953-80) Height 1953 M				
Months	*Min.*	*Max.*	*Avg.*	*Rainfall*
J	1.7	10.7	6.2	81.1
F	3.5	12.3	7.9	60.0
M	7.5	16.3	11.9	54.2
A	11.9	20.8	16.4	32.2
M	14.6	23.5	19.1	76.9
J	16.4	23.5	20.0	331.8
J	16.5	21.6	19.1	759.3
A	16.0	21.0	18.5	587.1
S	14.1	20.7	17.4	348.9
O	9.7	18.7	14.2	150.7
N	5.7	15.4	10.6	6.9
D	3.1	12.9	8.0	26.9
Dalhousie (1951-80) Height 1959 M				
Months	*Min.*	*Max.*	*Avg.*	*Rainfall*
J	2.1	11.4	6.8	139.1
F	3.2	13.1	8.2	126.4
M	6.9	17.7	12.3	142.5
A	11.1	22.0	16.6	94.7
M	14.5	25.5	20.0	68.3
J	16.8	27.2	22.0	137.1
J	16.0	23.8	19.9	615.7
A	15.8	22.8	19.3	603.1
S	14.3	23.0	18.7	284.8
O	11.2	22.2	16.7	80.4
N	7.2	18.3	12.8	40.0
D	4.7	14.8	9.8	75.1
Mukhim (1954-79) Height 1981 M				
Months	*Min.*	*Max.*	*Avg.*	*Rainfall*
J	3.5	12.5	8.0	62.6
F	4.4	13.5	9.0	70.4
M	7.9	18.0	13.0	76.3
A	12.0	22.7	17.4	46.7
M	15.0	25.1	20.1	68.3
J	16.8	25.9	21.4	167.6
J	17.1	23.8	20.5	416.3
A	16.9	23.4	20.2	405.7
S	15.3	23.5	19.4	212.2
O	11.8	21.9	16.9	41.6
N	7.9	18.3	13.1	11.8
D	5.3	14.9	10.1	30.0

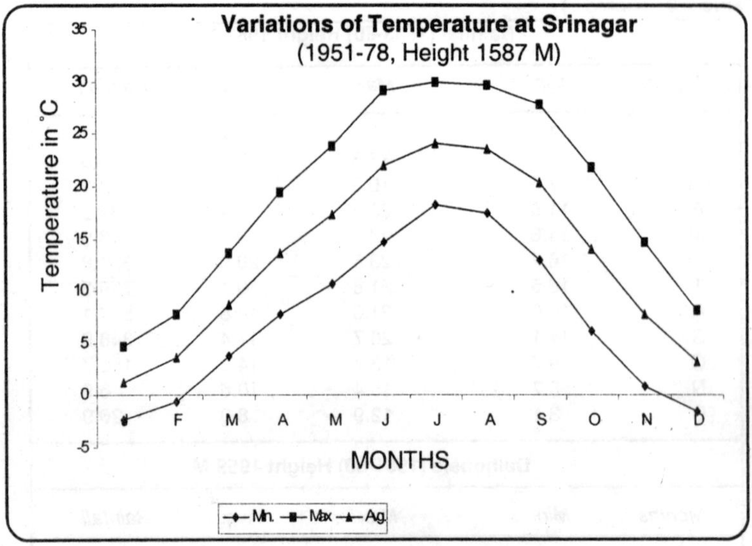

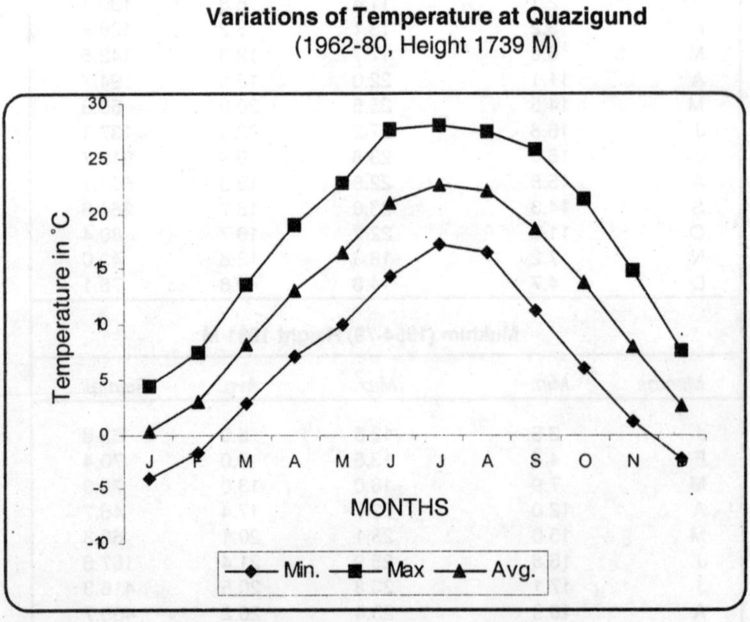

Fig. 2.7 : Variation of Temperature at soem Meteorological Stations

Fig. 2.7 : *(Contd.)*

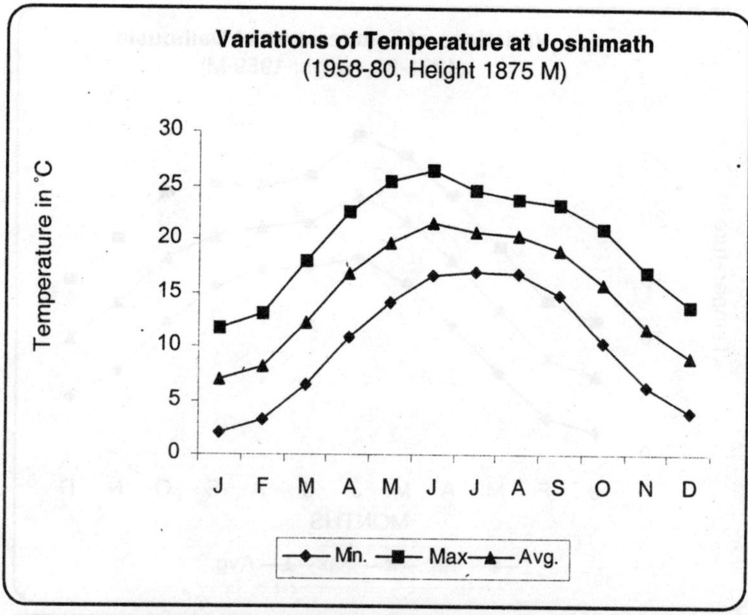
Variations of Temperature at Joshimath
(1958-80, Height 1875 M)
Temperature in °C
30
25
20
15
10
5
0
J F M A M J J A S O N D
Min. Max Avg.

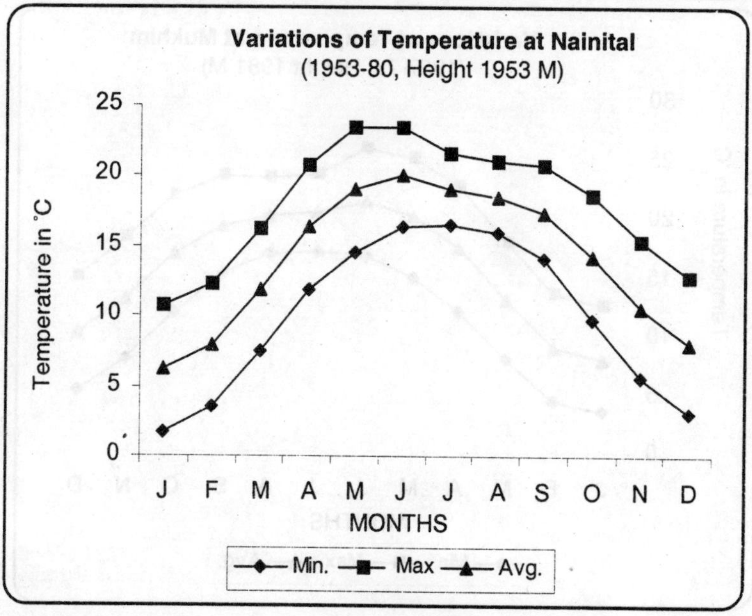
Variations of Temperature at Nainital
(1953-80, Height 1953 M)
Temperature in °C
25
20
15
10
5
0
J F M A M J J A S O N D
MONTHS
Min. Max Avg.

Fig. 2.7 : *(Contd.)*

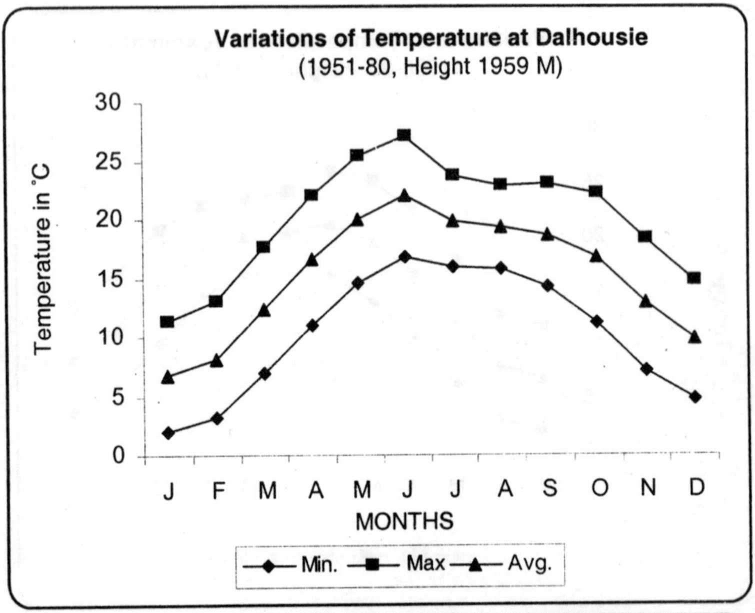
Variations of Temperature at Dalhousie
(1951-80, Height 1959 M)
Temperature in °C
30
25
20
15
10
5
0
J F M A M J J A S O N D
MONTHS
Min.
Max
Avg.

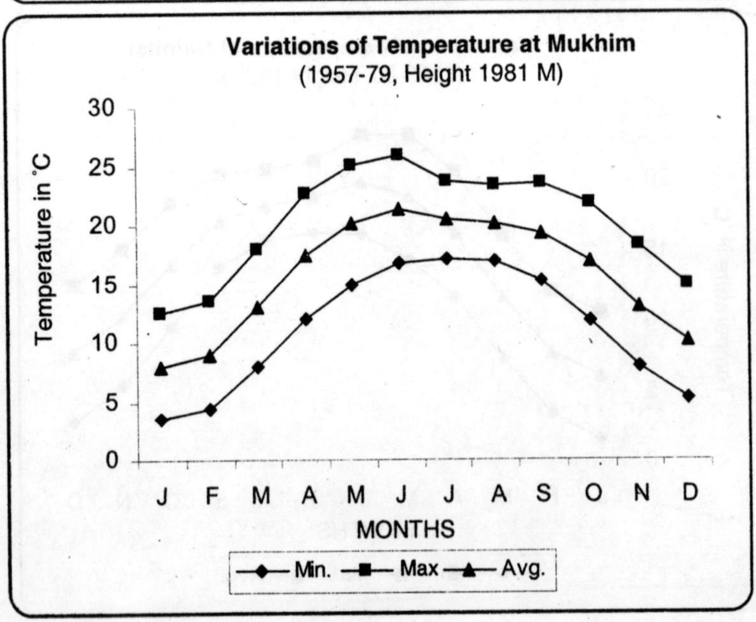
Variations of Temperature at Mukhim
(1957-79, Height 1981 M)
Temperature in °C
30
25
20
15
10
5
0
J F M A M J J A S O N D
MONTHS
Min.
Max
Avg.

In case of minimum temperature during this season it is observed that on seven stations it is found between 10°C – 20°C viz., Shimla, Dalhousie, Bhuntar, Nainital, Mussourie, Mukhim and Joshimath while on sometimes it goes more than 20°C at Jammu, Dharamshala, Tehri, Dehradun and Mandi. It also go below 10°C at some stations like Srinagar, Leh, Gulmarg, Manali, Quazigund and Mukteshwar.

Significance

It is envisaged from maximum temperatures during *Kharif* season on different stations in north-western Himalayan region that all the stations experiencing favourable temperature for agricultural practices and nowhere maximum temperature found close to limiting for any crop which affects the agriculture adversely and ceases its growth and development. But in case of minimum temperature some times it reaches close to limiting when it start to come down below 10°C. In such conditions the duration of very low temperature is very important to cease the crop growth. It is also observed that the maximum temperatures are not so limiting as the minimum temperature which can affect the crop at any stage of life cycle.

Variation of Temperature in Rabi Season

Rabi, is the second most important season of Indian agriculture. The general pattern of temperature is declining to its minimum and then increasing at its end. It starts after the Kharif season from October to March and most favourable for horticultural crops. Horticultural production which constitute a highly specialized branch of agriculture and the intensive market-oriented nature of industry is well expressed both by the British term **'Market Gardening'** and North American equivalent **'Truck Farming'**. With horticulture there is a particular incentive to make the optimum use of the climatic environment in order to produce the crops of high yield and quality. In areas which have a short growing season like north-western Himalaya, great émphasis is also attached to early cropping in order to avoid seasonal glut when supply may be in excess of demand. But the horticulture is often concentrated where early growth is associated with a milder winter climate.

Table 2.3 : Variations of Temperature and Rainfall at some Meteorological Stations of North-Western Hiamalaya

Manali (1968-80) Height 2039 M				
Months	*Min.*	*Max.*	*Avg.*	*Rainfall*
J	-1.8	10.1	4.2	145.0
F	-1.0	11.1	5.1	145.6
M	2.8	15.9	9.4	187.3
A	6.1	21.6	13.9	111.3
M	8.6	24.9	16.8	69.1
J	12.4	26.6	19.5	94.5
J	14.8	25.5	20.2	235.0
A	14.6	25.0	19.8	243.6
S	10.4	24.7	17.6	108.4
O	5.4	22.5	14.0	33.1
N	1.3	18.4	9.9	39.8
D	-0.3	14.0	6.9	46.5
Mussoorie (1951-80) Height 2042 M				
Months	*Min.*	*Max.*	*Avg.*	*Rainfall*
J	2.5	10.6	6.6	51.8
F	3.8	12.4	8.1	52.8
M	7.5	16.5	12.0	57.7
A	12.0	21.2	16.6	30.0
M	14.7	24.1	19.4	58.4
J	16.2	23.7	20.0	174.9
J	15.5	20.8	18.2	662.0
A	15.2	20.2	17.7	670.6
S	14.0	19.9	17.0	227.5
O	11.2	18.9	15.1	64.7
N	7.3	15.9	11.6	14.8
D	4.6	13.1	8.9	18.2

Table 2.3 :(Contd...)

Shimla (1951-80) Height 2202 M				
Months	*Min.*	*Max.*	*Avg.*	*Rainfall*
J	1.7	8.9	5.3	54.6
F	3.0	10.6	6.8	47.2
M	6.8	14.8	10.8	59.4
A	11.1	19.4	15.3	41.1
M	14.2	22.9	18.6	56.4
J	15.6	24.1	19.9	175.6
J	15.0	21.0	18.0	376.5
A	14.8	20.2	17.5	335.1
S	13.4	20.1	16.8	190.2
O	10.7	18.7	14.7	46.2
N	7.0	15.1	11.1	13.8
D	4.3	12.0	8.2	16.0
Mukteshwar (1951-80) Height 2311 M				
Months	*Min.*	*Max.*	*Avg.*	*Rainfall*
J	1.5	10.7	6.1	54.5
F	2.8	12.5	7.7	46.4
M	6.2	16.7	11.5	55.7
A	10.4	21.3	15.9	27.6
M	12.9	23.9	18.4	58.0
J	14.3	23.5	18.9	131.6
J	14.2	20.9	17.6	310.6
A	14.0	20.4	17.2	299.5
S	12.5	20.3	16.4	209.7
O	9.5	19.3	14.4	78.7
N	6.0	16.4	11.2	6.3
D	3.5	13.5	8.5	18.6
Gulmarg (1951-80) Height 2655 M				
Months	*Min.*	*Max.*	*Avg.*	*Rainfall*
J	-8.5	0.6	-4.0	161.2
F	-7.5	1.1	-3.2	274.6
M	-2.9	5.7	1.4	231.8
A	1.6	11.2	6.4	166.8
M	4.5	15.2	9.9	138.8
J	7.7	20.2	14.0	76.3
J	10.3	21.2	15.8	100.3
A	10.1	20.2	15.2	95.3
S	6.3	18.4	12.4	54.1
O	2.7	14.5	8.6	68.1
N	-2.0	9.1	3.6	19.2
D	-5.5	4.2	-0.7	67.3

Table 2.3 :(Contd...)

Leh (1951-80) Height 3514 M				
Months	*Min.*	*Max.*	*Avg.*	*Rainfall*
J	-14.4	-2.0	-8.2	9.5
F	-11.0	1.5	-4.8	8.1
M	-5.9	6.5	0.3	11.0
A	-1.1	12.3	5.6	9.1
M	3.2	16.2	9.7	9.0
J	7.4	21.8	14.6	3.5
J	10.5	25.0	17.8	15.2
A	10.0	25.3	17.7	15.4
S	5.8	21.7	13.8	9.0
O	-1.0	14.6	6.8	7.5
N	-6.7	7.9	0.6	3.6
D	-11.8	2.3	-4.8	4.6

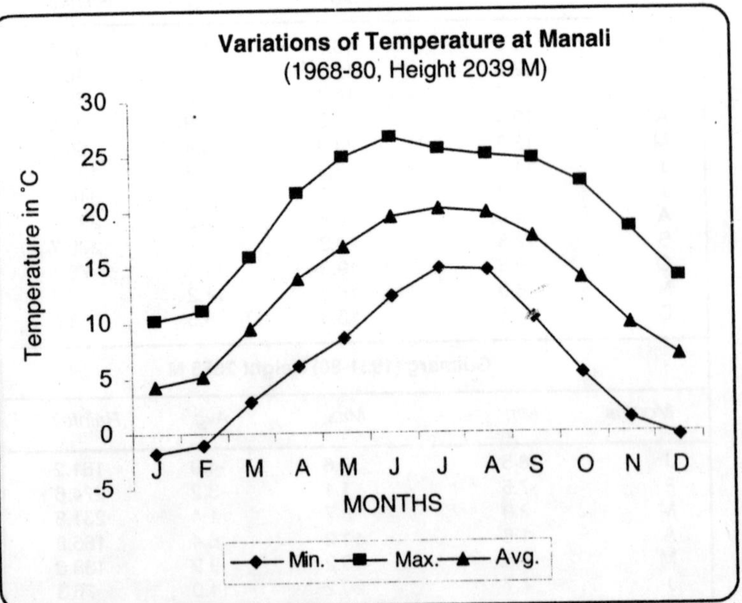

(*Contd...*)

Fig. 2.8 : Variation of Temperature at some Meteorological Stations

Fig. 2.8 : (*Contd...*)

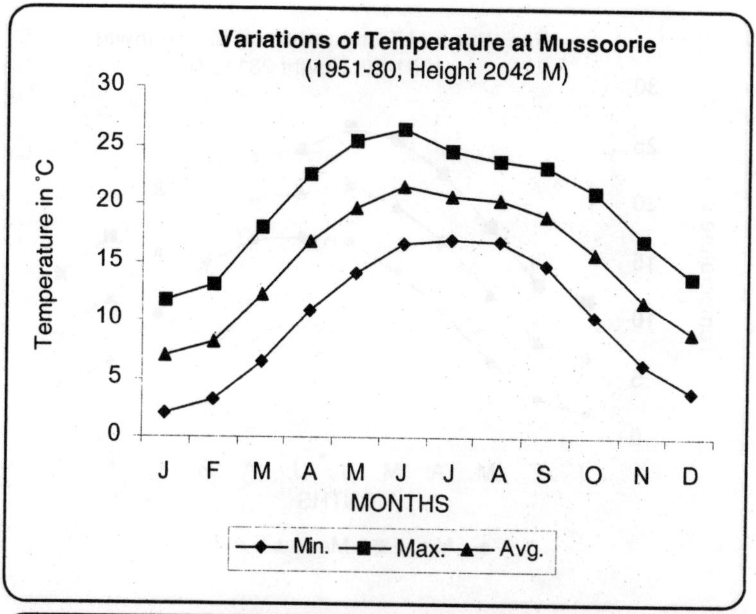
Variations of Temperature at Mussoorie
(1951-80, Height 2042 M)
Temperature in °C
30
25
20
15
10
5
0
J F M A M J J A S O N D
MONTHS
Min.
Max.
Avg.

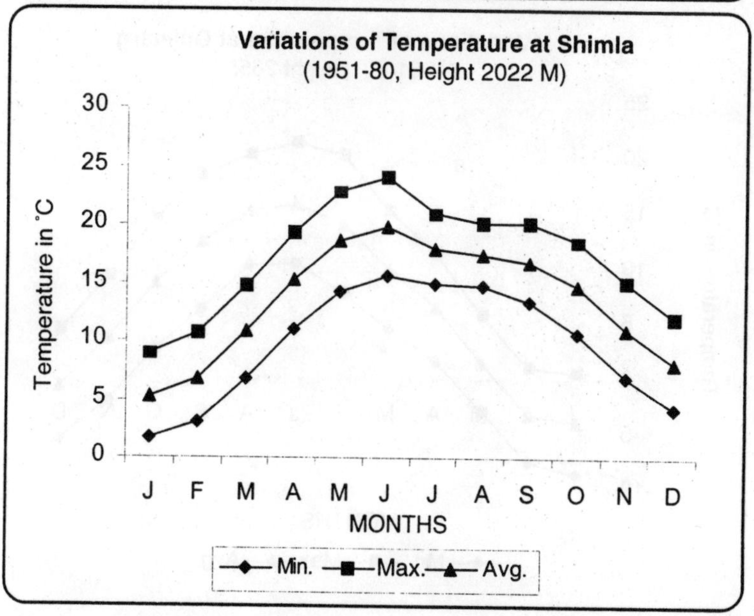
Variations of Temperature at Shimla
(1951-80, Height 2022 M)
Temperature in °C
30
25
20
15
10
5
0
J F M A M J J A S O N D
MONTHS
Min.
Max.
Avg.

Fig. 2.8 : *(Contd..)*

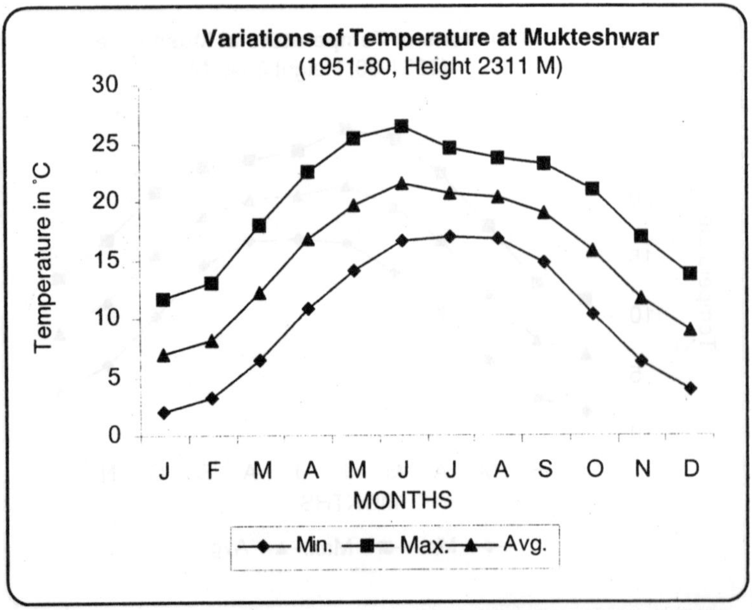
Variations of Temperature at Mukteshwar
(1951-80, Height 2311 M)
Temperature in °C
30
25
20
15
10
5
0
J F M A M J J A S O N D
MONTHS
Min. Max. Avg.

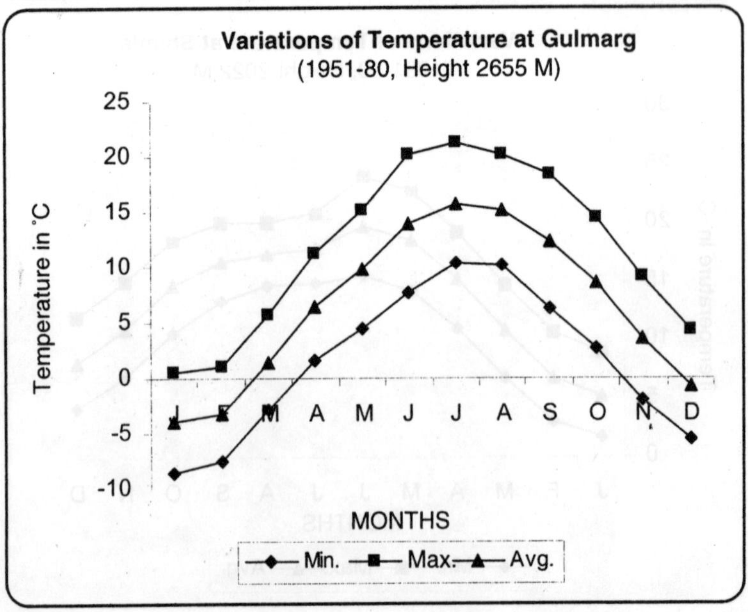
Variations of Temperature at Gulmarg
(1951-80, Height 2655 M)
Temperature in °C
25
20
15
10
5
0
-5
-10
J F M A M J J A S O N D
MONTHS
Min. Max. Avg.

Fig. 2.8 : *(Contd....)*

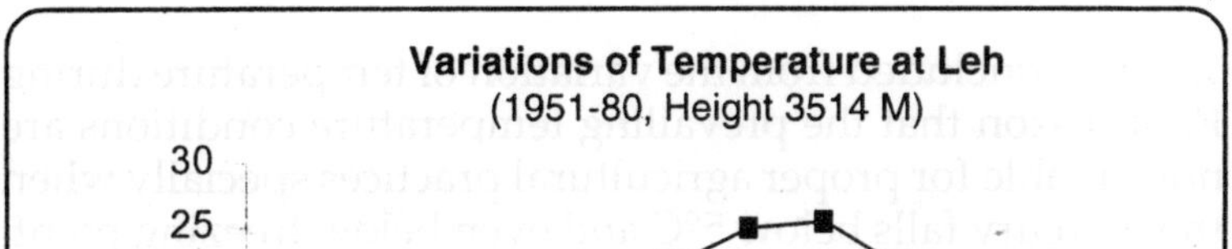

To analyse the temperature variation and distribution in the region during Rabi season, it is observed that maximum temperature is found between 5°C to 25°C at most of the stations i.e., Shimla, Srinagar, Dalhousie, Dharamshala, Nainital, Mussourie, Mukhim, Manali, Joshimath and Mukteshwar but the station at lower elevation sometimes recorded more than 25°C of temperature namely Jammu, Bhuntar, Tehri, Dehradun and Mandi, and stations of comparatively higher elevation sometimes record maximum temperature below 5°C viz. Leh, Gulmarg and Quazigund.

In case of minimum temperature, it is found that highest limit of minimum remains below 15°C with the only exception of Jammu where it is found between 0°C to 15°C namely, Shimla, Dalhousie, Dehradun, Dharamshala, Bhuntar, Tehri, Nainital, Mukteshwar, Mukhim, Manali, Mandi, and Joshimath. But there are also some stations where sometimes the temperature goes below the freezing point i.e., Srinagar, Leh, Gulmarg, Manali and Quazigund.

Significance

It is concluded from the variation of temperature during Rabi season that the prevailing temperature conditions are not suitable for proper agricultural practices specially when the mercury falls below 5°C and even below freezing point, it is limiting for most of the growth process of agricultural development. Horticultural crops, dairy farming and some short growing season crops are the alternate suggestions in such conditions.

Variation of Temperature in Zaid Season

Temperature characteristics during Zaid season (March to July) is generally increasing untill the monsoon rains check it. This season experiences the highest temperatures with low pressures. We can say that after Kharif crop's harvest, it is the extra season of Indian agriculture which is not very important because only a few hardy crop can grow in this water deficit season of highest temperature. A number of stations on high altitude experiences maximum temperature of more than 30°C viz. Jammu, Dehradun, Tehri, Dharamshala, Bhuntar and Mandi. On the other hand, 15°C to 30°C of temperature generally experiences Mukteshwar, Mussoorie, Nainital, Joshimath, Dalhousie, Mukhim, Shimla and Manali, while in Quazigund, Leh, Gulmarg and Srinagar, maximum temperature sometimes records below 15°C.

In case of minimum temperature during Zaid season, it is analyzed that the minimum temperature goes more than 20°C at Jammu, Dehradun, Tehri, Dharamshala and Mandi, while in Joshimath, Dalhousie, Mukhim, Mukteshwar, Mussoorie, Nainital, Bhuntar, and Shimla, it varies between 5°C to 20°C and at some stations it goes even below 5°C viz., Quazigund, Leh, Gulmarg, Srinagar and Manali.

Significance

It is concluded from the spatial distribution of

temperature during Zaid season that the temperature conditions are favourable for the proper growth of agricultural practices but some other factors such as water deficit situation, water thirsty environment and seasonal adaptability of crops make it difficult to grow a full-fledged agricultural season, which results in poor agricultural productivity and agricultural scenario during this Zaid or Extra-season of Indian agriculture.

NATURAL RESOURCES OF THE REGION

Soil Types

Earthen material that covers land surfaces and is formed by the action of natural physical, chemical and biotic forces on the unconsolidated residue (regolith) of rocks and minerals on the earth's surface is called as soil. Among its many important functions, soil serves as a substratum of plant and, thus further up the food chain of animal life.

The most fundamental process of soil formation and differentiation is weathering, which is the physical disintegration and chemical decomposition of rocks, minerals and immature soils. Rate of weathering vary, depending upon the material.

The process of soil formation create a soil profile that is composed of one of four biologically and chemical distinct master horizons (horizontal layers) conventionally designated from the surface down as 'A', 'B', 'C' and the unconsolidated rock 'R' of variable depths. 'A' horizons are the most weathered and leached and generally contain the most organic matter and plant nutrient; 'B' horizons are moderately weathered and accumulate the products of removed from 'A' horizons; 'C' horizons contain parent material from which the above horizons are formed; and 'R' layers, not part of the soil proper, exhibit properties much different from the overlying 'C' horizons.

Snow and Snowfield

The perpetual snowfields are not actually soils but barren areas of the Himalayas. The present glaciers zone in the Himalayas lies approximately between 3350 m and 7000 m above mean sea level and found in districts of Pithoragarh, Chamoli, Uttarkashi, Kinnaur, Lahaul and Spiti and a number of peaks in the Ladakh region. The snow keeps the Himalayan rivers perennial from the ecological point of view and such areas need special care and preservation.

Skeletal Soils

These shallow skeletal soils have their genesis over sloping land of high plateaux and small valleys. They are formed from variety of parent materials. These lithic and light textured soils are rich in humus and have mobile form of sesquioxides. The other peculiarity of these soils is an altered 'B' horizon. These soils are largely found in Ladakh and Kargil region with few patches in Kinnaur, Lahaul and Spiti, Uttarkashi, Chamoli and Pithoragarh. These soil are suitable for wildlife, forestry and grazing.

Mountain Meadow Soils

The mountain meadow soils are developed in depression pockets in the high altitude valleys. The presence of mollic epipedon with a variety of horizons, semi-decomposed organic debris with dark grey colour in upper horizon, structureless second horizon with slight blueish decomposed rocks, sometime brown in colour and coarse sandy to loamy texture are the general characteristic of these soils. These soils have medium nitrogen, low phosphorus and medium potassium contents. These soils are found in a continuous stretch from north-western limit of Kashmir to Pithoragarh amidst the snowfields and submontane soils. Also found in

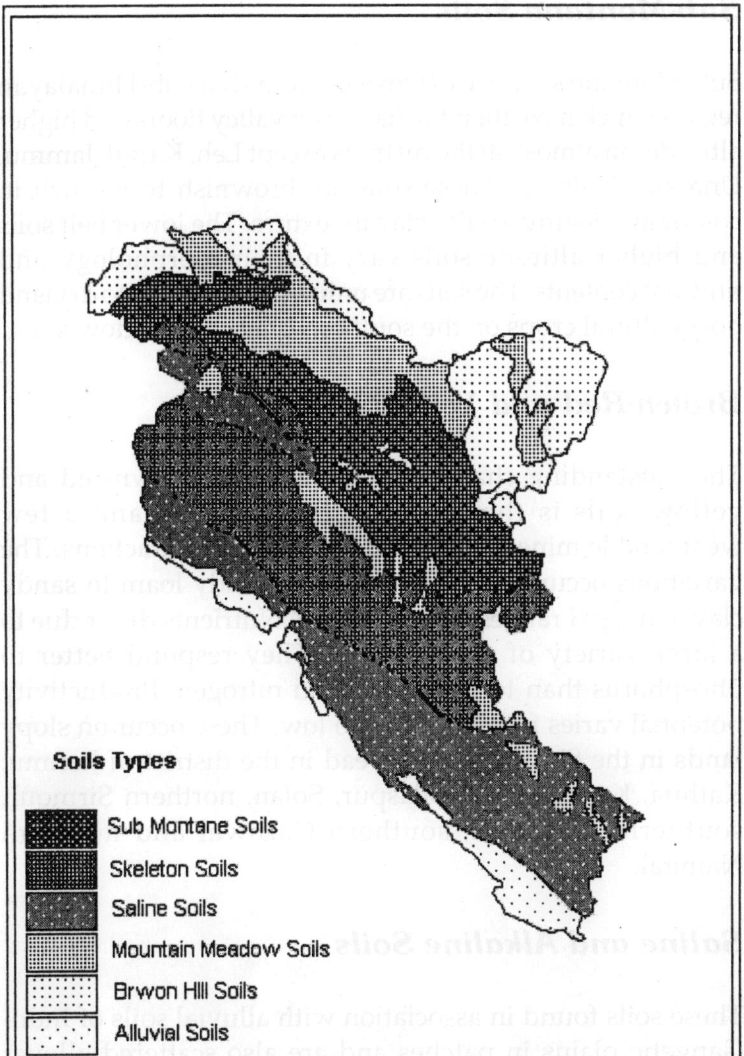

Fig. 2.9 : North-western Himalaya : Soil Types

patches in Nainital, Anantnag, Budgam, Leh, etc. These are best grazing grounds and climate as well as science beauty provide a wide scope for tourism development.

Sub-Montane Soils

Sub-Montane soils are extensively found in sub-Himalayan region which have their formation on valley floors and higher altitudes in almost all the districts except Leh, Kargil, Jammu, Una and Nainital. These soils are brownish to reddish in colour and loamy to silty clay in texture. The lower belt soils and higher altitude soils vary in their morphology and nutrient contents. The soils are unirrigated and put to dryland horticultural crops on the southern limits of meadow soils.

Brown-Red and Yellow Soils

The outstanding characteristic feature of brown-red and yellow soils is the thick agrillic horizons and a few weatherable minerals in the 20 to 200 micron fractions. The variations occur in texture from fine sandy loam to sandy clay loam, pH ranges from 5.5 to 8.2. Nutrients differ due to a large variety of parent rocks. They respond better to phosphorus than to potassium and nitrogen. Productivity potential varies from medium to low. These occur on slopy lands in the Siwaliks and spread in the district of Jammu, Kathua, Kangra, Una, Bilaspur, Solan, northern Sirmour, southern Dehradun, southern Garhwal and northern Nainital.

Saline and Alkaline Soils

These soils found in association with alluvial soils of Indo-Gangetic plains in patches and are also scattered allover India. These soils have hard and impervious salt-pans consisting of various salts which restrict the downward and upward movements of water within the root-zone. It is but natural that no crop or vegetation can be grown on such pans, and they remain barren. It has happened due to other reasons, however the basic reason is waterlogging.

LIVESTOCK RESOURCES AND AGRO-ECOLOGY

People in hills have agriculture as their main occupation. Livestocks play a vital role in agricultural development as it improve soil fertility through the Farm Yard Manures and draught power for cultivation. The livestock population in north-western Himalaya increased from 10.58 million in 1961 to 18.12 million in 1992. Grazing by livestock put pressure on grasslands and forests. High growth rate of livestock population causing land degradation. Overgrazing impedes regeneration, retard growth of vegetation and leads to extinction of good palatable grasses which are replaced by less palatable and inferior grasses. Extensive areas have been invaded by bushes which are not browsed (eat as fodder), excessive trampling makes the soil compact and impervious and prevent circulation of air, water thus exposing the soil to erosion by wind and water.

Sheep and goat constitute the important species of livestock in India contributing greatly to food, fibre, rural employment and gross domestic product. These species are of economic importance in arid, semi-arid and hilly areas where crop or dairy farming is not economical and there is a large percentage of small and marginal farmers and landless labourer (Charya, 1984).

Around 73 per cent of the rural household owns one or more kinds of livestock which contribute 15-20 per cent of household income. Livestock provide economic security and social status to the family which is still highly regarded by farming communities in most part of the country (Taneja, 1998).

Overgrazing by livestock on private, communal and state owned farms is perhaps the single leading cause of removal of vegetative cover. As the human beings and livestock population increase so will the pressure to overexploit already marginal grazing lands (I.P. Abrol, 1994).

In the Himalayan region, most livestock species are being used raised under mixed cropping system. Cattle are the most common (38.74%), followed by sheep (22.61%), goats (21.61%) and buffaloes (8.80%) and rest others (Livestock Census,1992). Under the mixed crop farming systems in the north-western Himalayan region, livestock pressure is increasing. (Table 2.4.) For example in the low hills, the average number of livestock per household were 11 animals, cattle accounting for 53 per cent and sheep and goat for 36 per cent. In the mid hills, the numbers were eight animals in mixed farming systems and nine animals in vegetables based farming system (Mahendra Dev, 1994).

Temporal changes in the livestock population and composition between 1961 and 1992 shows that the proportion of cattle population is declining. In 1961, the ratio of cattle population in total livestock was 48.31per cent, it slightly increased in 1972 and its share became 49.28 per cent, but in 1982 and 1992, it continuously declined to 42.28 per cent and 38.74 per cent of total livestock population.

Buffaloes also following the same pattern of fluctuation in 1961, its share was 11.29 per cent, which increased to 13.32 per cent in 1972, but in 1982 and 1992, continuously declined to 13.05 per cent and 8.80 per cent of total livestock population.

Sheep are harmful for the ecology of any region because they graze the grasses not only from above the ground but from the depth with-roots, so there is no chance of regeneration once grazed completely and with minor fluctuation its share remains about 20 per cent of total livestock population. It is evident from the analysis that the goats and sheep constitute more than 40 per cent of livestock but there growth pattern is just reverse of cattle and buffaloes temporally.

Growth Pattern of Livestock

Increasing pressure of livestock is a matter of great concern from the agro-ecological viewpoint of the region and it is noted that the pasture lands are not increasing at par to feed the growing livestock. As a result, overgrazing, become unavoidable causing land degradation, retard growth of vegetation, areas invaded by bushes which are not browsed, soil become compact and impervious, and prevents circulation of air and water thus exposing the soil to erosion by wind and water.

Analysis of livestock growth is done in two ways :

- Growth in terms of livestock population.
- Growth in terms of percentage share.

Cattle, buffaloes, sheep and goats are taken into account to analyse the growth pattern of livestock. It is evident from the Table 2.5, that the growth of cattle population was very high during 96 to 972, i.e., 24 per cent, while during 1972-82 and 1982-92, it is only 1.10 per cent and 10.7 per cent respectively. But the growth of cattles was not at par with livestock population. On the other hand, the percentage share of cattle population to total livestock is somewhat different, as the growth in percentage share of cattle during 1961-72 was just 2.0 per cent while it is –12.8 per cent and –19.8 per cent during 1972-82 and 1982-92 respectively. The total growth in the cattle population from 1961 to 1992 is 38.9 per cent while at the same time growth in the percentage share cattle to the total livestock is –19.8 per cent. It is clear from the above analysis that the growth in number is not necessarily the same with percentage share.

Table 2.4 : Livestock Population in north-western Himalaya

	1961	*1972*	*1982*	*1992*
Cattles	51,100,28	63,411,82	64,108,52	70,983,51
Buffaloes	11,941,82	17,138,64	19,471,44	16,122,05
Sheeps	22,088,10	24,604,91	34,065,88	41,431,02
Goats	19,231,87	22,124,65	29,199,03	39,594,71

Table 2.5 : Growth in Livestock Population in north-western Himalaya (Perentage)

	1961-1972	*1972-1982*	*1982-92*	*1961-1992*
Cattles	24.1	1.1	10.7	38.9
Buffaloes	43.5	13.6	-17.2	35.0
Sheeps	11.4	38.5	21.6	87.6
Goats	15.0	32.0	35.6	105.9

Table 2.6 : Percentage Share of livestock in north-western Himalaya

	1961	*1972*	*1982*	*1992*
Cattles	48.31	49.28	42.96	38.74
Buffaloes	11.29	13.32	13.05	8.80
Sheeps	20.88	19.12	22.83	22.61
Goats	18.18	17.19	19.57	21.61

Table 2.7 : Growth in Percentage Share of Livestock in north-western Himalaya

	1961	*1972*	*1982*	*1992*
Cattles	2.0	-12.8	-9.8	-19.8
Buffaloes	18.0	-2.0	-32.6	-22.1
Sheeps	-8.4	19.4	-1.0	8.3
Goats	-5.4	13.8	10.4	18.9

Growth analysis in case of buffaloes shows that there is tremendous increase of 43.5 per cent in the number of buffaloes during 1961-72, while during 1972-82, it is only 13.6 per cent, but the case is worst during 1982-92 where growth shows severe decline in buffaloes population and the growth rate is found –17.2 per cent. If we analyse the growth in percentage share of buffaloes in the total livestock, it is found to be 18.0 per cent during 1961-72, –2.0 per cent during 1972-82 and further declines to –32.6 per cent during 1982-92. Overall growth in livestock population during 1982-92 is estimated to be 35.0 per cent, while growth in percentage share of buffaloes to total livestock population is –22.1 per cent which shows that other livestock grow on a faster rate than the buffaloes in this region.

Changing Land-use Pattern and its Ecological Limitations

The term 'Land-use' is employed for varied surfacial and dynamic uses of the land and soil surveys e.g. land under cultivation, pasture, barren, orchard, fellow, waste, culturable waste, settlement, forests, water bodies, etc. on varied terrain conditions and soil types. In other words, land-use is the surface utilization of all developed and vacant land on a specific point, at a given time and space.

To know difference between land-use and land utilization is important. Land-use is the use actually made of any parcel of land, house, apartment and industrial location are land-use categories, whereas the term residential, industrial and agricultural refers to a system of land utilization implying roads, neighbourhood, retail and service activities as well as location of industries and the carrying of agricultural pursuits.

Of all the environmental resources land is the most valuable for mankind as it is the basic means of production. It is the most fundamental geographical element in the development and growth of other bio-physical resources of the earth. All human needs are fulfilled directly by the land. The land is a very comprehensive term, includes the physical, biological and geological system of the earth. Land utilization statistics provide the area figure, showing distribution of total geographical area of the country in its various uses.

The existing land-use pattern of the region has been controlled and regulated by a set of physical and human factors operating in the region. Surface configuration and climatic complexes have largely conditioned the land utilization system of the region on the one hand whereas on the other, the process of land transformation has to a great extent, been determined and directed by other factors such as increasing pressure of population, transfer of technology to rural areas and almost complete absence of other means of livelihood and on the other hand agricuture which is extremely inefficient in terms of feeding the regional population.

For the convenience of land-use analysis and its change, the reporting area of the region is classified in four categories:

(a) Forests
(b) Uncultivable lands
(c) Cultivable lands
(d) Cultivated lands

Reporting area is the area for which the information on landuse by classification of area is available. Generally land utilization figures are based on the returns prepared by the village staff for revenue purposes. Geographical area is the total area under the political or administrative limits including both surveyed and unsurveyed area, and the 'reporting area is only that area whose survey has been done out of total geographical area.'

Forests

Area under forests include area actually forested or lands classed or administered as forests whether state-owned or private and whether wooden or maintained as potential forest land. The area of crops raised in the forests and grazing lands open for grazing within the forests are also included under the forest area.

Changing Pattern of Forest Cover during 1961-63 to 1992-95

It seems that the forest cover is increasing during the last three decades. The area under forest cover during 1961-63 was 3678433 hectares or 41.95 per cent of the total reporting area which increased to 4645065 hectares (42.24%), 4952172 hectares (43.34%), and 5135858 hectares (44.67%) during 1971-74,

1982-85 and 1992-95 respectively. But the variation in forest cover is found at macro, meso and micro level analysis. During 1961-63, it is evident from the table below that out of 41.95 per cent forest cover in north-western Himalaya, highest forest cover found in U.P. Himalayan region i.e. 46.63 per cent followed by Jammu region (37.41 per cent), Himachal region (33.35 per cent), while in Kashmir region it is almost negligible i.e., 0.73 per cent of total reporting area (Table 2.8). At districts level only Dehradun have forest cover necessary for the standard requirement of ecological balance in mountainous area and rest of the districts found below the standard requirement.

Table 2.8 : Changing Pattern of Forest Cover (Region-wise) *(Percentage)*

Regions	*1961-63*	*1971-74*	*1982-85*	*1992-95*
U.P. Himalayan region	46.63	61.84	64.16	60.67
Himachal region	33.35	21.16	26.74	30.68
Jammu region	37.41	36.82	36.78	36.15
Kashmir region	0.73	0.86	0.95	1.06
Total NW-Himalaya	41.95	42.24	45.34	44.67

During 1971-74, the forest cover of the region increased to become 42.24 per cent in the region as a whole, but at regional level it is evident from the table above that the forest cover in the U.P. Himalaya was 61.84 per cent, means it fulfil the ecological requirement, a little decline in Jammu region which become 36.82 per cent, while severe decline in Himachal region where it become only 21.16 per cent from 33.35 per cent of 1961-63. Kashmir region have negligible forest cover i.e., 0.86 per cent, but it increased from the previous 0.73 per cent. At district level Uttarkashi shown highest forest cover with 72.30 per cent followed by Garhwal (65.80%), Dehradun (63.40%), Tehri Garhwal (62.63%) and Nainital (60.85%) are the districts, having forest cover necessary for the balance of ecological requirement. On the other extreme Anantnag (0.53%), Baramula (0.79%), and

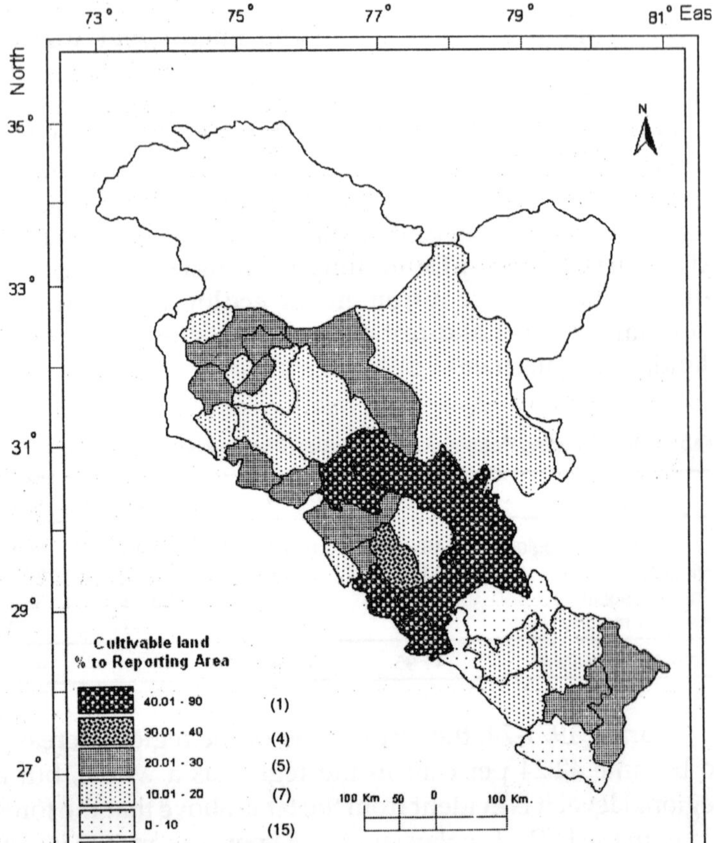

Fig. 2.10 : North-western Hiamalaya Forest Cover (1982-85)

Srinagar (1.55%) in Kashmir region and Kinnaur (0.10%), Kullu (0.00%) and Lahaul and Spiti (0.00%) of Himachal region have none or negligible forest cover which is a havoc for the agro-ecology of the region.

During 1982-85, forest cover of the region become 45.34 per cent from 42.24 per cent (1971-74). But at regional level the disparities exists e.g., U.P. Himalaya with 64.16 per cent forest cover is on the top followed by Jammu region (36.78%),

Himachal region (26.74%) and Kashmir region (0.95%) forest. It is observed that during this period the forest cover in all the regions of N-W-Himalaya is increased from the various percentages. At district level, it is found that only five districts of U.P. Himalaya i.e., Uttarkashi (88.76%), Tehri Garhwal (69.40%), Dehradun (69.38%), Garhwal (62.81%) and Chamoli (60.12%) have forest cover more than the ecological requirement i.e., 60 per cent, while in rest of the districts i.e., Nainital, Almora and Pithoragarh the forest cover ranges between 50-60 per cent of the reporting area. On the other side, Kullu in Himachal Pradesh shows complete lack of forest cover, while Anantanag (0.40%), Pulwama (0.70%), Budgam (0.89%), Baramula (0.96%), Kupwara (0.58%) and Srinagar (3.31%) of Kashmir region have negligible forest cover. Doda district of Jammu region is the only district where forest cover found more than 50 per cent.

In 1992-95, there observed a slight decline in the forest cover of N-W-Himalaya and the forest cover decreased to 44.67 per cent against the 45.34 per cent (of 1982-85) with 1.47 per cent rate of decline. At regional level U.P. Himalaya found to be declined from 64.16 per cent to 60.67 per cent with 5.44 per cent rate of decline, and in Jammu region it become 36.15 per cent with a declining rate of -1.71 per cent. In Himachal region forest cover with 14.73 per cent growth rate become 30.68 per cent, while in Kashmir region forest cover with 14.73 per cent growth rate become 30.68 per cent, while in Kashmir region forest cover just come out of fraction to become 1.06 per cent with a growth of 11.58 per cent during the meantime. At district level, Uttarkashi with 88.84 per cent forest cover remain on the top, followed by Dehradun (69.14%), Tehri Garhwal (69.12%), and Chamoli (61.90%), while in Nainital and Pithoragarh forest cover found between 50-60 per cent. Almora showing severe decline of 38.72 per cent, as a result the forest cover shrink to 38.71 per cent from 53.70 per cent (1982-85). Kullu with complete lack of forest now have a fraction of 0.15 per cent forest cover. On the other hand, Anantnag (0.42 per cent), Pulwama (0.84%), Budgam

(1.09%), Baramula (0.82%), Kupwara (1.15%) and Srinagar (3.53%) show a very little forest cover which is almost nil in terms of ecological need.

As a whole, if we analyse the forest cover in the ecological terms it is envisaged that the forest cover in NW-Himalaya is less than the ecological requirement for the mountainous region, but in regional level only U.P. Himalaya has more than 60 per cent forest cover since 1971-74. The forest cover in Jammu region shows meagre fluctuations and remained

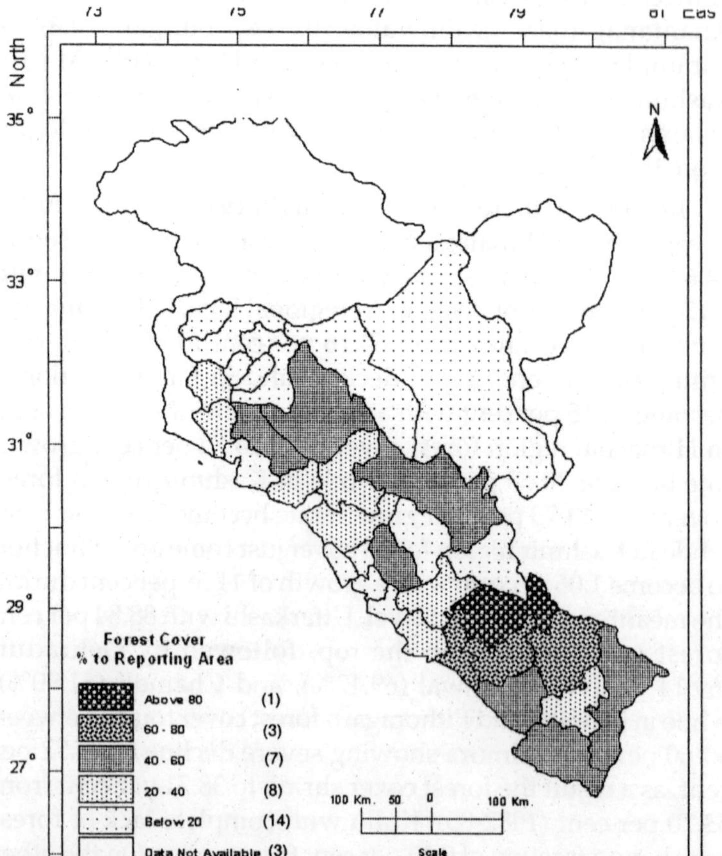

Fig. 2.11 : North-western Himalaya : Forest Cover (1992-95)

almost 36 per cent from 1961-63 to 1992-95. While in Himachal region, it increased continuously from 21.16 per cent in 1971-74 to 30.68 per cent in 1992-95. The Kashmir Valley is the poorest in forest cover but show a continuous increase from 0.73 per cent to 0.86 per cent, 0.95 per cent and 1.06 per cent during 1961-63, 1971-74, 1982-85 and 1992-95 respectively. But the variations are noted at district level (Table 2.9)

Table 2.9 : Changing Pattern of Forest Cover (in %) (District-wise)

Districts	*1961-63*	*1971-74*	*1982-85*	*1992-95*
Bilaspur	10.89	10.29	9.92	9.95
Chamba	47.07	12.12	37.45	39.22
Hamirpur	N.A.	17.57	14.62	18.06
Kangra	N.A.	36.96	37.97	38.82
Kinnaur	2.35	0.10	8.07	10.02
Kullu	N.A.	0.00	0.00	0.15
Lahaul and Spiti	N.A.	0.00	11.51	51.03
Mandi	39.95	37.27	38.52	43.37
Shimla	37.38	15.47	17.16	25.77
Sirmour	31.44	21.48	21.61	21.45
Solan	N.A.	10.84	10.87	11.19
Una	N.A.	20.58	18.58	18.58
Anantnag	0.53	0.53	0.40	0.42
Pulwama	N.A.	N.A.	0.70	0.84
Srinagar	1.24	1.55	3.31	3.53
Budgam	N.A.	N.A.	0.89	1.09
Baramula	0.65	0.79	0.96	0.82
Kupwara	N.A.	N.A.	0.58	1.15
Leh	0.11	0.11	0.00	0.00
Kargil	N.A.	N.A.	0.33	0.36
Doda	53.10	53.53	53.27	53.21
Udhampur	43.94	43.52	44.49	44.06
Jammu	14.91	14.94	14.23	12.73
Kathua	27.29	26.33	29.34	27.01
Rajouri	N.A.	40.85	36.17	37.04
Poonch	38.95	28.50	30.12	29.66
Nainital	56.47	60.85	57.52	57.30
Almora	20.23	57.56	53.70	38.71
Pithoragarh	20.23	53.93	52.32	51.94
Chamoli	57.13	55.73	60.12	61.90
Uttarkashi	N.A.	72.30	88.76	88.84
Tehri Garhwal	56.38	62.63	69.40	69.12
Garhwal	57.13	65.80	62.81	59.32
Dehradun	60.83	63.40	69.38	69.14
Total	41.95	42.24	45.34	44.67

Note : N.A = Not Available.

Changing Pattern of Uncultivable Lands

The uncultivable lands are those areas which cannot be cultivated either because of absolutely barren, consists of mountain peaks, areas with very high gradient, rocky surfaces or the areas covered by buildings, water bodies, roads, railways and other non-agricultural uses. From the land use statistics, area under non-agricultural uses and barren and unculturable wasteland combined is considered as uncultivable land. Under this heading of uncultivable land, the changing pattern is shown in Table 2.10.

Table 2.10 : Changing Pattern of Uncultivable Land (Region-wise) (In %)

District	*1961-63*	*1971-74*	*1982-85*	*1992-95*
U.P. Himalayan region	32.02	6.90	7.71	12.72
Himachal region	9.64	11.12	10.63	10.22
Jammu region	18.96	30.95	25.02	25.57
Kashmir region	18.50	18.02	17.28	16.55
Total N-W-Himalaya	26.41	11.81	12.34	14.2

During 1961-63, the extent of uncultivable land was 22.98 per cent, such a large extent of area under non-agricultural uses or barren land is a bad indicator for the agro-ecology of the region but the condition is improved during 1971-74 and this proportion of uncultivable land declines to 11.81 per cent, and again it is inching towards unbalanced ecology and its proportion increased to 12.34 per cent and 14.42 per cent during 1982-85 and 1992-95 respectively. But the variation can be observed at regional as well as district level. It is evident from the Table 2.10 that the U.P. Himalaya have the highest percentage of uncultivable lands during 1961-63, i.e., 32.02 per cent followed by Jammu region (18.96%), Kashmir region (18.50%) and Himachal region (9.64%). The districts having high percentage share of uncultivable land includes Tehri Garhwal (31.83%), Jammu (29.20%), Kathua (28.41%), Ladakh (50.67%) and Kinnaur (32.30%) where it is inimical for the agro-ecology of the region.

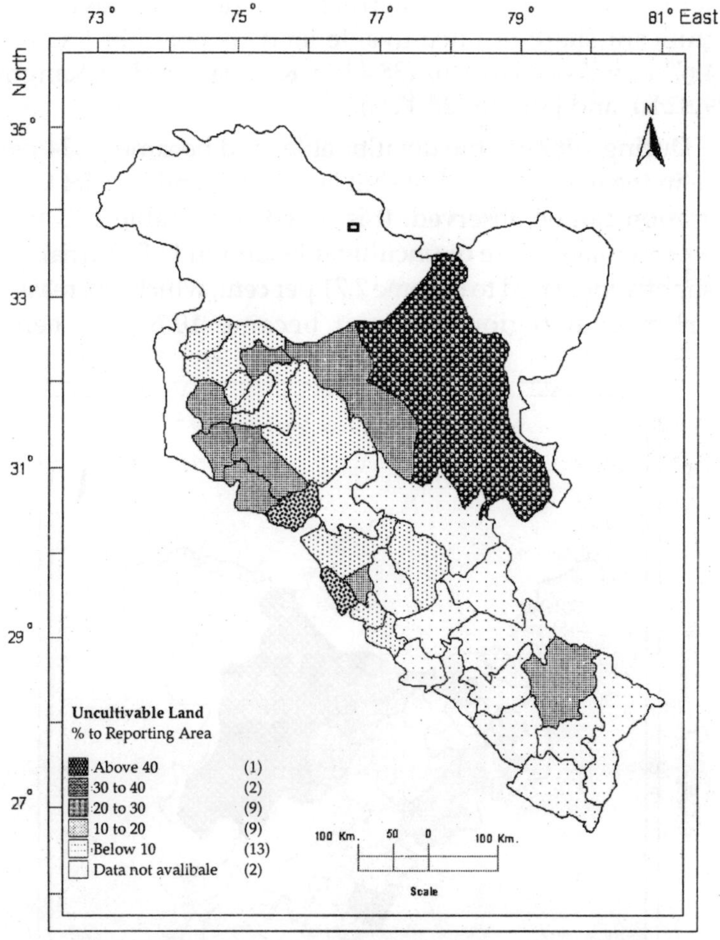

Fig. 2.12 : North-western Himalaya : Uncultivable Land (1982-85)

During 1971-74, the pattern of uncultivable lands is such that, its percentage share is reduced to almost half of 1961-63 i.e., 11.81 per cent. At regional level, it is estimated that Jammu region shows an increase of 63.24 per cent and become 30.95 per cent, while U.P. Himalaya with a decline of 78.45 per cent to become 6.90 per cent of the reporting area. Himachal and Kashmir region didn't show any significant change with 11.12 per cent and 18.02 per cent uncultivable lands respectively. At district level the significant share of uncultivable land is noted at Ladakh (54.67%), followed by Una (38.42%), Kathua (29.98%), Kangra (29.82%), and Jammu (28.42%).

During 1982-85, the uncultivable land became 12.34 per cent in the region as a whole but at regional and district level variation can be observed. It is noted from Table 2.11 that the percentage share of uncultivable land in U.P. Himalaya is slightly increased to become 7.71 per cent, which decreased in Himachal region where it became 10.63 per cent.

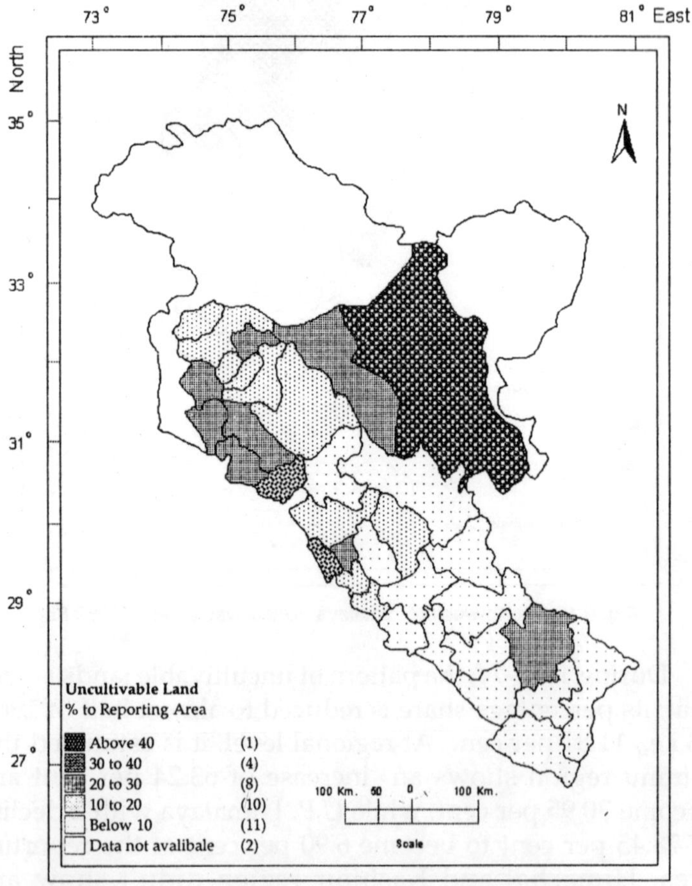

Fig. 2.13 : North-western Himalaya : Uncultivable Land (1992-95)

Jammu region recorded the highest share of uncultivable land is inimical for the agro-ecology of the region. A large number of districts such as Hamirpur, Una, Leh, Kargil, Udhampur, Kathua, Jammu and Rajouri have more than 25 per cent of its land cover as uncultivable land which is unecological and creates several serious problems of ecological nature.

Table 2.11 : Changing Pattern of Uncultivable Land (District-wise)

Districts	*1961-63*	*1971-74*	*1982-85*	*1992-95*
Bilaspur	13.49	17.88	18.59	18.40
Chamba	4.61	1.95	2.41	2.40
Hamirpur	N.A.	29.82	28.23	30.72
Kangra	N.A.	18.99	20.13	14.13
Kinnaur	32.30	11.96	6.34	17.48
Kullu	N.A.	19.15	15.22	14.92
Lahaul and Spiti	N.A.	14.68	1.24	2.79
Mandi	6.32	5.69	5.84	5.84
Shimla	4.51	9.17	7.04	5.65
Sirmour	5.17	7.17	7.32	7.91
Solan	N.A.	17.43	10.60	11.91
Una	N.A.	38.42	32.94	32.60
Anantnag	19.59	18.21	19.06	17.06
Pulwama	N.A.	N.A.	17.24	16.86
Srinagar	19.02	18.39	24.05	24.20
Budgam	N.A.	N.A.	13.00	12.25
Baramula	17.09	17.63	16.52	17.26
Kupwara	N.A.	N.A.	15.62	12.79
Leh	50.67	54.67	59.89	61.41
Kargil	N.A.	N.A.	27.55	28.37
Doda	17.27	14.00	16.67	23.58
Udhampur	22.61	24.66	25.47	24.91
Jammu	29.20	28.04	28.40	22.81
Kathua	28.41	29.98	30.79	31.26
Rajouri	N.A.	24.40	29.00	28.29
Poonch	19.02	23.16	21.84	23.43
Nainital	4.01	5.36	5.17	5.58
Almora	68.29	16.08	6.76	33.28
Pithoragarh	68.29	6.67	6.46	6.41
Chamoli	14.27	2.88	22.40	21.64
Uttarkashi	N.A.	0.27	1.77	3.25
Tehri Garhwal	31.83	6.77	2.93	4.00
Garhwal	14.27	12.86	5.70	6.91
Dehradun	6.85	6.52	5.65	6.00
Total	26.41	11.81	12.34	14.42

Note : N.A. Not Available

During 1992-95, the area of uncultivable land increases to 14.42 per cent from 12.34 per cent with a growth of 16.85 per cent in the region as a whole. The highest growth in uncultivable land is observed in U.P. Himalayas where it became 12.72 per cent from 7.71 per cent or a growth of 64.98 per cent while the highest share of it, is recorded in Jammu region (25.57 %) followed by Kashmir region (16.55%), U.P. Himalaya (12.72%) and Himachal region (10.22%). At district level, Hamirpur (30.72%), Una (32.62%), Leh (61.41%), Kargil (28.37%), Kathua (31.26%), Rajouri (28.29%) and Almora (33.28%) have significant share of uncultivable land which is inimical for the agro-ecology of the region and an alternate planning is needed for these non-agricultural and barren lands.

Cultivable Lands

Here the cultivable lands are those lands which can be cultivable but at present because of some reason not under plough. These lands are also considered as potential lands because they have potential of cultivation and presently provide fuel and fodder in rural areas. In this study cultivable land includes:

- Permanent pastures and other grazing lands
- Land under miscellaneous tree crops and groves
- Culturable wastelands
- Other fallow lands
- Current fallow lands

Permanent pastures and other grazing lands, covers all grazing lands whether they are permanent pastures and meadows or not and village common grazing lands. The areas lying between agricultural land and reserved forests and which are not suitable for cultivation or from where forests have been cleared for the purposes and the agricultural land which have been left fallow for several years have been categorized as pasture lands. These areas are left open for grazing by the village cattle.

Land under Miscellaneous Trees and Groves not included in the net sown area which includes all cultivable land, not included under sown area but put to other agricultural uses. Land under Casuarina trees and other groves for fuel etc. which are not included under orchards are also included in this category. Land under social forestry and silviculture to maintain the ecological balance are also included in this category.

Culturable wasteland includes all lands available for cultivation but not taken up for cultivation or taken up for cultivation once but not cultivated during the year and the last five years or more in succession.

Table 2.12 : Changing Pattern of Cultivable Land (Region-wise)

Region	*1961-63*	*1971-74*	*1982-85*	*1992-95*
U.P. Himachal region	6.34	13.29	14.93	14.80
Himachal region	44.50	48.13	43.91	42.31
Jammu region	22.06	28.12	18.33	17.59
Kashmir region	22.13	20.01	19.19	21.32
Total N-W- Himalaya	20.60	25.54	23.46	23.74

Fallow land implies all lands which were taken up for cultivation but are temporarily out of cultivation for a period of not less than one year and not more than five years. Land lying fallow for a period of one year, are included under '***current fallow***', those lying for more than one year but less than five years, are included under '***other fallow land***' while those fallow beyond a period of five years are included under '***culturable wastes***' or under miscellaneous tree crops and groves.

Changing Pattern of Cultivable Land during 1961-63 to 1992-95

Changing pattern of cultivable land shows that it increased from 1961-63 to 1971-74 in the region as a whole from 19.08 per cent to 25.54 per cent of the reporting area respectively.

But at regional level Kashmir region showed a decline while the U.P. Himalayan region showed an increase from 6.34 per cent to 13.29 per cent or a growth of 109.62 per cent during this period. In Himachal region and Jammu region, it increased from 44.50 per cent to 48.13 per cent and 22.06 per cent to 28.12 per cent respectively. At district level during 1961-63, it is noted that Kinnaur with 60.67 per cent cultivable land followed by Bilaspur (50.33%) and all other districts in Himachal Pradesh have more than 40 per cent cultivable land which has the potential of cultivation to fulfil the growing demands of food and increase the economic condition of the region, if proper management and planning is chalked out, to harness its resources.

During 1971-74, it is noted that the Chamba with 81.43 per cent cultivable land is on the top while Bilaspur (45.18%), Mandi (35.55%), Shimla (52.26%), Sirmour (52.37%), Solan (48.27%) and Chamoli (35.96%) also have significant share of cultivable land which can support a large animal and human population.

Table 2.13 : Changing Pattern of Cultivable Land (District-wise)

Districts	*1961-63*	*1971-74*	*1982-85*	*1992-95*
Bilaspur	50.33	45.18	44.14	43.72
Chamba	42.77	81.43	53.69	52.25
Hamirpur	N.A.	17.27	21.54	18.05
Kangra	N.A.	19.23	21.09	26.43
Kinnaur	60.67	29.83	80.00	69.44
Kullu	N.A.	11.35	12.57	12.73
Lahaul and Spiti	N.A.	40.00	85.18	44.72
Mandi	28.82	35.55	31.80	26.90
Shimla	43.28	52.26	53.29	51.94
Sirmour	47.47	52.37	51.48	51.94
Solan	N.A.	48.27	53.84	54.95
Una	N.A.	13.83	18.59	21.95
Anantnag	19.73	21.81	15.31	14.83
Pulwama	N.A.	N.A.	23.53	27.11
Srinagar	31.65	17.62	23.95	26.91
Budgam	N.A.	N.A.	15.56	19.50
Baramula	19.09	19.63	21.29	22.53
Kupwara	N.A.	N.A.	15.93	19.50
Leh	21.00	18.05	18.55	16.93
Kargil	N.A.	N.A.	26.67	25.67
Doda	17.25	19.28	15.49	8.74

(Contd...)

Table 2.13 : *(Contd...)*

Udhampur	19.73	18.39	14.49	15.35
Jammu	24.28	29.90	24.31	30.26
Kathua	20.83	19.15	20.82	20.07
Rajouri	N.A.	18.39	16.27	14.94
Poonch	29.50	27.25	24.91	22.57
Nainital	10.01	3.66	8.06	8.57
Almora	4.24	7.74	23.11	17.49
Pithoragarh	4.24	24.65	29.01	30.43
Chamoli	4.67	35.96	12.18	11.04
Uttarkashi	N.A.	1.54	5.67	4.23
Tehri Garhwal	8.37	13.88	15.03	14.45
Garhwal	4.67	7.10	17.93	22.52
Dehradun	9.04	9.33	7.05	7.78
Total	20.60	25.54	23.46	23.74

Note : N.A. Not Anailable

During 1982-85, the proportion of cultivable land in N-W-Himalaya decreased to become 23.46 per cent, the decreasing pattern is also noted in Kashmir, Jammu and Himachal region but a slight increase in U.P. Himalayan region where it become 14.93 per cent from 13.29 per cent in this time period of 1971-74 to 1982-85. At district level, Bilaspur (44.14%), Chamba (53.69%), Kinnaur (80.00%), Lahaul and Spiti (85.18%), Shimla (53.29%), Sirmour (51.48%) and Solan (53.84%) have significant share of cultivable land, on the other hand, Nainital (8.06%), Uttarkashi (5.67%) and Dehradun (7.05%) have very little share of cultivable land.

During 1992-95, the pattern of cultivable land is such that it is slightly increased from 1982-85 at the growth rate of 1.19 per cent to be 23.74 per cent out of total reporting area. At regional level, U.P. Himalaya, Himachal and Jammu region showed a decline from 14.93 per cent to 14.80 per cent, 43.91 per cent to 42.31 per cent and 18.33 per cent to 17.59 per cent respectively while Kashmir region recorded an increase from 19.19 per cent to 21.32 per cent. At district level, it is found that Chamba (52.25%), Kinnaur (69.44%), Shimla (51.54%), Sirmour (51.94%) and Solan (54.95%) have more than 50% of their reporting area as cultivable while Bilaspur (43.72%), Mandi (26.90%), Pulwama (27.11%), Srinagar (26.91%), Kargil (25.67%) Jammu (30.26%) and Pithoragarh (30.43%) have

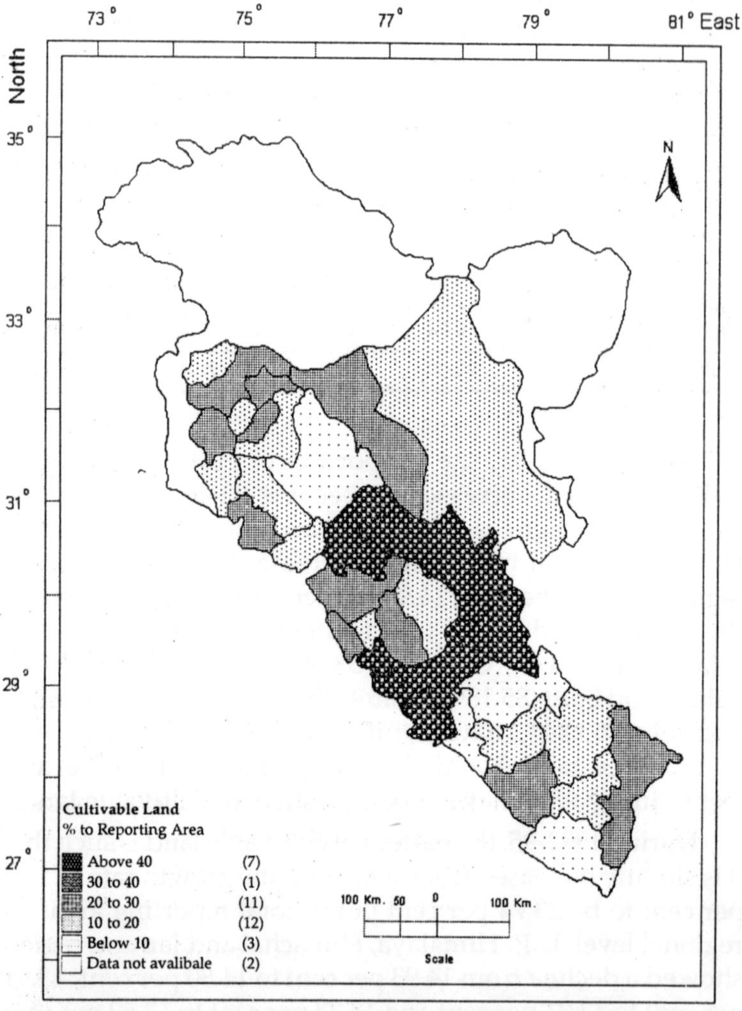

Fig. 2.14 : North-westerm Himalaya Cultivable Land (1982-85)

more than 25 per cent of their area under this category. Only four districts namely Doda (8.74%), Nainital (8.57%), Uttarkashi (4.23%) and Dehradun (7.78%) have less than 10 per cent share under this category of potential lands.

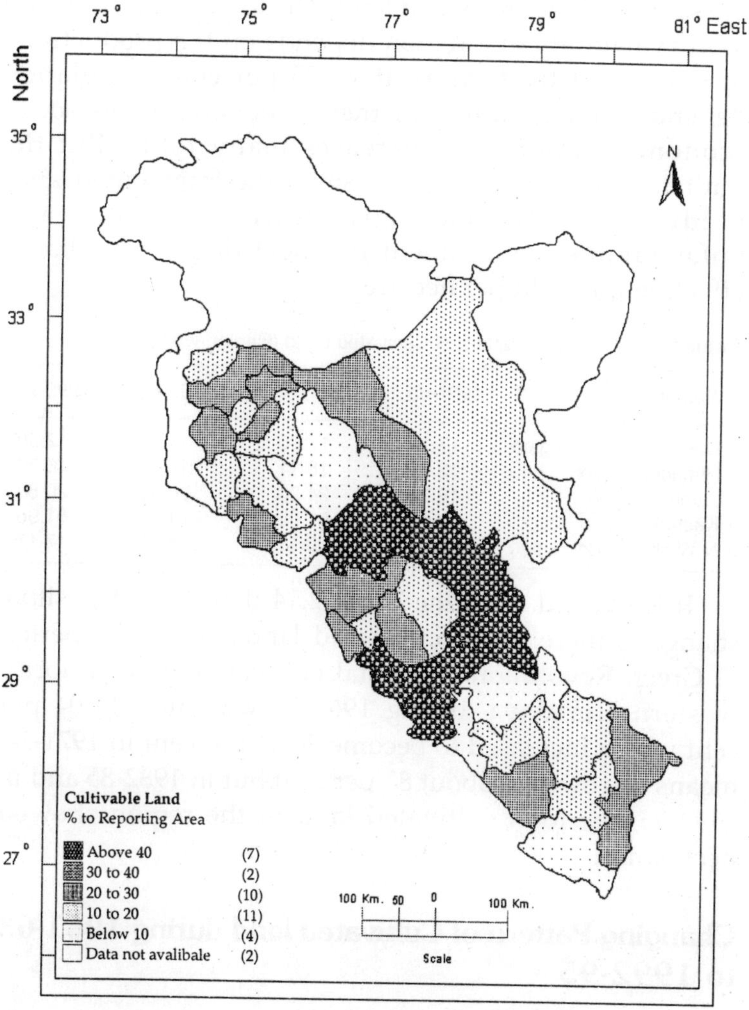

Fig. 2.15 : North-westerm Himalaya : Cultivable Land (1992-95)

Cultivated Lands

Cultivated lands are those areas which fall under the heading of 'Net Area Sown' in the statistics of landuse. All those areas which are under cultivation or under plough at a given time are said to be the cultivated land. The food production and food availability of a region depends on the productivity of cultivated land. More than 75 per cent population depends on agriculture for their livelihood, so its socio-economic importance increases many folds. But the limitations of slope, soil cover, small and fragmented land-holdings, traditional agricultural practices and unfavourable environment for most of the time badly affecting the yield per hectare.

Table 2.14 : Changing Pattern of Cultivated Land (Region-wise)

Region	*1961-63*	*1971-74*	*1982-85*	*1992-95*
U.P. Himalayan region	17.15	17.95	13.21	12.30
Himachal region	12.68	18.11	19.21	16.77
Jammu region	18.74	24.20	20.45	20.68
Kashmir region	61.68	61.08	61.75	61.06
N-W-Himalayan region	11.04	20.41	18.86	17.17

It is evident from the Table 2.14 that a great positive change is there in the cultivated land during the period of Green Revolution. The total cultivated land in north-western Himalaya during 1961-63 was only 11.04 per cent which increased to become 20.41 per cent in 1971-74, means a growth of about 85 per cent but in 1982-85 and in 1992-95, the total cultivated land in the region showed declination:

Changing Pattern of Cultivated land during 1961-63 to 1992-95

It is evident from the Table 2.14 and Table 2.15, that during 1961-63, the area under cultivation was 11.04 per cent of the total reporting area but the variation can be seen at regional and district level. Region-wise it is found that Himachal region have only 12.68 per cent cultivated land while U.P. Himalaya, Jammu, and Kashmir region occupies 17.15 per cent, 18.74 per cent and 61.68 per cent of the cultivated land respectively. There are wide variations at district level e.g., all the district of Kashmir region occupy more than 60 per cent land under cultivation while in Nainital, Jammu and Bilaspur, the cultivated land occupied more than 25 per cent area. Tehri Garhwal (3.42%), Chamba (5.53%) and Kinnaur (2.34%) have insignificantly low share under cultivation to sustain the existing population.

Table 2.15 : Changing pattern of Cultivated Land (District-wise)

Districts	*1961-63*	*1971-74*	*1982-85*	*1992-95*
Bilaspur	25.30	26.64	27.34	27.94
Chamba	5.53	4.51	6.16	6.14
Hamirpur	N.A.	35.34	35.62	33.17
Kangra	N.A.	24.81	20.82	20.62
Kinnaur	2.34	58.11	5.59	3.06
Kullu	N.A.	69.49	71.96	72.20
Lahaul and Spiti	N.A.	45.48	1.56	1.46
Mandi	24.90	21.44	23.83	2.88
Shimla	17.39	2.36	22.50	17.04
Sirmour	15.91	18.98	19.58	18.70
Solan	N.A.	23.38	24.69	21.95
Una	N.A.	27.17	47.19	27.27
Anantnag	60.15	59.45	61.48	67.68
Pulwama	N.A.	N.A.	58.53	55.19
Srinagar	59.74	62.45	48.69	45.37
Budgam	N.A.	N.A.	70.55	67.16
Baramula	63.31	61.95	61.23	59.39
Kupwara	N.A.	N.A.	67.88	66.56
Leh	28.22	27.19	21.56	21.66
Kargil	N.A.	N.A.	45.43	45.59
Doda	12.37	13.19	14.56	14.46
Udhampur	13.52	13.43	15.47	15.68
Jammu	31.60	27.13	33.28	34.20
Kathua	23.46	24.55	22.83	21.66
Rajouri	N.A.	16.35	18.56	19.63
Poonch	17.39	21.09	23.13	24.34
Nainital	29.50	30.12	29.34	28.74
Almora	15.07	18.62	16.43	10.52
Pithoragarh	15.07	13.75	12.22	11.23
Chamoli	23.93	5.44	5.31	5.43
Uttarkashi	N.A.	25.89	3.80	3.68
Tehri Garhwal	3.42	16.72	12.64	12.43
Garhwal	23.93	14.24	13.56	11.11
Dehradun	22.03	20.75	17.91	17.08
Total	11.04	20.41	18.86	17.17

Note : N. A. Not Available

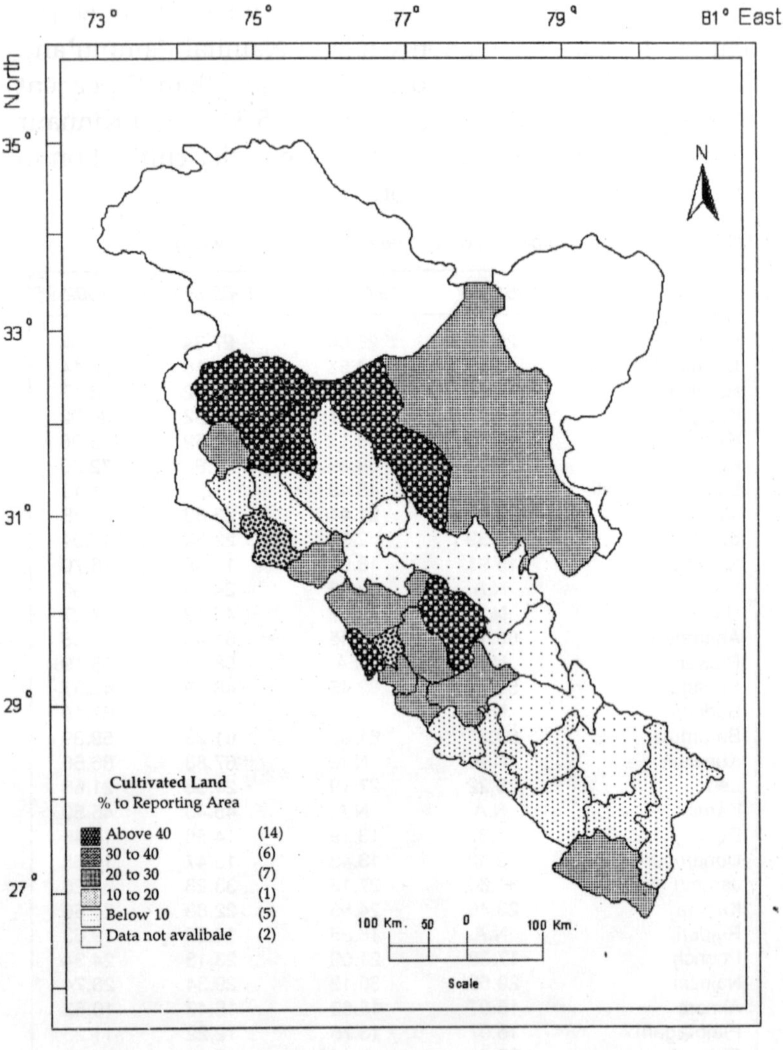

Fig. 2.16 : North-western Himalaya : Cultivated Land (1982-85)

During 1971-74, there is minor increase in the spread of cultivated land in U.P. Himalaya region where it became 17.95 per cent against the 17.15 per cent of 1961-63. On the other hand, minor decrease is recorded in Kashmir region where it declined to 61.08 per cent from 61.68 per cent. Himachal and Jammu region noted with significant growth in cultivated area where it raised 18.11 per cent from 12.68 per cent and 24.20 to 18.74 per cent respectively. At district level, it is found that Anantnag (59.45%), Baramula (61.95%), Srinagar (62.45%), Kinnaur (58.11%), and Kullu (69.49%) have more than 50 per cent area under cultivation while Nainital (30.12%), Uttarkashi (25.89%), Jammu (27.13%), Bilaspur (26.64%), Hamirpur (35.34%), Lahaul and Spiti (45.48%) and Una (27.17%) have 25 to 30 per cent area under Cultivation while Chamoli (5.44%), Chamba (4.51%) and Shimla (2.36%) have very low area under cultivation to sustain the existing population.

During 1982-85, the total cultivated land in the region as a whole declined to 18.86 per cent from 20.41 per cent of 1971-74. The pattern of decline is not same at regional level. The rate of decline in U.P. Himalaya was 26.40 per cent where the cultivated land shrank to 13.21 per cent from 17.95 per cent while in Jammu region, it is decreased to 20.45 per cent from 24.20 per cent. Minor increase is also noted in Himachal region and Kashmir region where it became 19.21 per cent from 18.11 per cent and 61.75 per cent from 60.08 per cent in the same time period respectively. At district level, it is found that Kullu (71.96%), Anantnag (61.48%), Budgam (70.55%), Baramula (61.23%), Kupwara (67.88%) have more than 60 per cent area under cultivation while in Hamirpur (35.62%), Una (47.19%), Pulwama (58.53%), Srinagar (48.69%), Kargil (45.43%) and in Jammu (33.23%) the cultivated land found between 30.60%. On the other extreme in Chamba (6.16%), Kinnaur (5.59%), Lahaul and Spiti (1.56%), Chamoli (5.31%), and Uttarkashi (3.80%) have very low share of cultivated land to sustain the existing population.

During 1992-95, the cultivated land further declined to 17.17 per cent from 18.86 per cent of 1982-85 in the region as a whole. At regional level, it is noted that in U.P. Himalaya

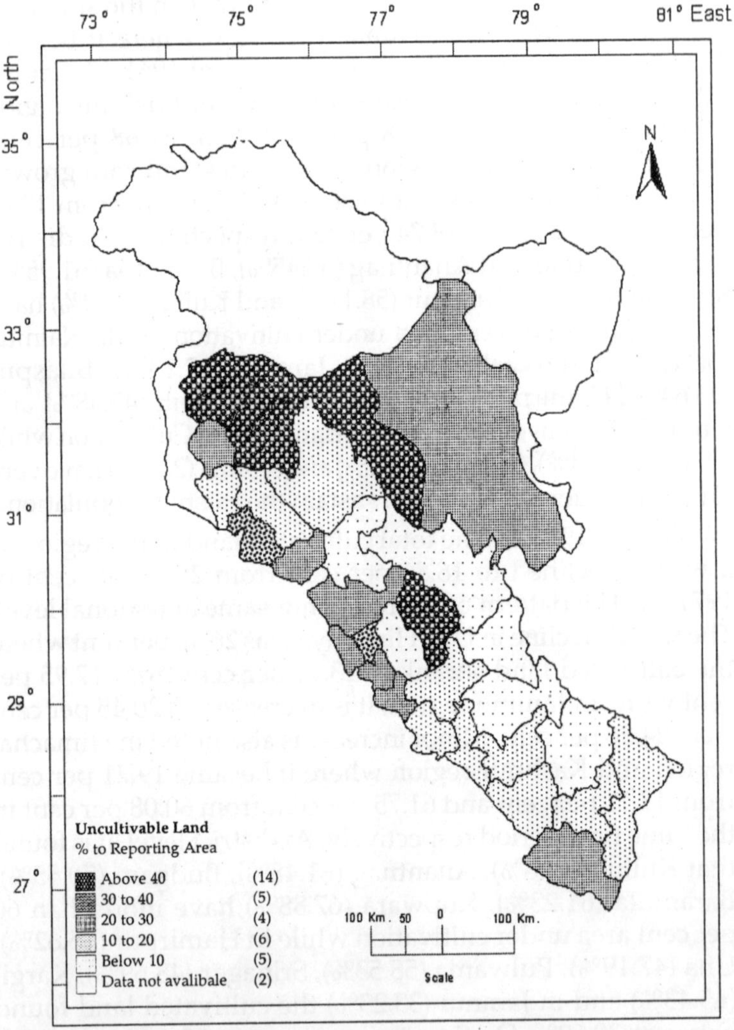

Fig. 2.17 : North-western Himalaya : Cultivated Land (1992-95)

region, Himachal region and Kashmir region the pattern of cultivated land is declined to 12.30 per cent, 16.77 per cent and 61.06 per cent respectively of the previous point of time while Jammu region show a little increase from 20.45 per cent to 20.68 per cent. At district level, the variation in cultivated land is such that in Kullu (72.20%), Anantnag (67.68%), Pulwama (55.19%), Budgam (67.16%), Baramula (59.39%) and Kupwara (66.58%) i.e., more than 50 per cent share of reporting area, while in Bilaspur (27.94%), Hamirpur (33.17%), Una (27.27%), Srinagar (45.37%), Kargil (45.59%), Jammu (34.20%) and Nainital (28.74%), it is found between 25 per cent to 50 per cent. On the other hand in Chamba (6.14%), Kinnaur (3.06%), Lahaul and Spiti (1.46%), Mandi (2.88%), Chamoli (5.43%) and Uttarkashi (3.68%), the share of cultivated land recorded below 10 per cent of the reporting area.

Changing Cropping Pattern (1961-63 to 1992-95)

Cropping pattern is the percentage share of area under different crops in a region at a given time. The fluctuations and variations occurred in these percentages with time is known as changing cropping pattern. The unscientific changes in the agricultural transformation processes now operating in the ecologically sensitive Himalayan region, particularly, the extension of cultivation to upslope, marginal, submarginal lands and forest areas under the increasing pressure of population and absence of other means of livelihood alarming an ecological havoc in the region.

Rice, wheat, maize, barley, total pulses, total fruits, total vegetables and total oilseeds are taken into account to know the pattern of cropping in the end the changes occurred in it. Among these eight crop groups, it is revealed from the analysis that rice, wheat and maize are the most dominant crops of the region and collectively occupied more than 80% of the total cropped area all the time from 1961-63 to 1992-95. Barley and pulses are continuously declining in cropping pattern while oilseeds, fruits and vegetables show promising return in the same time period. (Table 2.16).

Table 2.16 : Changing Cropping Pattern

	1961-63	*1971-74*	*1982-85*	*1992-95*
Wheat	27	36	36	35
Rice	24	21	24	30
Maize	31	24	23	20
Barley	4	5	3	2
Pulses	6	6	5	4
Oilseeds	4	3	4	4
Fruits	1	2	3	3
Vegetables	3	2	2	2
Total	100	100	100	100

Changing Wheat in the Cropping Pattern

Undoubtedly, India has more than trebled foodgrains production from almost the same area as existed three decades ago, but this increase has negated by the increase in population. Wheat production in India is increasing at a greater pace than the population. From a mere 79 grams in 1961 availability has increased to about 185 grams of wheat per day in 1996.

Wheat is the most dominating crop in this region since the pre-green revolution and in all the regions of north-western Himalaya except in the Kashmir region where it covers only 7 per cent during 1961-63 and declined to 6 per cent, 1 per cent and negligible during 1971-74, 1982-85 and in 1992-95 respectively. Almost half of the cropping pattern in U.P. Himalaya is covered by wheat which spread on 53 per cent in 1971-74, 47 per cent in 1982-85 and 54 per cent in 1992-95. In Himachal region wheat occupied 37 per cent in 1961-63 as well as in 1971-74.

Changing Rice in the Cropping Pattern

Contributing consistently to around 45 per cent of India's cereal production rice continues to hold the key to sustained food sufficiency in the country. The real challenge, however, lies in accomplishing the goal on a sustainable basis. It is important to ensure that all future research and development strategies are environmental friendly, not cause further damage to our already fragile natural endowments on the one hand and economically viable and attractive to the grower on the other hand.

Keeping in view the twin objectives of progressive prouctivity growth and growth on sustainable basis. It is evident from Table 2.16, that the share of rice in cropping pattern of north-western Himalaya is increasing and occupying the place only next to wheat. It is found that the per cent share of area under rice recorded a declining trend between 1961-63 to 1971-74 i.e., from 24 per cent to 21 per cent, but later on it increased to 24 per cent and 30 per cent in 1982-85 and 1992-95 respectively. Rice is the most dominant crop of Kashmir valley in terms of area as well as production while in other region it occupies either 2nd or 3rd position.

Changing Maize in the Cropping Pattern

Agro-ecology of the region is most favourable for maize crop than other parts of the country. As a result it responded best and is the most dominant crop of the region in terms of productivity. The productivity of maize in the region as a whole increased by 178.6 per cent from 1954-55 to 1995-96 while area increased by 164.66 per cent in the same period of time.

It is evident from Table 2.16, that the maize was the most dominant crop of the region during pre-green revolution period but later on other crops acquired its area with the help of green revolution technology particularly rice and wheat and its share going decreasing gradually: It is found that the per cent share occupied by maize in the region was 31 per cent during 1961-63 and later it declined to 24 per cent, 23 per cent and 21 per cent during 1971-74, 1982-85 and 1992-95 respectively. A large area under maize is still unaware of green revolutionary measures and is being grown in traditional manner.

Changing Barley in the Cropping Pattern

Barley is an insignificant crop but hardy in nature which can survive in harsh climatic condition also. Its share in area remained almost stagnant with a slight decline from 1954-55 to 1995-96 i.e., 4.25 per cent and 4.07 per cent respectively but per cent share production decreased significantly i.e., from 3.70 per cent to 1.19 per cent.

Barley is a minor crop of the cropping pattern in the region but an important component of the agro-ecology to maintain the ecological balance. It is seen from Table 2.16, that the percentage share of area under barley is declined from 4 per cent in 1961-64 to 2 per cent in 1992-95, which shows its negligence towards this crop. It is also observed in the region that because it is a hardy crop and grows in the areas which are not favourable for other major crops either economically or ecologically. As a result its share is continuously declined which is a matter of concern and if proper steps are not taken immediately it will be removed from the agricultural scenario of the region.

Changing Pulses in the Cropping Pattern

Pulses are an important part of Indian diet, besides the rich source of protein. They also maintain the soil fertility through the biological nitrogen fixation by bacteria prevalent in their root nodules and thus play a vital role on furthering sustainable agriculture. But it is seen in general that the pulses have been cultivated in rainfed conditions where fertility and water retentive capacity of soil is poor. As a result the per capita availability of pulses at national level has declined from 64 gm/head/day (1951-56) to less than 40/gms/head/day (1995-96), which is far below the FAO/WHO's recommendations of 80 gms/head/day.

Declining pattern of pulses is not only seen at national level but it is also found in most of the region as well as in this mountainous region too. Percentage share of pulses in the cropping pattern is gradually declining and it is evident from the Table 2.16, that during 1961-63, pulses occupied 6 per cent of the area and maintained the same position in 1971-74 but in 1982-85 and 1992-95, it declined to 5 per cent and 4 per cent respectively. Pulses are an integral part of balanced diet and for sustainable agricutlure as well.

Changing Oilseeds in the Cropping Pattern

India is the third largest edible oil economy in the world after U.S. and China. It occupies a distinct position not only in terms of area under oilseeds but also in terms of diversity in oilseeds. Its agro-ecological conditions favourable for growing nine major oilseeds i.e., groundnut, rapeseed, mustard, soyabean, sunflower, safflower, sesame and niger and two non-edible sources namely caster seeds and linseeds and oil bearing tree species.

The abrupt change of Indian oilseeds economy from a net-importer status in the eighties to a self-sufficient and more so a net-exporter status during the early nineties has been popularly termed as *"Yellow Revolution."*

The percentage share of area under oilseeds in north-western Himalaya is found stagnant from 1961-63 to 1992-95. It is revealed from Table 2.16 that the area under oilseeds during 1961-63 was 4 per cent which declined to 3 per cent in 1971-74 but it again improved its position of 4 per cent in 1982-85 which remain existed in 1992-95.

Changing Fruits and Vegetables in the Cropping Pattern

India is the second largest producer of fruits and vegetables, the production of which has been tripled during the last 50 years. In the present 21st century a major shift is expected in the consumption pattern in favour of fresh and processed fruits and vegetables as a result greater modern technology should be adopted.

Recent developments have made it possible to grow a wide range of crops including fruits and vegetables not only during summer but also at sub-zero (-30°C) in north-western Himalayan region. A number of fruit crops and over 56 vegetables of different kinds can be grown in most harsh climate of Ladakh region under the polythene cover trenches of 4.67M × 1.52M × 0.91M.

Though the share of fruits and vegetables is very low but the condition of fruits is seen promising, which increased from 1% to 3% during 1961-63 to 1992-95. While on the other hand the conditions of vegetables is discouraging with the declination in the areal spread from 3 per cent to 2 per cent in the same period of time. Priority should be given to increase the area as well as production of fruits and vegetables, which play an important role to fulfil the calorific requirements of malnourished people of backward regions.

3

Ecological Parameters and Agricultural Development

In 1968, an eminent agricultural scientist, Dr. M.S. Swaminathan, warned about the ecological implications of a purely exploitative system of farming even at the very beginning of what was subsequently called Green Revolution in the following words :

> "Intensive cultivation of land without conservation of soil fertility and soil structure would lead....., it may lead us only in an era of agricultural disaster in the long run, rather than to an era of agricultural prosperity."

The interactions of the various constituent elements of an ecosystem are highly complex and often beyond the comprehension of the human brain. It is because of such complexity that human intervention and meddling with balanced ecosystem with objective of producing some desirable effects, also often brings forth some additional, unexpected and undesirable side effects.

There are several examples to show that through the lack of understanding of ecological principles, the efforts of well intentioned people are frequently misdirected leading to the upset of natural balances. A few such cases are given below:

MAN'S INTERVENTION IN NATURAL ECOLOGICAL BALANCE

(a) The Killing of Koyotes in North America

There is a story about some sheep ranchers, who became convinced that Koyotes (wolves of N. America) were robbing them of their young sheep. The community slaughtered all the Koyotes that could be located for a mile around. Soon after, the rabbits, the field mice and other rodents increased tremendously and ate most of the grass of the pastures. The sheep-man then stop killing of Koyotes and started poisoning rodents. The Koyotes filtered in from the surrounding areas but finding their food scarce, turned to young sheep as their only source of food (Dhaliwal, 1995).

(b) Construction of Aswan High Dam in U.A.R.

Another important example of disruption of nature by man is of Aswan Dam in U.A.R. The entire economy of U.A.R. is tied to the ecology of river Nile. The river enters from the south carrying water, silt and nutrients from the Ethiopian high lands over 1500 miles. The Nile valley is the centre of agriculture and population in the Sahara desert, where population is largely depended on agricultural production. The annual flooding was primarily beneficial as it restored soil, water supply and flushed salts from lower layer and kept the Nile valley one of the most fertile land areas on earth in spite of continuous cultivation over thousands of years.

The Nile was dammed at several points. Aswan High Dam is the fourth in the series and created Nassar lake. No doubt it added about a million acres to agricultural land but Aswan dam which was intended to become a living monument to man's ignorance of ecological effects of massive environmental intervention. The major consequences as given by Southwick in 1976 are :

(i) There is no silt deposition during flood periods, all the silt goes to lake, resulting in increased requirement of artificial fertilizers. Crop yields are coming down. The farmers of Nile valley have to spent 100 million US dollars to buy fertilizers.

(ii) The flood water now does not flush the soil in delta and there is waterlogging.

(iii) The silt free water below the dam flows faster and tends to reacquire normal silt load leading to erosion.

(iv) The expansion of canal irrigation resulting in water-logging has increased incidence of several diseases e.g. malaria and schistosomiasis.

(v) There is decline on fish harvest in the sea coast to the tune of 95 per cent.

(c) The Killing of Deer's Natural Predator in Arizona

Man's intervention often upsets the delicately balanced nutritional factors on an environment, many of which are subtle and escape initial of casual investigation. The Kaibab plateau of Arizona supported a population of approximately 4000 deers in 1918, an aggregation that was far below the carrying capacity of the area, since it was estimated that 30,000 deers could have inhabited the region. However, many of the deer's natural predators, including Wolves and Pumas were killed to the same extent. This reduction in predation pressure resulted in a maximum population of 100,000 deers during the winter of 1924-25. An excess number of ungulates, exceeding the carrying capacity of the areas depleted the food supply and caused the starvation of 60,000 deers during the next two years. Surviving members of the population were undernourished and more susceptible to disease. The depletion of the initial food supply resulted in a continued attrition until only 10,000 deers were estimated to be inhabited the area in 1940 (Klages, 1965, p.194).

(d) Shooting of Otter by Fishermen

A good example of such an unexpected side effect came to light following the shooting of Otter by fisherman, which prey on fish in rivers. The idea in shooting the Otter was that the fisherman's rival will be destroyed and thus the fisherman could catch more of fishes, but in actual fact it turned out that the fish population itself seriously dwindled. Investigation showed that the Otter was catching mostly sick or old fishes. When the Otter was killed, there were more sick fishes to spread disease and as a result, the fish population dwindled. Thus, by shooting the Otter, the fisherman had unwittingly destroyed nature's safeguard against epidemics and degeneration.

In the same way man disturbed the natural ecological balance by the modern techniques of farming like heavy dozes of chemical fertilizers, insecticides, pesticides, irrigation, clearing of forest for agriculture and by changing cropping pattern to get maximum output. It is obvious that all the crop plants can be grown in varieties of environment but responded best only in optimal environmental conditions or gives highest return in terms of productivity which is called eco-friendly condition for that crop plant and the study of such relationships comes under the discipline of 'Agro-ecology.' In its true sense agro-ecology is the study of physical characteristics of environment, climate and soils in relation to the optimal agricultural production with least deterioration.

Cardinal Points and Plant Responses

All the plant activities operates within certain more or less well-defined limits. A seed cannot germinate and a young seedling is incapable of development unless the environment supplies certain definite requirements of temperature, moisture, oxygen, carbon dioxide and mineral nutrients. There are three stages for the requirements of any component for the life and growth of any plant i.e., minimum, maximum and optimum which are said to be the cardinal points of vital activity. The minimum and maximum are the range of tolerance for any crop, below and above of which a crop plant cannot survive. Tolerance is a kind of resistance characterized by the host's ability to withstand or repair the effects of infections of a pest that would cause economically significant damage to a susceptible host. Somewhere between minimum and maximum limits a crop shows the highest rate of activity or responded best, such as point is known as optimum condition for that crop. Pfeffer[1] in 1903, recognized that "the cardinal points can never be determined with more than the approximate accuracy, since their position is involved by the external conditions, by the duration of exposure, by the age of the plant and by its previous treatments."

The activities of separate environmental factors such as temperature above the maximum will sooner or later result in cessation of all manifestations of life. Beneck and Jost[2] in 1923, showed the effect of the length of exposure to given temperatures on the location of the optimum condition. The rate of activity of *Lepidium sativum* exposed to the temperatures indicated for intervals of 3.5, 7 and 14 hours was measured by the increase in the length of the roots. A short exposure of 3.5 hours, showed an optimum temperature at 30°C, with the doubling of time-interval the optimum temperature was found at 29°C and when the period of exposure was increased to 14 hours the highest rate of activity was in evidence at 27.2°C.

Ecological Laws of Crop Production

In short the organisms are influenced by their environment in a number of ways. These environmental factors govern the survival and growth of their life cycle. The distribution and abundance of plants and animal are also influenced by the physical factors of the environment. The responses of the organisms to their environment are governed by the following ecological laws :

Liebig's Law of Minimum

A German Botanist Justin Liebig in 1840, proposed that "*the essential material available in minimum amounts, most closely approaching the critical minimum would tend to be limiting,*" a concept has come to be known as Liebig's Law of Minimum. It showed that the growth of crop plants is dependent on the amount of nutrient that is available in minimum quantity. Therefore, he came to the conclusion that the growth of the plants is limited by the essential nutrient which is in short supply in relation to the needs of the plant. In other words, if any essential nutrient is present only in minimal quantity in relation to the needs of the plants, the growth rate of the plant would be minimum.

Since, Liebig was a botanist, he visualized primarily in terms of light, temperature, nutrients and essential elements. He tried to explain the absence of some plants in shaded areas or lack of vegetation above certain altitudes in the Alps. He gave justifications of plants distribution in terms of inadequate light, temperature or nutrients. His concept that the crop yields are not often limited by nutrient that are in abundant supply such as carbon dioxide and water, but by others which are needed in minute quantities and are in short supply such as zinc, also holds goods in modern agriculture. Although the general concept of Liebig's Law has been well accepted, but criticized as, he didn't develop the concept at length that there might be too much of these factors as well as too little. It is strictly applicable in steady-state conditions only. Action of some factors other than the minimum one may modify the rate of availability and utilization. In some plants, zinc may act as a limiting factor only when grown in sunlight, since these plants require less zinc when grown in shade than in full sun light.

Schimper's Optima

Schimper in (1903) coined three terms which can be used for

the purpose of illustrating the reactions of plants to increasing intensities of a particular factor of the environment. These terms are the *'absolute'*, *'harmonic'* and *'ecological optima'*. These optima may be taken in a broader sense than the previously considered cardinal points and their meanings can be expressed as follows :

The absolute optimum corresponds to the highest degree of activity of any one function of a plant such as transpiration or respiration. With increasing temperatures, the plant up to a certain point, transpires at given intervals ever-increasing amount of water. Beyond this particular point, because of interference or the breakdown of certain of the intricate portions of the organism, the rate of activity decreases sharply. The absolute optimum point of activity may, therefore be defined as that point where limiting factors or checks come into activity.

The harmonic optimum corresponds to the most favourable intensity of any one function in relation to the other functions of the plant. Transpiration increases up to a certain point with increasing temperatures. While transpiration is a necessary function of the plant, an excessive activity of this particular or any other function would soon lead to the destruction of the plant. The plant reaches its highest activity at that particular point where the rates of activities of the various functions are in harmony with each other or in other words, at that point where they are properly coordinated.

The theoretical ecological optimum consists of the summation of the various harmonic optima. It is difficult to give the exact location of the summation of the various harmonic optima or the exact location of the ecological optimum.

Shelford's Law of Tolerance

V.E. Shelford in 1913, extended the concept of limiting factors to include the limiting effect of maximum as well as minimum on organisms. He proposed that a factor may be limiting not only at low quantities but too high quantities of the factor may also be determined to the growth and development of the organism. Thus, *"any environmental factor which is below the critical minimum or well above the critical maximum requirement of an organism, would certainly limit the growth of organism in a given area."* This is known as Shelford's Law of Tolerance.

Organisms have ecological maximum and minimum requirements for each environmental factor. These are the limits of tolerance of the organisms for that factor. It may, however, be pointed out here that the various environmental factors do not influence the organisms independently. The effect of one factor is modified by the other and hence the organisms respond to the totality of the environment. According to the Law of tolerance, every environmental factor has two zones viz. zone of tolerance and zone of intolerance.

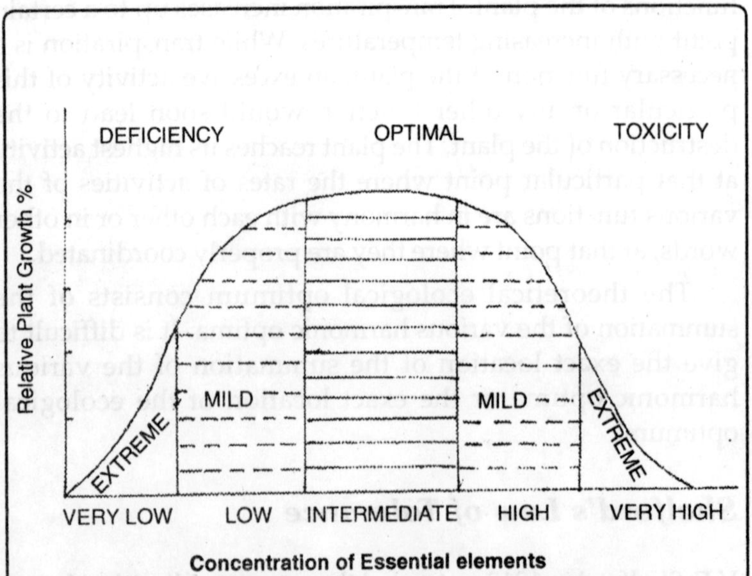

Fig. 3.1 : Graphical Representation of Law of Tolerance

Zone of Tolerance

This zone is favourable for the growth and development of the organisms. It is composed of three parts:

(i) *Optimum Zone* : This zone is the most favourable to the organisms where their growth and development are the maximum.
(ii) *Critical Minimum Zone* : This is the minimum limit of any environmental factor below which the growth and development of any organism ceased.
(iii) *Critical Maximum Zone* : This is the maximum limit of any environmental factor beyond which the organisms usually cease their normal activities.

Zone of Intolerance

This zone is well below the critical minimum and above the critical maximum zones. This zone is unfavourable for the growth and development of the organisms and they cannot survive and tolerate for a long.

Odum stated some subsidiary principles to the law of Tolerance in 1971 which are as follows :

(i) Organisms may have a narrow range of tolerance for one factor and a wide range of tolerance for another.
(ii) Organisms with wide range of tolerance for all factors have more chances of fairly wide distribution.
(iii) When conditions are not optimum for an organism with respect to one environmental factor, the limits of tolerance for other factors may also change, e.g., the resistance of grass to the drought is reduced when soil nitrogen is limiting i.e., less water is required to prevent witting at high nitrogen levels than at low levels.

(iv) In nature, organisms are not frequently found to actually live at the optimum range of a particular factor which is determined experimentally. This is because of greater significance of some other factors e.g., certain tropical orchids if kept cool, grow well in full sunlight than in shade. However, in nature they grow only in shade as they cannot tolerate the heat of the direct sunlight.

(v) Environmental factors are the most limiting during the period of reproduction. Therefore, the non-reproducing adult plants or animals have wider limits of tolerance than for reproductive individuals, eggs, embryos, seeds and larvae e.g., the geographical distribution of birds is determined by the effect of climate on eggs or the young rather than an adult. Similarly, an adult Cypress tree can grow continually on dry upland or submerged in water, but fails to reproduce unless there is moist unflooded field for seedling development.

In ecology, the prefixes 'steno' and 'eury' are used to indicate narrow and wide range respectively i.e., 'stenothermal' and 'eurythermal' means those organisms having tolerance to narrow and wide range temperature respectively similarly stenohydriceuryhydric (for water), stenohaline-euryhaline (for salinity), stenoecious-euryecious (for habitat selection).

ECOLOGICAL PARAMETERS OF SOME IMPORTANT CROPS

It is clear from the Liebig's law of minimum and Shelford's Law of Tolerance, that all the crops have some minimum and maximum requirements of ecological parameters for their sustenance, while below and above of these cardinal points of requirement, the crop suffers badly. In the same time somewhere between these cardinal points, a situation exists where the crops responded best or optimum. Since so many parameters are involved in the growth of a plant, it is very difficult or almost impossible to find out the cardinal point of one parameter because all the parameters depend on each other. But rainfall, temperature and characteristics of soil are of immense importance and required in different quantities at different stages of their life cycle. A brief account of requirements of ecological parameters for some crops are given below :

Ecological Requirements of Maize Crop

Maize is a warm weather crop. It grows from sea level up to 3000 metres of altitude. It is a kharif (monsoon) season crop mainly in northern India. It requires considerable moisture and warmth from germination to flowering. The most suitable temperature of germination is 21°C and for growth 32°C extremely high temperature and low humidity during flowering damage the crop, resulting in poor grain formation. Rainfall of 50 cms. to 75 cms. with good drainage is conducive for proper growth. It is very sensitive to stagnant water particularly in early stages of life cycle. Maize is best adapted to well drained sandy loam to silty loam soils. It will not thrive on heavy clays. It can be grown successfully in soils whose pH ranges from 5.5 to 7.5.

Ecological Requirements of Rice Crop

Rice growing seasons vary in different parts of India, depending upon temperature, rainfall and other climatic conditions. It is cultivated from sea level to as high as 3000 meters of altitude. Rice crop need hot and humid climate. The average temperature required throughout the life period of the crop ranges from 21°C to 37°C. At the time of tillering, the crop requires a higher temperature than for growth. At the time of flowering, the temperature requirement is in the range of 26.5°C to 29.5°C, while at the time of ripening the temperature should be between 20°C to 25°C. A well distributed rainfall of 25cms. to 150 cms. is necessary for the growth of rice crop. Soils having good water retention capacity with good amount of clay and organic matter are ideal for rice cultivation. Clay or clay loam are most suited for rice cultivation. Rice being a semi-aquatic crop, grows best in submerged conditions. It grows well in soils having a pH range between 5.5 to 6.5.

Ecological Requirements of Wheat Crop

Wheat is mainly a rabi (winter) season crop in India. It is cultivated from sea level up to even 3000 metres of altitude. It require about 14°C to 15°C optimum average temperature at the time of ripening. During the growth and flowering stages, excessive high or low temperatures and drought are harmful. Cloudy weather, with high humidity and low temperature is conducive for rust attack. A rainfall of 25 cm. to 75 cms. is favourable for the proper growth of this crop. Soils with a clay loam having moderate water holding capacity are ideal for wheat cultivation. Very porous and excessively drained soils should be avoided. Heavy soils with good drainage are suitable for wheat cultivation under dry conditions. Heavy soils with poor drainage are not suitable, as wheat is sensitive to waterlogging.

Ecological Requirements of Barley Crop

Barley requires cool weather during early growth and warm and dry weather at maturity. It can be grow from sea level up to 4000 metres of altitude. This crop requires 40 cms. to 90 cms. of rainfall for optimal growth. But it is drought resistant, so it suits the areas of scanty rainfall too. The optimal temperature requirement for this crop is 15°C to 18°C where it gives economic returns. Barley thrives best on well drained, fertile deep-loam soils. Being a salt tolerant crop, it can do well even in salt affected soils. It gives higher yields than wheat and other cereal crops on moderately saline-alkaline soils.

Ecological Requirements of Gram Crop

Gram is a winter season crop. It is best suited to the areas having 60 cms. to 90 cms. of rainfall per annum. It is grown from sea level to an altitude of 2000 metres. Severe cold or frost are injurious to it and badly affected at the time of flowering. Excessive rain soon after the sowing or at flowering and fruiting causes heavy loss. The best type of soils for gram is one that is well drained and not too heavy. It is not suited to soils having a pH higher than 8.5.

Ecological Requirements of Lentil Crop

Lentil requires cold climate. It can tolerate severe winter and frost to a great extent. It can be grown up to an altitude of 3000 metres. It remains unaffected by rain at any stage of its growth, including flowering and fruiting. The optimum temperature for its growth is 18°C to 30°C. Well drained loam soils are best for lentil cultivation.

Ecological Requirements of Pea crop

Pea requires cold climate. Moderate temperatures are essential throughout the growing season, 22°C of temperature is considered favourable for germination. High temperatures are more injurious to pea crop than frost. High humidity is harmful to pea crop as it favours incidence of disease. The optimum temperature suitable for its growth is 13°C to 18°C. Well drained loam soil is considered best for pea cultivation. Optimum soil pH for this crop is 6.5.

Ecological Requirements of Urd Crop

Urd as an important pulse crop require a hot and humid growing season. It can be grown successfully from sea level up to an elevation of 1800 metres. The annual rainfall of 60 cms. to 75 cms. is optimum. Heavy rain during flowering stage is harmful and adversely affect the production. It can be grown on all types of soils ranging from sandy loam to heavy clay, except alkali and saline soils. Loam or slightly heavy soils with neutral pH are best suited for Urd cultivation.

Ecological Requirements of Tomato Crop

Tomato shows a wide climatic tolerance and can be grown in the open wherever there is more than three months of frost free weather. It is a kharif season crop which can be easily transplanted. The range of temperature they can tolerate varies from minimum of –19°C to a maximum of 15-35°C. These are more successful where there are long sunny periods with light evenly distributed, rainfall and night temperatures between 6.5°C to 20°C. Very wet weather in the tropics with low sunshine incidence and high night temperatures results in excessive growth at the expense of fruiting and also leads to diseases. Tomato can be grown at sea level in the tropics, but usually do better at higher altitudes. A light, free-draining and fertile loam with a soil pH of 5-7 is the best for the crop, but they can grow on a variety of soils.

Ecological Requirements of Cauliflower, Cabbage, Turnip and Radish (*Cruciferous crops*)

Most of the cruciferous crop plants originated in temperate countries and grow best in cool moist climates and at higher altitudes in the tropics. They are not suited to the low humid tropics. In Trinidad (10°N) with a minimum monthly temperature ranging from 20-22°C and maximum tempeatures of 30-33°C fair yields of cabbages and cauliflowers are obtained with irrigation in the dry season which is coolest time of the year. They can grow from sea level to 9000 feet. Average annual rainfall of about 85 centimetres and a mean relative humidity of 91 per cent at 8:30 a.m. and 56 per cent at 2:30 p.m. gives good yields. The ideal soil is rich sandy loam.

Ecological Requirements of Potato Crop

Potato is cool season crop. It thrives best in cool regions where there is sufficient moisture and fertile soils. Satisfactory tuber growth occurs if soil temperatures are between 17°C–19°C. Higher soil temperatures adversely affect the tuber development. Tuber development virtually stops if temperature rises above 30°C. At higher temperatures, the respiration rate increases, and the carbohydrates produced by photosynthesis are consumed rather than stored in the tuber. High temperatures at any part of the growing period affect the size of the leaflets, thereby reducing the tuber formation. It grows best under long day conditions. Sunshine along with cooler nights are essential for reducing the spread of diseases. Potato can be grown on a wide range of soils ranging from sandy loam, silt loam, loam and clay soil. Soils for potato should be friable, well aerated, fairly deep and well supplied with organic matter. Well drained sandy loam and medium loam soils, rich in humus are most suitable for potato. Soil structure and texture has a marked effect on the quality of the tuber. Light soils are preferred because they tend to promote more uniform soil temperatures and make harvesting of the crop easier. The root growth is usually restricted to top layers, at a depth of about 20-25 centimeters from soil surface. In rich soils roots of some varieties may reach upto depth of 90-100 centimetres.

Ecological Requirements of Sweet Potato Crop

Sweet potato can be grown from sea level up to the elevation of 9000 feet but grow best where the average temperature is 24°C or more with well distributed annual rainfall of 75 to 125 centimetres and abundance of sunshine. A frost free growing period of 4–6 months is essential. In high rainfall areas they are often planted towards the end of the wet season. They will not persist in long periods of drought without irrigation. This crop can be grown on wide range of soils, but a well-drained sandy loam with a clay subsoil is considered ideal. It cannot bear waterlogging and is usually grown on mounds of ridges.

Ecological Requirements of Sugar beet Crop

Sugar beet is a crop of temperate and cold climate. It is highly versatile and can be raised successfully in any region with temperature ranging from 12–45°C during the crop season (October to May). The favourable soil temperature for germination of sugarbeet seed is around 15°C. For proper growth of plants and sugar accumulation, an average temperature of about 20–22°C is ideal. Temperatures above 30°C retards sugar accumulation. For sugars beet well drained fertile subsoil is most preferable. Sugarbeet can thrive very well in saline-alkaline soils with a pH value as high as 9.5. This should have enough slope and drainage to prevent flooding of the bed.

Ecological Requirements of Sugarcane

Sugarcane is a tropical crop but can be successfully grown in sub-tropics too. Temperature above 50°C ceases its growth, those below 20°C slow down it. Low temperature reduces its tillering. An average mean temperature of 26°C to 32°C is best suited for the growth of sugarcane. The crop does best in the regions receiving a rainfall of 75 to 120 centimetres per annum along with fairly high humidity. Sugarcane requires a long growing season of ten to twelve months because certain number of heat units is required to bring the plant to maturity. Besides temperature and rainfall, light also plays a very important role in proper growth and development of sugarcane. Under bright sunlight conditions the stems are thicker but shorter and leaves are broader and greener, while under low sunshine, the stem are slender and long with narrower and yellowish leaves. Day length considerably influences the tillering. Short day length decreases number of tillers per plant and ultimately the tonnage. Plants grown under long day conditions produce more dry matter. Day length also determines the flowering in cąne. Thus climate plays a very important role in yield of cane. Sugarcane can be grown on all types of soils ranging from sandy loam to clay loam. However, it thrives best in well drained loamy soils. It can also be raised successfully on lighter soils provided there is adequate irrigation facilities and on heavy clays with proper drainage and addition of organic matter. Saline, alkaline and acidic soils are not at all suitable for sugarcane. Sugarcane usually flower at the age of ten to twelve months.

Ecological Requirements of Soybeans

Soybean grows well in warm and moist climate. The climatic requirements for soybean are almost the same as for maize. A temperature of 26.5°C to 30°C appears to be the optimum for most of the varieties. Soil temperatures of 15.5°C or above favour rapid germination and vigorous seedling growth. The minimum temperature for effective growth is about 10°C. A lower temperature tends to delay the flowering. Day length is the key factor in most of the soybean varieties as they are short day plant and are sensitive to photo-periods. Most of the varieties will flower and mature quickly if grown under condition where the day length is less than 14 hours provided that temperatures are also favourable. Well drained and fertile loam soils with a pH between 6.0 and 7.5 are most suitable for the cultivation of Soybean. Sodic and Saline soils inhibits germination of seeds. In acidic soils liming has to be done to raise the pH to about seven. Waterlogging is injurious to the crop.

Ecological Requirements of Safflower

Safflower is considered a cool season crop; optimum temperature for germination being about 15.5°C. The crop is grown during rabi primarily as a rain-fed crop, but in some areas it is raised under irrigation. Seedlings are tolerate to low temperatures. High temperatures at the time of flowering is harmful to the crop. The day temperature in the range of 24-32°C at flowering are moderate for higher yields. When flower buds are being formed or flowering has just commenced, temperature below 0°C may cause considerable damage in the form of sterile heads. Frost is harmful to the crop. At all the stages of growth of the plant, excessive rainfall or humidity increases damage from fungal diseases. Water-logging due to poor drainage or prolonged rains can cause substantial reduction in yield. The crop is not fit for tracts of heavy rainfall, the general rainfall suited to the crop being between 60 to 90 centimetres. It is cultivated on all types of soils, including sandy soils but the crop is best suited to deep, well drained fertile soil with a high water holding capacity and neutral pH. Though, it is fairly tolerant to salinity but under such soils oil content of seed is reduced.

Ecological Requirements of Sunflower

The crop require a cool climate during germination and seedling growth. Seedlings tolerate frosts moderately well untill they reach the four to six leaf stage of development. It requires warm weather from the seedling stage up to flowering stage and warm sunny days during flowering to maturity. High humidity accompained with cloudy weather and rainfall at the time of flowering results in poor seed set. The amount of linoleic acid decreases with high temperatures at maturity. It is a photo insensitive crop, therefore, it can be grown successfully in any season viz. kharif, rabi and spring. It takes 80-90 days in kharif, 105-130 days in rabi and 100-110 days in spring season. It can be grown on a wide range of soils and tolerates a moderate pH range and some salinity. It thrives best on deep loam soils with good drainage and irrigation facilities. The optimum range of soil pH for this crop is 6.5 to 8.5.

COMPONENTS OF AGRO-ECOLOGY AND AGRICULTURAL DEVELOPMENT

The agricultural development of any region is influenced by their environment in a number of ways. These environmental factors govern the survival and growth of individuals at all stages of their life cycle. Not only the productivity but distribution and abundance of plants in any region are also influenced by the physical factors of the environment. In the words of Morgan (1908)[3], "between an organism and its environment there takes place a constant interchange of energy and material. This, in general, is also true of all bodies whether living or lifeless, but in living organism this is a peculiar one; first because the plant or animal is so constructed that it is suited to a particular set of physical conditions, and second because it may so respond to a change in the outer world that it further adjust itself to changing conditions, i.e., the response may be of such a kind that it better insures the existence of the individual, or of the race. The two ideas contained in the foregoing statement cover, in a general way, what we mean by adaptation of living things."

The potential crop producing ability of a given area is

dependent primarily upon the existing climatic and soil conditions under which the crop is grown. The responses of various factors on agriculture can be summarized in following paragraphs.

Response of Temperature on Agriculture

Temperature provides a working condition for nearly all plant functions. Life is known to exist within a range of 300°C, i.e., from -200°C to 100°C (Dhalwal, 1995). Most of the activities are actually restricted to a much narrow band, and for most of the organisms, the favourable temperatures lies between 10°C and 35°C. However, certain bacteria can tolerate up to 90°C in hot springs and a few blue-green algae may also occur in hot water around 80°C. The upper temperature tolerance for insects and the most tolerant fish is 50°C. Certain bacteria, blue-green algae and Lichens can also tolerate sub-zero temperatures. The spores and seeds of certain plants can also tolerate both sub-freezing and very high temperatures for a short duration without the loss of viability.

The organisms live in a rather limited range of temperature which is different for different species and outside this range they fail to reproduce and grow. It is also noticed that the upper limits are more critical than the lower ones. The organisms have narrow range of tolerance to temperature are called ***stenothermal*** while those having wide range of tolerance to temperature are known as ***eurythermal***. However, many organisms appear to work more efficiently towards the upper limits of their range of tolerance. The range of temperature variation tends to be more on land than in water. That is why, terrestrial organisms have broader ranges of tolerance than the aquatic organisms.

The organisms adapt themselves to unfavourable temperatures in various ways. Most cold blooded animals undergo hibernation, dormancy or diapause to hide away

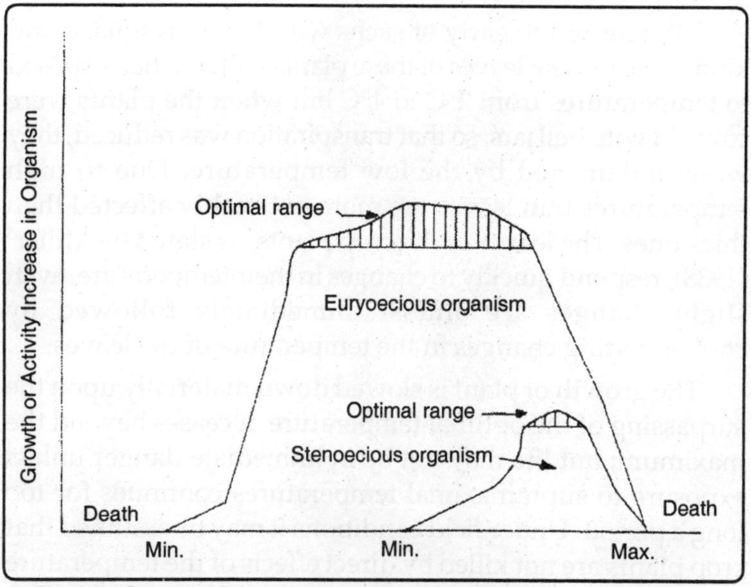

Fig. 3.2 : Graphical Representation of optimum range of Euryoecious and Stenoecious organism

the effect of low temperatures. While plants, in cold climates have evolved adaptation (cold hardiness) to cope with low temperatures.

Organisms can also extend their limits of tolerance by adapting to extreme conditions of temperature, like that of four species of cotton which were perennial shrubs confined to frost free tropical countries about 600 years ago (Hutchinson, 1965). The early fruiting cottons were selected and planted in temperate climates with cold winters and hot summers. The annual growth habit was imposed on them by winter frost and now annual cultivated cottons are grown in cold winter and semi-arid climates. Thus, during the Course of about 600 generations, the cotton plant has adapted itself to climate. The chilling of plants, that is, exposure to the temperatures that are low but above the freezing point, has detrimental effects on certain plants, thus serves to limit their distribution. The chilling of plants not only has retarding effects but may leave some species definitely injured. Molisch

(1896)[4], referred to early of sachs with tobacco, squash and kidney beans. The leaves of these plants wilted when exposed to temperatures from 2°C to 4°C but when the plants were covered with bell jars, so that transpiration was reduced, they were undamaged by the low temperature. Due to high temperatures thin leaves are more noticeably affected than thick ones. The leaves of the crop plants, as stated by Miller[5] (1938), respond quickly to changes in their temperature, even slight changes are almost immediately followed by corresponding changes in the temperature of the leaves.

The growth of plant is slowed down materially upon the surpassing of the optimal temperature, it ceases beyond the maximum, but life may not be in immediate danger unless exposure to supramaximal temperatures continues for too long a period. Under field conditions it may be assumed that crop plants are not killed by direct effects of the temperature as such, but rather by secondary effects induced the high temperatures such as inability of the plant to re-establish the necessary water balance, the dehydration of protoplasm, or sometimes by a partial precipitation of the cell proteins.

There are few optimal temperature crops which lie between the minimum and maximum ranges of tolerance and suitable at different stages of their life cycle and are summrised above in respective heading of these crops.

Response of Precipitation on Agriculture

Precipitation is a major limiting factor affecting the distribution and abundance of both animal communities and agricultural development particularly in terrestrial environment. Precipitation can be found in about fifty different forms, but available to the organisms mainly in the form of moisture, snow and dew, etc. The availability of the amount of precipitation and its distribution over time is highly variable in different regions of the earth surface. In topics and sub-tropics, rainfall tends to be unevenly distributed over seasons as against the temperature and climates where rainfall is generally evenly distributed throughout the year.

The evaporative power of the air is an important ecological factor particularly for terrestrial plants. The land plants are adapted to conserve or loss moisture through transpiration. Moisture has more pronounced effect on agriculture than does any other single factor on the environment. Schimper in 1903, emphasized strongly the many fold influences of water on the expressions and appearances of plant life by stating that "no factor affecting plant life is so thoroughly clear as the influence of water." Warming in 1909, stated that "the ecological importance of water to the plant is fundamental and almost surpasses that of light or heat." Robbins in 1927, also stated this, in these words, "water is the chief limiting factor in the growth of most crops."

For the majority of crops, there is ample sunshine and an abundance of oxygen and carbon dioxides in the air, the temperature of the air and soil is seldom seriously unfavourable, as a rule, there are sufficient nutrients in the soil, but the farmer, except in the most rainy sections of the country, is usually confronted at some time during the season with a shortage of water. This is particularly true in arid and semi-arid regions.

Moisture as an Edaphic factor

The amount of water present in soil at any given time has a direct influence on the concentration of the soil solution and constitutes one of the main factors determining the ease with which water and soluble nutrients can be absorbed by the roots of the plants. In this respect, soil moisture becomes an edaphic factor. Soil moisture in relation to its numerous direct and indirect influences can, without doubt, be designated as one of the most significant factors determining the subterranean habitats of plants.

Ecological Classification of Plants based on Water

Warming's classification of vegetation types into three groups, the hydrophytes, mesophytes and xerophytes is generally accepted. Plants growing in 'fresh water or very humid habitats' are known as hydrophytes. The mesophytes or typical land plants, grow under medium climatic and soil conditions. Plants capable of enduring without injury for a prolonged period of drought are considered as xerophytes. Schimper used the term hygrophytes, tropophytes and xerophytes for the designation of vegetation types, of habitats in increasing degrees of dryness. Maximov (1929), points out that "the limits of these groups are naturally ill defined, and in practice it is sometimes difficult to decide to which group a given plant shall be assigned." The mesophytes or common land plants take an intermediate position between the hydrophytes and true xerophytes. Rice is the only cereal that may be classified as a hydrophyte while the sorghums are the only crop plants with characteristics approaching the true xerophytes.

Xerophytes are defined as 'plants capable of enduring without injury a prolonged period of drought'. Delf (1900), defined xerophyte plants as "those which with the help of certain structural modifications can continue to perform their normal functions when exposed to climatic conditions involving atmospheric or edaphic drought or both." Ecologically suitable rainfall requirement for few crops are given above in their respective headings.

Response of Light on Agricultural Development

Without light there would be no production of chlorophyll, no assimilation of carbon dioxide and no life can exist on the earth surface. Schimper in 1903, designate light as next to moisture, the most important environmental factor determining the structure of plants. Both water and light provide actual material essential to the building up of the structures of higher plants, while temperature, as has been pointed out, provides the necessary working conditions.

Schimper then pointed out that the light factor is of less importance than water and temperature factors as it relates to the geographical distribution of plants, notwithstanding its prime importance to nutritional and structural effects. This is because the differences in both the quantity and quality of light in relation to the needs of plants are not so pronounced in different climatic regions as are differences in the moisture and temperature factors.

Light is essential for photosynthesis on which all other organisms depend for their energy supply. Light affects the daily and seasonal rhythms of both animals and plants in a number of ways. These effects depend upon the intensity, quality and duration of light.

(a) *Intensity of Light*

The intensity of light (the energy inputs), control the entire ecosystem through its effects on the primary production. The rate of photosynthesis of plant leaves increases linearly with increasing light intensity up to the optimum i.e., light saturation level. In most of the cases, there is decrease in photosynthesis beyond this level. At high intensities, synthesis is reduced by photo-oxidation of enzymes and photosynthate is used up by rapid respiration.

Under most conditions the quantity of light present is sufficient for the normal requirement of crop plants. The actual amount of radiant energy which may be utilized by plants is highly dependent on other factors of the environment either favourable or unfavourable to the establishment or maintenance of a proper physiological balance. Thus a plant well supplied with moisture and necessary nutrients is able to utilize more radiant energy than one growing in an unfavourable environment. On the basis of tolerance of intensity of light, plant can be grouped into two i.e., shade tolerant plants (Sciophytes) and shade intolerant plants (heliophytes). The shade tolerant plants maintain a high rate of photosynthesis at low intensities of light, whereas the shade intolerant plants are adversely affected by the shade.

(b) *Quality of Light*

The term 'quality of light' is used here in reference to the composition of light in relation to its effects on plants. When a beam of light is dispersed by refraction through a prism and rays arrange themselves in a series according to their wave lengths. The relationship of these rays to plants behaviour and to the question of optimum intensities of light is pointed out by MacDougal (1901), in these words, "not all of the rays of the spectrum are concerned in the various influences exerted by light upon living matter, but only waves of certain wavelengths are active in each. It is not possible therefore to fix upon a minimum, optimum and maximum intensity of light which is common, to all of the relations, between the plant and light." Shirley (1929), consider the measurement of separate ray of secondary importance to the evaluation of the total light intensity. The principle wavelengths absorbed by the plants are blue (400-500 nm) and red (600-700 nm), the rate of photosynthesis is maximum on these wavelengths. The green (500-600 nm) light of the spectrum is less strongly absorbed. The infra-red containing more heat is also very weakly absorbed and hence the plants are saved from the deadly effect of high temperatures.

(c) *Duration of Light*

Owing to spheroidal shape of the earth and the inclination of its axis, the duration of the light period (photoperiod) that is, the length of the day varies from 12 hours at the equator to increasing lengths at the higher latitude to finally contain sunlight throughout the 24 hour period at the poles. Thus in the tropics the plants are exposed to sunlight half of each day and exposure increases in higher latitudes. This results in actually greater amounts of insolation at the higher latitudes than in the tropics at the summer solstices, June 21 and December 21 for the northern and southern hemisphere respectively. Therefore, the length of the day remain constant or nearly so for all season of the year while in higher latitudes the length of the day increases up to the summer solstice and than decreases.

Light may influence the plant behaviour by its intensity, its composition and by its duration. These variables of light together with the temperature and other environmental conditions determine not only photosynthetic material formed but also the utilization of these materials.

The response of plants to the relative length of day and night is referred to as photoperiodism and term was first used by Garner and Allard in 1920. The photoperiod influences the behaviour and activities of many organisms. Photoperiodicity coupled with organism's biological clock is a time mechanism which maintains rhythmicity of many organisms. Photoperiod affects the moulting, coloration of plumage, migration and breeding in birds, fishes, reptiles and mammals and onset of the diapause in insects.

Similarly, photoperiod brings about growth and flowering in many plants. The flowering of plants takes place only when the daily cycle in light of a specific duration called the critical photoperiod. On the basis of their photoperiodic requirements, plants can be grouped into four categories:

(i) *Short-day Plants* : They come to flowering when the day length is less than the critical photoperiod, like paddy.

(ii) *Long-day Plants* : They require photoperiod longer than the critical day length to flower, like wheat.

(iii) *Intermediate Plants* : They require neither too short nor too long photoperiods to flower.

(iv) *Day-neutral Plants* : They do not have any specific photoperiod requirement and flower after a period of vegetative growth, irrespective of photoperiod.

Response of Soil on Agricultural Development

The soil is not a static body but should be regulated as a living and highly dynamic entity with natural provisions for continued development and renewal. Soil differs from parent material entering into its formation in colour, structure, texture, physical constitution, chemical composition and biological characteristics, etc. In relation to its genesis and development of its characteristics, soil is regarded by Kellogg (1936) as a function of climate, vegetation, relief, age and parent material. Soils are the base of the terrestrial ecosystem. It is the site of decomposition of organic matter and the return of minerals of the nutrient. It is anchoring medium for plants from where they derive their water and nutrient supply. The abiotic and biotic components are especially intricate in soils. Cold climate, high elevation and steep slopes are the determining factors in the soil characteristics of north-western Himalayan region. The following characteristics of the soil are important from ecological viewpoint:

(a) *Soil Texture*

The physical properties of a soil may be approached from the standpoint of its texture and structure. The depth of the soil also is of great importance to its economic utilization. The close relation of these factors to the water economy of plants is evident in such a way that, they determine both the ease with which water may penetrate and the amount of water holding capacity of the soils.

Soil texture refers to the proportion of mineral particles of various sizes or diameters viz., gravel (> 2.0 mm), coarse-sand (2.0–0.2 mm), fine-sand (0.2–0.02 mm), silt (0.02–0.002 mm) and clay (<0.002 mm). Sandy soils can not support good vegetation due to little water holding capacity. Clayey soils have good water holding capacity and provide nutrients in available form to plants. However, clay particles become sticky on wetting and do not separate on drying leading to formation of cracks which injure the root system. Therefore, loamy soils (mixture of sand and clay) are better suited for plant growth.

(b) *Organic matter*

The decomposition of dead plants and animal remains provide organic matter, which is a major source of nutrients for plants. Deficiencies of plant nutrients and lack of proper balance between the essential elements have decided depressing effects on crop yield. And an abundant supply of nutrients is especially important during the grand period of growth.

(c) *Micro-organism*

Micro-organisms in the soils aid in decomposition of organic matter and release of nutrients. Many species of bacteria and algae help in fixation of atmospheric nitrogen. The nitrogen contents of the soil are more or less associated with its fertility. The various factors entering soil genesis, especially the climate and temperature factors, come definitely into play in determining the nitrogen level. The relative availability of nitrogen determine not only the type and luxuriance of vegetation produced but also its rate of decomposition upon its return to the soil. On the other hand, several species of bacteria and algae produce toxic substance and cause diseases in plants.

(d) *Nutrients*

The growth of plants is regulated by nutrients present in the soils. Deficiencies of plant nutrients and lack of proper balance between the essential elements have depressing effects on crop yields. Deficiencies may be and often are supplied to meet specific requirements, either by the inclusion of such crops as legumes or green manure crops in the course of rotation or by means of commercial fertilizers. The excess of several nutrients like chloride, sulphates and nitrates makes the soils saline and such soils are suitable only for some halophytes. The availability of nutrients in balanced amount is regulated by the soil pH. Highly alkaline soils (excess of carbonate and bicarbonates of sodium and magnesium) are not suitable for plant growth.

These edaphic factors may affect the plants in a number of ways, viz., ability of seed to germinate, size and erectness of plants, vigour of vegetative organs, woodness of stems, depth of roots, susceptibility to drought and frost, number of flower per plant and time of appearance of flower.

Response of Biogenic Salts on Agricultural Development

Biogenic salts are the dissolved nutrients vital to life. Among the known elements, sixteen (16) are found to be essential for the growth of plants. Although nitrogen and phosphorus are of major significance, phosphorus is of highest importance from ecological point of view. In fact, too much of nitrogen and phosphorus is as limiting as too little. The interaction between nitrogen and phosphorus is quite important. In an average biomass, N/P ratio is 16:1. Nitrogen and phosphorus are the principal macronutrients (the nutrients required in relatively large quantities by all living organisms) followed by potassium, calcium, sulphur and magnesium. Magnesium is a necessary component of chlorophyll.

Micro-nutrients or trace elements are required in minute

quantities, but still plants are fail to grow without them. Thus micro-nutrients are important limiting factors. The micro-nutrients which are specially important for plants are boron, chloride, cobalt, copper, iron, manganese, molybdenum, silicon, vanadium and zinc. Manganese, iron, chloride, zinc and vanadium are important for photosynthesis, molybdenum, boron, cobalt and iron are for nitrogen metabolism while manganese, boron, cobalt, copper and silicon are important for other metabolic functions.

The growing plants have three sources from which they get necessary nutrients, the air, water and the soil. Carbon, hydrogen and oxygen are getting from air and water, while rest are getting from soil. Among the nutrients getting from soil, nitrogen, phosphate and potassium are considered as major macro-nutrients of first order while calcium, magnesium and sulphur are macro-nutriens of secondary order. Iron, manganese, boron, molybdenum, copper, zinc, and chlorine are required in minute quantities and known as micro-nutrients.

Carbon and moisture (oxygen and hydrogen) constitute the bulk of the weight of a plant. Field crops obtain most of their carbon and oxygen directly from the air by photosynthesis.

Functions of Essential Plant-nutrients

Nitrogen is an essential constituent of all living matter. It is a part of proteins and chlorophyll. It promotes leaf and stem growth, and increases protein contents of food and fodder crops.

Phosphorus is a constituent of nucleic acid, phytin and phospholipids. It stimulates early development of root growth and help to establish seedlings quickly and brings about early maturity of crops particularly the cereals. It stimulates blooming and aids in seed formation and improve the quality of the foodgrains and other crops. It help in formation of root nodules and nitrogen fixation.

Potassium does not enter into the composition of any of the important plant constituents such as protein, chlorophyll, fats and carbohydrates concerned in plant metabolism. It is essential in the formation and transfer of starch and sugar. Thus it is required in large quantities for potatoes, sweet potatoes, turnip and bananas, etc. It regulates water condition within the plant cell and water loss by transpiration and acts as an accelerator of enzymes action.

Calcium is a constituent of cell wall. It increases stiffness of straw and promote root development and growth. It is essential to activate growing points, especially root tips and encourages seed production.

Magnesium is a constituent of chlorophyll and essential for all green plants. It helps in maintaining a dark green colour in leaves. It promotes the formation of oils and fats in seeds. It helps in translocation of starch and regulates the uptake of other nutrients.

Sulphur is not a constituent of chlorophyll but it help in chlorophyll formation and encourages vegetative plant growth. It stimulates seed formation and promotes nodule formation on legumes.

Boron is concerned with the uptake of calcium by the roots and with its efficient use in the plants. It helps in absorption of nitrogen and act as a regulator of potassium-calcium ratio. *Manganese* helps in chlorophyll formation and acts as a catalyst in oxidation and reduction reaction within plant tissues.

Iron is not a constituent of chlorophyll but it helps in its formation. A deficiency of iron causes chlorosis. It helps in enzyme systems which bring about oxidation-reduction reactions in plants. It is essential for the synthesis of protein contained in the chloroplasts.

Zinc is essential for enzyme systems necessary for plant metabolism. It is considered to be helpful in the formation of some growth harmones. It is helpful in the reproduction of certain plants.

Molybdenum acts in enzyme systems which bring about oxidation-reduction reaction in plants. It increases efficiency of legumes in fixing atmospheric nitrogen.

Copper acts as an "electron carrier" in enzymes which bring about oxidation-reduction reactions in plants. These reactions are essential to plant development and reproduction. It regulates respiration and also helps in utilization of iron.

Chlorine has been proved to be essential plant nutrients for proper plant development especially in sugar beet, carrot, lettuce, barley, wheat, cotton and clover.

Response of Altitude on Agricultural Development

The relief is an important element of the ecological setting directly influencing land cultivation and farming by affecting the ease of cultivation and the degree of accessibility. Very diverse climatic conditions are met within the same area due to the variations in altitude. For example in Bhagirathi catchment following climatic zones (Table 3.1) can be identified:

Table 3.1 : Climatic Zones

Altitude (Meters AMSL)	Climatic Zones	Temperature °C		
		Mean Annual	Mean June	Mean January
300-900	Tropical Zone	18.90	27.20	11.10
900-1800	Warm Temperate Zone	13.90	21.10	6.10
1800-2400	Cool Temperate Zone	10.30	17.20	2.80
2400-3000	Cold Zones	4.50	13.30	1.70
3000-4000	Alpine Zones	3.00	5.60	below zero six months
4000-4800	Glacial Zones	Ten months below zero snow melts in two months between July and August 2.2 to 3.9.		
Above 4800	Perpetual Frozen Zones	Cold desert with no vegetation.		

Source : Soil and Landuse Survey, New Delhi.

The effects of altitude is reflected in the occurrence of natural vegetation. At higher altitudes conifers viz. Chir, Deodar, Kail and Fir are found whereas at low altitude broad leave trees viz. Oak, Sal, Jamun, etc. are found.

The effect of altitude is also evident from the changes in agricultural landuse pattern at different altitudes. In general it is observed that up to 7000 feet the double cropping is done on a large scale and the climate support the agriculture in both rabi and kharif season while between 7000 to 9000 feet the mixed agriculture i.e., cultivation of field in one season and rearing of animals in another season is practicable. But beyond the 9000 feet of altitude due to severe cold, high wind velocity, snow covered areas and very short season available for growing crop, the cultivation of field is not possible and only animal rearing supports the human life on such elevated remote areas.

Response of Slope on Agricultural Development

Slope is an important and only single factor which has got predominant infeuence on the problem of soil erosion. The effect of slope on agriculture may both direct and indirect. A direct influence of slope is in the form of the restraint on cultivation and accessibility. The indirect effect of slope manifests itself in pedological and climatic modifications including the position and depth of water table, development of slope and drainage pattern. It is a fact that with the increase in steepness of slope, use of even very simple farm machinery become difficult.

Cultivation is being practiced on steeper slopes in hilly parts of the region is a fact that the farmers use terraces for cultivation is a testimony of their consciousness of the problem of erosion. These terraces are generally constructed on slope steeper than 33 degrees, the present deadline for arable farming, though this limit may be increased except where the conditions are highly unfavourable. The formation of benches or terraces on sloppy lands is necessary if the cultivation is to be made possible. But the condition of soil depth will be of great importance in this case. If the soil layer is shallow, the formation of terraces would be more harmful than the fruitful. It is therefore, the combination of slope and soil depth that determines the economy of cultivation on higher slopes.

On steep slopes, the run-off and erosion are active. Landslides are common in such mountainous areas particularly in the Kharif (rainy) season. Construction of terraces on steep slopes is not proper and do not meet the technical requirement as they are constructed on gently to moderate slopes. From the run-off and erosion point of view, the condition of the area is pitiable excluding the river terraces which have slight to moderate erosion. It is emphasized that in agricultural lands having steep slopes, terraces with inward slopes should be constructed. And to check the soil erosion proper afforestation and grass cover should be planted.

It manifests, that as the gradient of slope increases, the soil erosion and amount of surface run-off also increases which affects adversely the cultivation on the slopes.

Response of Fire on Agricultural Development

Fire which is thought to be the devastating for agriculture but the truth is not so. Fire is also a major factor of environment which plays a significant role in shaping the vegetation in terrestrial ecosystem. Therefore, properly managed fire can be an ecological tool of great significance. The fire can be managed for elimination of competitors, harmful animals and release of nutrients in forests and grassland ecosystems.

Fire can be grouped into two viz.:

(a) Crown fires or wild fires

(b) Surface fires

Wild fires destroys all types of vegetation and the organic matter. These are limiting to almost all types of organisms. After wild fires, the biotic communities starts on a bare area and it takes many years to establish the normal life and regenerate leads to the area becoming productive again.

Surface fires which selectively destroy the organisms i.e., they are more limiting to certain types of organisms and enhance the development of other organisms which are tolerant to fire. Light surface fires promote bacterial activity which helps in breaking down the bodies of plants and releasing the nutrients for fresh growth. Nitrogen fixing legumes are known to perform well after a light fire, which promote their germination. In Eucalyptus and Pinus, auxiliary buds sprout and exhibit rapid growth after the fire. In several grasses like Andropogan scoparius, the crude protein and nutrients are known to increase after fire (Gopal and Bhardwaj, 1979).[6] Fire is of immense use in grassland agriculture. Under dry conditions, fire helps to maintain grassland against the invasion of desert shrubs, whereas under moist conditions fire favours grass over trees.

NOTES

1. Pfeffer (1903), *The Physiology of Plants*, vol. 2, Clarendon Press, Oxford.
2. Beneck, W. and L. Jost (1923), *Pflanzenphysiologie*, vol. 2, Gustav Fischer, Jena.
3. Morgan, T.H., (1908), *Evolution and Adaptation*, Macmillan, New York.
4. Molisch, H., (1896), Das Erfrienren Von Pflanzen bei Temperaturen uber den Eispunkt, Sitzber Akad. *Wiss. Wien, Math. Naturw.* Kl., 105(1):82-95.
5. Miller, E.C. (1938), *Plant Physiology with special reference to Green Plants*, McGraw Hills, New York.
6. Gopal, B. and N. Bhardwaj (1979), *Elements of Ecology*, Vikas Publishing House Pvt. Ltd., New Delhi.

4

Trend of Agricultural Development

Trend of Agricultural Development can be assessed by the adoption of Modern Agriculture inputs on the one hand and its response on the agricultural productivity on the other. Besides ecological condition is a deciding factor in agricultural development. Weather is also a key variable in agricultural production, and variability of rainfall also play its role in the entire production schedule.

Though the environmental, technological and institutional factors are all important in shaping the pattern of agriculture, the environmental condition hold the greatest influence not only on the existing agriculture but also to measure the agricultural potential for future planning. The implementation of technological measures are also conditioned by the environment.

In this investigation first, the trend of agricultural development has been analyzed on the basis of agricultural inputs applied for the improvement of agricultural development. Secondly, the trend of agricultural development has been analyzed on the basis of agricultural productivity of wheat, rice, maize, barley, pulses and foodgrain for some selected districts of north-western Himalaya.

TRENDS OF AGRICULTURAL DEVELOPMENT BASED ON AGRICULTURAL MACHINERY

Trend of Wooden Plough Development

On the basis of availability of data from 1951 to 1992, the trend line and rate of change in Wooden Plough development have been find out for 17 districts of the region and also for the region as a whole.

It is very clear from Fig. 4.1 that the trend and rate of change in wooden plough is negative in four districts out of 17 namely Baramula, Srinagar, Leh and Dehradun with the rate of change equal to -4.28 per cent, -12.30 per cent, -2.05 per cent, and -2.29 per cent respectively. In general these districts are agriculturally advanced and negative trend of wooden plough is a positive factor for agricultural development if it is overtaken by iron plough. While the overall region show a positive trend with the rate of change equal to 31.22 per cent. Positive trend with highest rate of change in wooden plough is noted in Doda (31.22%), Poonch (25.31%), Pithoragarh (23.37%), and Udhampur (22.74%). Wooden plough is an indicator of traditional form of agriculture dominance in these districts. The actual number of wooden plough in the region as a whole increased from 11,03,993 to 18,56,775 in the given period of time and in terms of growth it is 68.19 per cent. Log values of wooden plough development are given in Appendix-I.

Trend of Iron Plough Development

It is evident from the trend line drawn and rate of change calculated on the basis of values given in Appendix-II that the overall trend of iron plough utilization in the region is positive and its rate of change from 1951 to 1992 is 205.49 per cent (Table 4.1). It is clear from Fig. 4.2, that out of

Table 4.1 : Rate of Change in various Technological Parameters of Agricultural Development (in %)

	Wooden Plough	*Iron Plough*	*Pumping Sets*	*Sugarcane Crushers*	*Tractors*	*Wheat Threshers*	*Rice Threshers*	*Total Shellers and Threshers*	*Intensity of land use*	*Irrigation*
	1961 to 1992	*1961 to 1992*	*1961 to 1992*	*1961 to 1992*	*1961 to 1992*	*1972 to 1992*	*1972 to 1992*	*1972 to 1992*	*1961-63 to 1992-95*	*1961-63 to 1992-95*
Bilaspur	8.94	237.55	282.82	16.78	336.52	1413.56	108.93	92.75	3.24	10.39
Chamba	11.94	390.91	N.A.	N.A.	N.A.	1102.26	242.77	84.08	-2.06	-9.00
Hamirpur	N.A.	N.A.	N.A.	-31.61	N.A.	2851.21	530.96	94.98	N.A.	N.A.
Kangra	N.A.	N.A.	137.14	-30.82	N.A.	389.78	443.25	81.97	N.A.	7.03
Kinnaur	N.A.	N.A.	N.A.	N.A.	N.A.	N.A.	N.A.	N.A.	N.A.	-7.01
Kullu	N.A.	N.A.	-80.28	N.A.	181.84	751.14	73.78	73.78	N.A.	338.31
Lahaul and Spiti	N.A.	N.A.	N.A.	N.A.	N.A.	N.A.	N.A.	41.25	N.A.	N.A.
Mandi	20.78	726.04	86.44	69.04	263.08	545.65	216.23	84.08	1.85	-2.97
Shimla	1.39	572.73	0.23	N.A.	N.A.	781.05	175.42	51.36	-2.97	-12.70
Sirmour	10.66	315.86	177.33	3.51	225.63	888.55	455.90	92.75	3.25	11.36
Solan	N.A.	N.A.	172.27	-45.05	126.46	293.55	227.34	94.98	N.A.	48.90
Una	N.A.	N.A.	54.88	-41.79	242.77	302.72	65.96	92.75	N.A.	185.50
Anantnag	2.80	724.14	-67.27	N.A.	143.58	N.A.	N.A.	N.A.	7.50	5.84
Baramula	-4.28	N.A.	-28.39	N.A.	246.71	N.A.	3.51	N.A.	N.A.	3.48
Doda	31.22	N.A.	N.A.	N.A.	N.A.	N.A.	N.A.	N.A.	0.69	-7.90
Jammu	6.41	189.73	46.22	3.28	103.41	172.27	316.87	99.53	4.89	18.03
Kathua	19.12	79.06	21.06	8.14	230.56	154.10	N.A.	90.55	14.55	20.56
Leh	-2.05	-14.89	73.78	N.A.	N.A.	N.A.	N.A.	N.A.	-1.37	N.A.
Poonch	25.31	768.96	N.A.	40.45	129.25	N.A.	N.A.	N.A.	N.A.	N.A.
Rajouri	N.A.	N.A.	N.A.	N.A.	N.A.	N.A.	N.A.	94.98	N.A.	N.A.
Srinagar	-12.30	206.90	-58.31	N.A.	N.A.	N.A.	N.A.	N.A.	4.42	7.27

(Contd...)

Table 4.1 *(Contd...)*

Udhampur	22.74	269.83	N.A.	10.87	N.A.	N.A.	N.A.	N.A.	N.A.	-9.87
Almora	N.A.	151.18	57.96	0.46	N.A.	N.A.	-46.30	99.53	15.29	N.A.
Chamoli	N.A.	338.31	N.A.	N.A.	N.A.	N.A.	N.A.	N.A.	10.39	N.A.
Dehradun	-2.29	86.11	199.05	-45.29	122.87	231.13	983.93	99.53	3.95	N.A.
Garhwal	2.32	248.26	N.A.	N.A.	509.63	145.47	N.A.	N.A.	10.15	N.A.
Nainital	N.A.	N.A.	1413.56	-56.85	110.12	N.A.	N.A.	79.89	5.84	N.A.
Pithoragarh	23.37	107.28	N.A.	-75.11	N.A.	N.A.	N.A.	N.A.	18.29	N.A.
Tehri Garhwal -	N.A.	N.A.	N.A.	N.A.	N.A.	N.A.	N.A.	N.A.	-0.46	N.A.
Total	31.22	205.49	78.65	27.94	139.65	530.96	129.09	88.36	7.26	25.18

Note : N.A. Not Available.

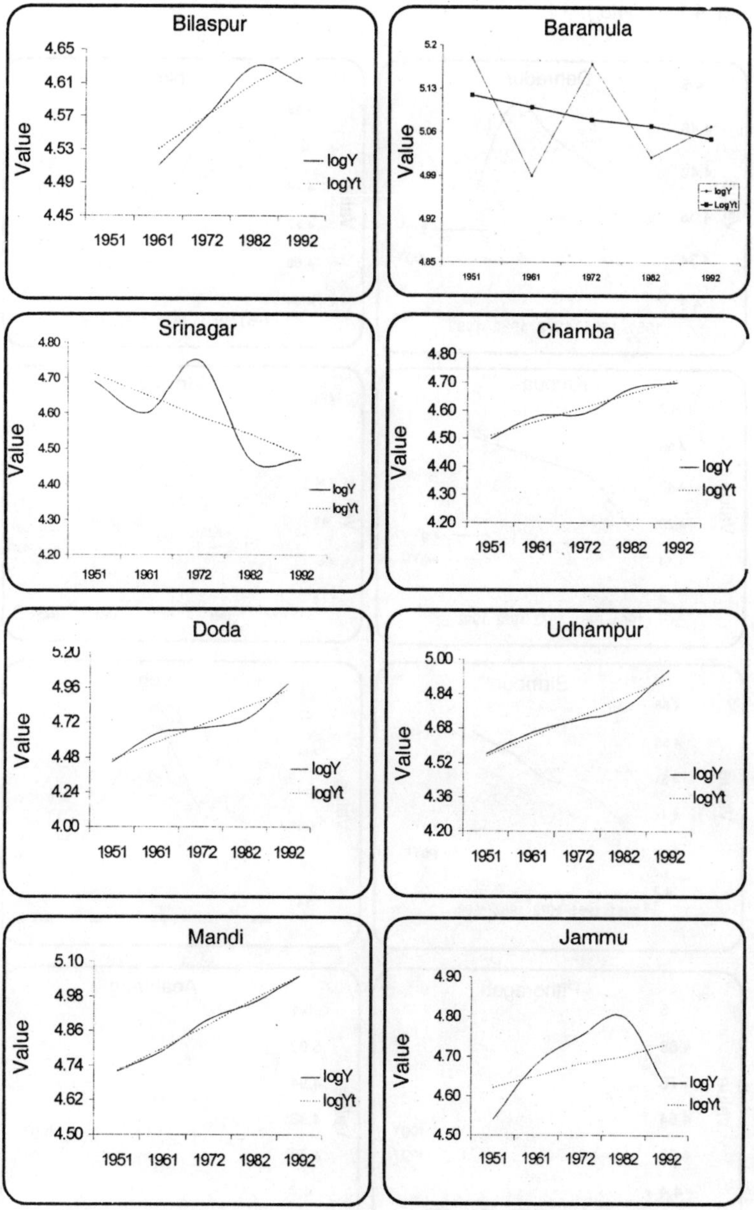

Fig. 4.1 : Trend line of Wooden Plough Development in Selected Districts of North Western Himalaya

Fig. 4.1 : *(contd.)*

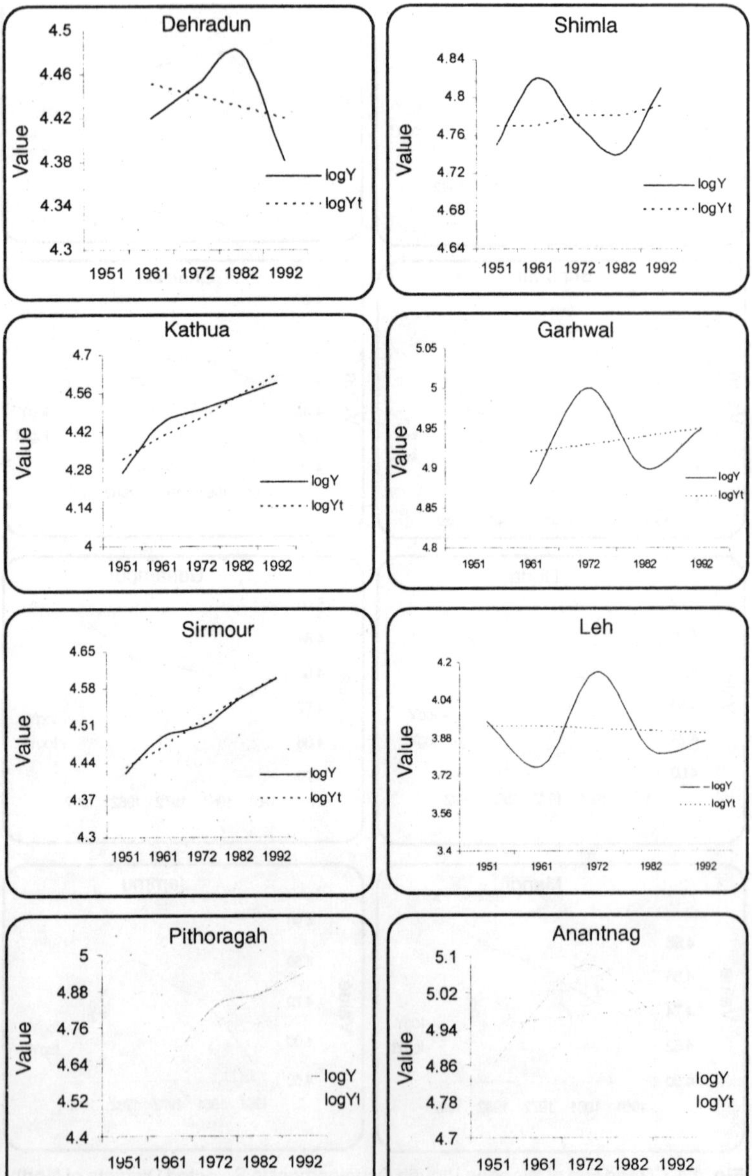

(Contd....)

Fig. 4.1 *(Contd....)*

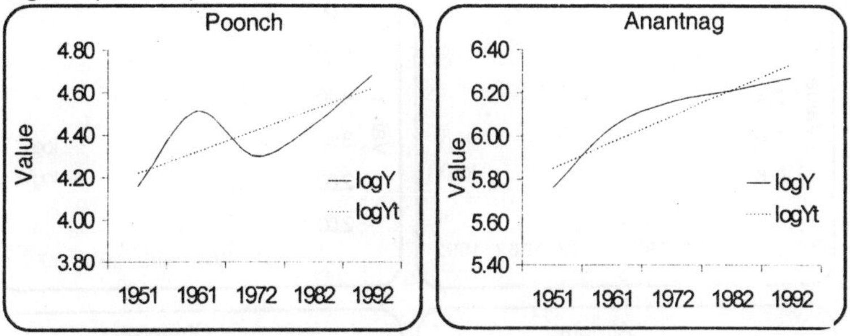

17 district taken into account for trend line curves 16 shows positive trend and rate of change except Leh district which is a cold arid district mostly shown uncertain level of agricultural development because only scientific agricultural practices can be possible here on a limited land which shows high fluctuation with a little development, in any parameter.

In general, it is revealed that there is a shift towards the adoption of iron plough instead of traditional plough because its effciency and durability is more as compared to the wooden plough. In some of the districts, the rate of change is found abnormally high like in Mandi (726.04%), Shimla (572.73%) and Poonch (768.96%). Reason for such a high growth is that the condition of iron plough utilization is very low in these districts and a little increase in its magnitude cause high rate of growth in percentage. As a result only Kathua (79.06%) and Dehradun (86.11%) are noted with less than 100 per cent rate of change in case of iron plough development.

Trend of Pumping Set Development

It is evident from the trend line drawn (Fig. 4.3) and rate of change calculated (Table 4.1) on the basis of value given in Appendix-III, that overall trend of pumping set development in the region is positive. Value of its rate of change from 1951 to 1992 is found to be 78.65 per cent. Development in the

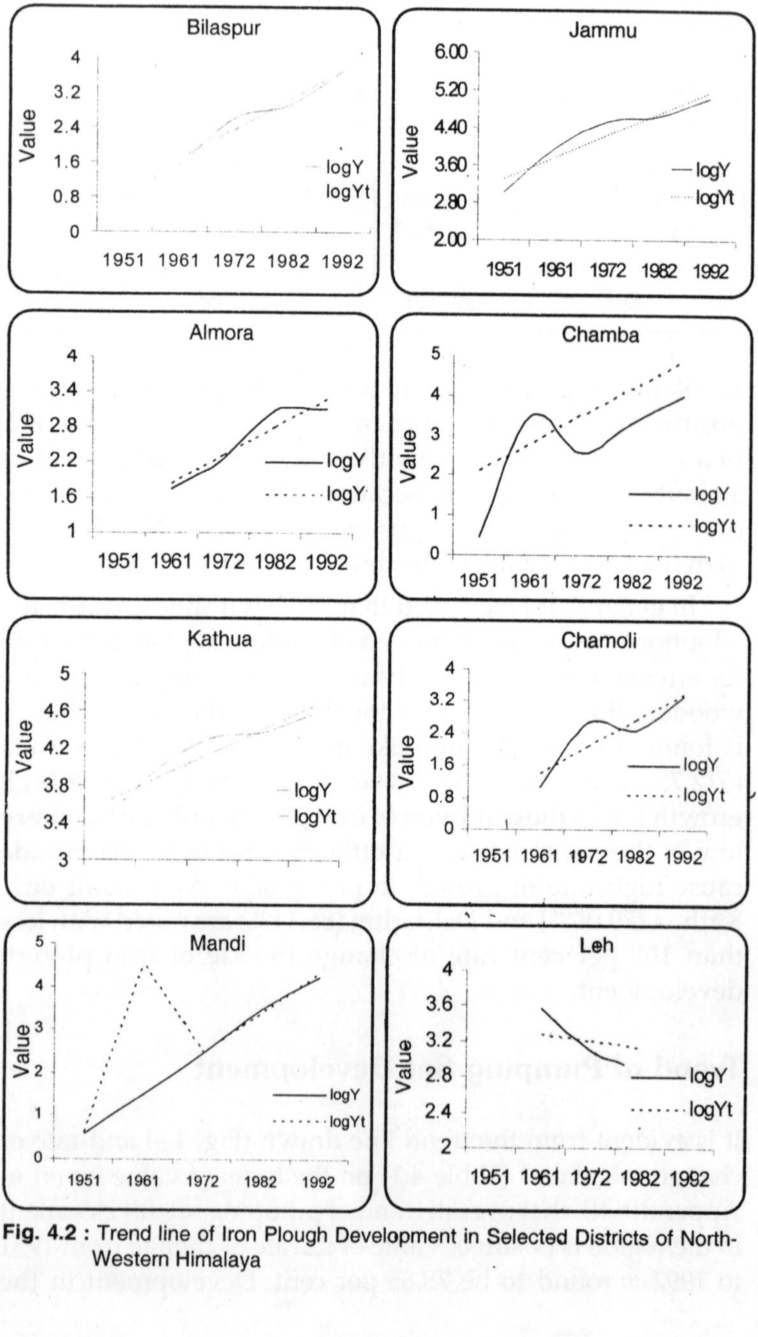

Fig. 4.2 : Trend line of Iron Plough Development in Selected Districts of North-Western Himalaya

Fig. 4.2 : *(Contd.)*

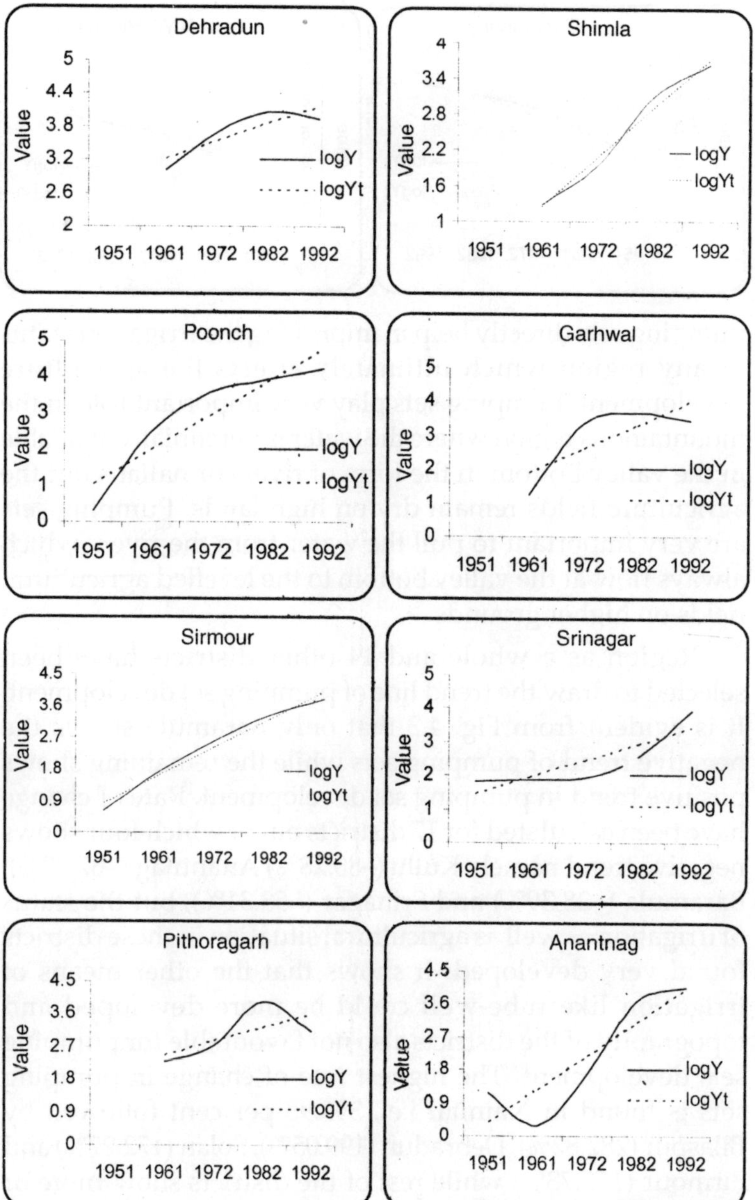

(Contd....)

Fig. 4.2 : *(Contd.)*

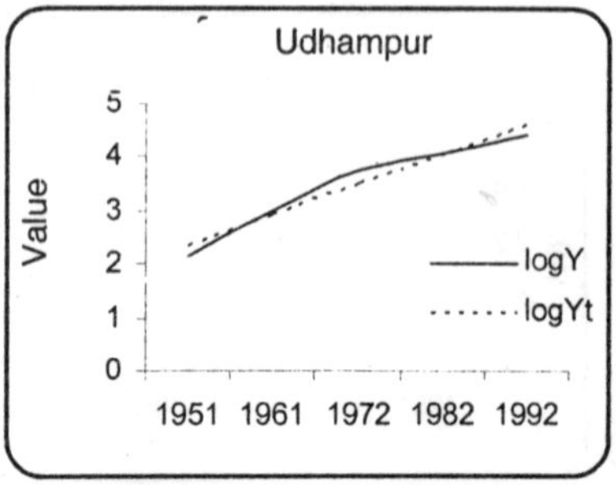

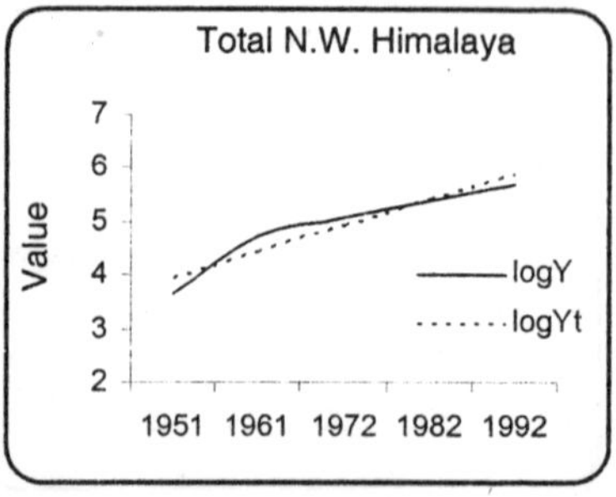

pumping sets directly help in improving the irrigation status of any region which ultimately effects the agriculture development. Pumping sets play very important role in the mountainous region where the water is perennially available in the valley bottom in the form of rivers or nallahs but the agriculture fields remain dry on high lands. Pumping sets are very important to pull the water from the rivers which always flow at the valley bottom to the levelled agricultural fields on higher grounds.

Region as a whole and 14 other districts have been selected to draw the trend line of pumping set development. It is evident from Fig. 4.3 that only Baramula shows the negative trend of pumping sets while the remaining shows positive trend in pumping set development. Rate of change have been calculated for 17 districts out of which four shows negative trend namely Kullu (–80.28%) Anantnag (–67.27%), Baramula (–28.39%) and Srinagar (–58.31%), but the status of irrigation as well as agricultural situation in these districts found very developed. It shows that the other means of irrigation like tube-well could be more developed and topography of the districts also not favourable for pumping sets development. The highest rate of change in pumping sets is found in Nainital i.e., 356.98 per cent followed by Bilaspur (282.82%), Dehradun (199.05%), Solan (172.27%) and Sirmour (177.73%) while rest of the districts show more or less average rate of change in pumping set utilization and development.

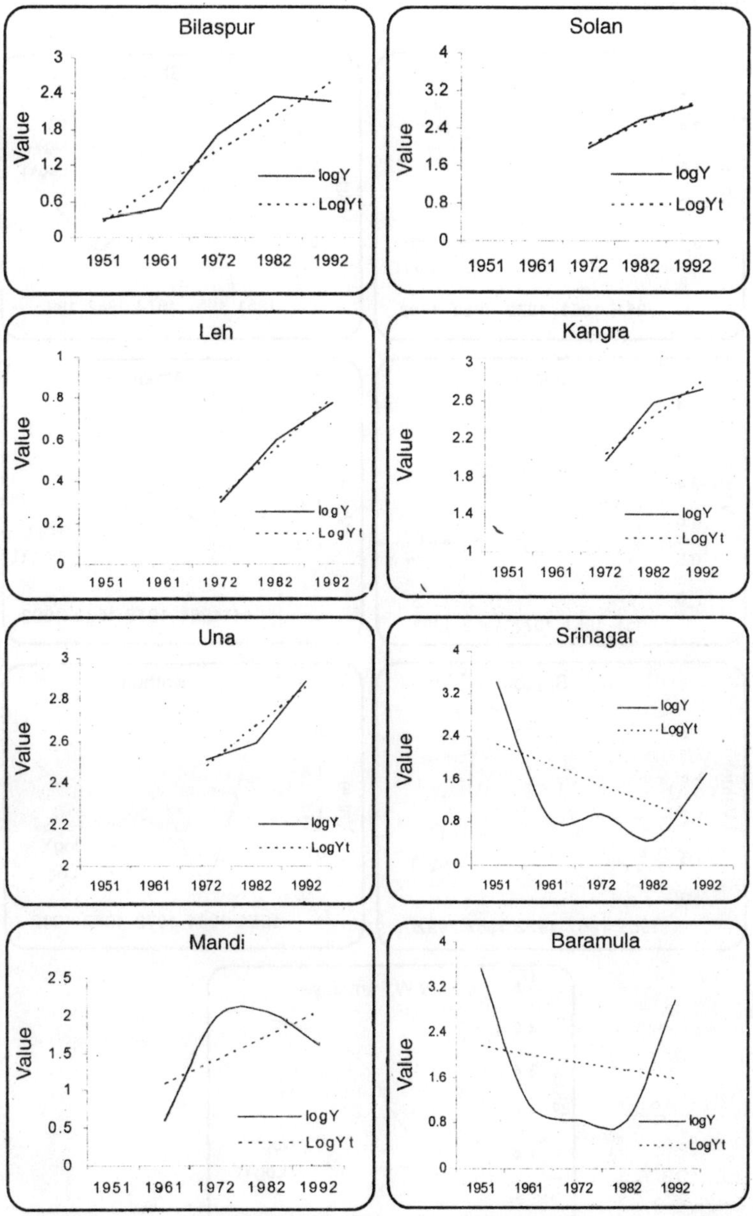

Fig. 4.3 : Trend line of Pumping Sets Development in Selected Districts of North-Western Himalaya

Fig. 4.3 : *(Contd.)*

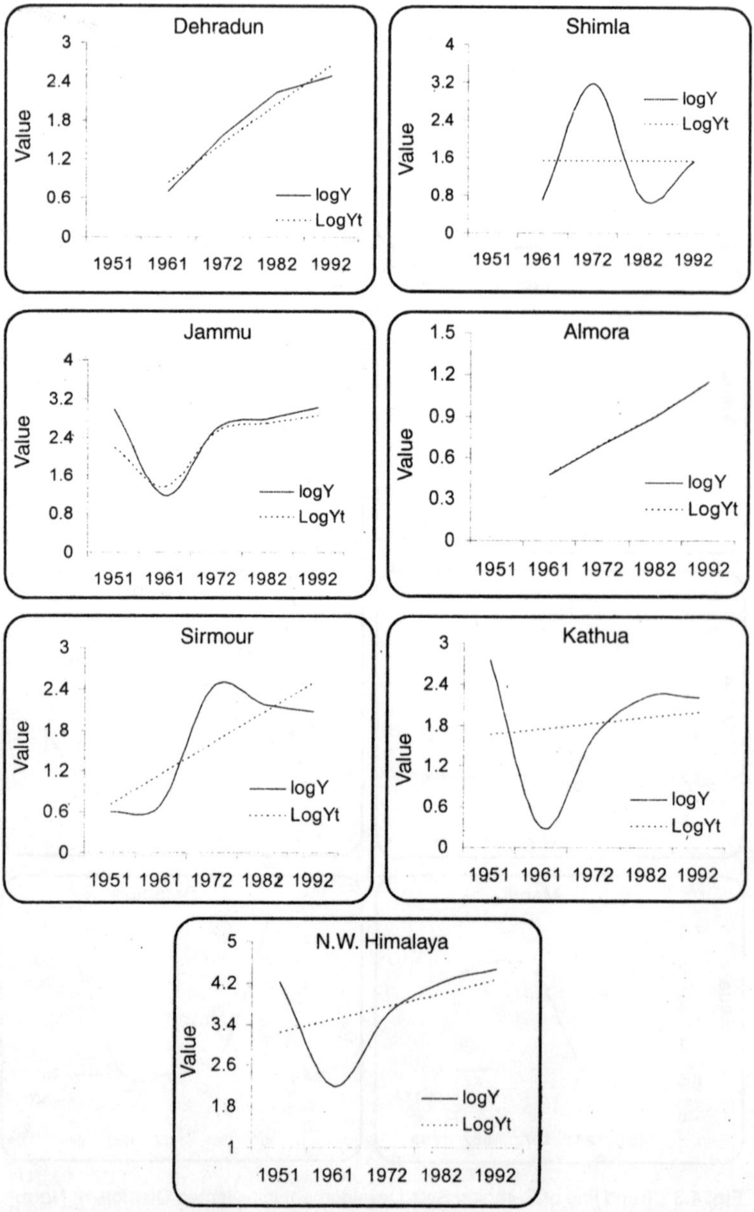
Dehradun
Value
3
2.4
1.8
1.2
0.6
0
1951 1961 1972 1982 1992
logY
LogYt
Shimla
Value
4
3.2
2.4
1.6
0.8
0
1951 1961 1972 1982 1992
logY
LogYt
Jammu
Value
4
3.2
2.4
1.6
0.8
0
1951 1961 1972 1982 1992
logY
LogYt
Almora
Value
1.5
1.2
0.9
0.6
0.3
0
1951 1961 1972 1982 1992
logY
LogYt
Sirmour
Value
3
2.4
1.8
1.2
0.6
0
1951 1961 1972 1982 1992
logY
LogYt
Kathua
Value
3
2.4
1.8
1.2
0.6
0
1951 1961 1972 1982 1992
logY
LogYt
N.W. Himalaya
Value
5
4.2
3.4
2.6
1.8
1
1951 1961 1972 1982 1992
logY
LogYt

Trend of Tractorization

Tractor is an important indicator of agricultural development in a region. It is utilized in different form at all stages of agricultural practices e.g., in the beginning it is used to plough the fields, then levelling the seed beds, to irrigate the field, sometimes its motor has been used to pull the water through pumping sets, it is also used to pump the water from tube-well in those areas where electric supply is not proper, to the post harvest scene its motor is used to thresh out the crops produce, up to this its work is not complete because in the final stage it is also used to distribute and transport the agricultural produce for different markets and destination. In the afforestation programme it is also used in sowing of plant. It is multipurpose agriculture machinery.

It is revealed from Fig. 4.4 that the trend line drawn for all 14 districts and the region as a whole shows a positive trend of tractorization in the region. Rate of change in tractorization has been calculated for 14 districts on the basis of values given in Appendix-IV. It is revealed from Table 4.1 that the rate of change in tractorization for the region as a whole is 139.65 per cent which shows considerable development in this agricultural machine. However, the highest rate of change is recorded in Garhwal district i.e., 509.63 per cent from 1961 to 1992. While Bilaspur, Mandi, Sirmour, Baramula, and Kathua shows more than 200 per cent besides Kullu, Solan, Anantnag, Jammu, Poonch, Dehradun and Nainital where the rate of change lies between 100-200 per cent. There is no district to show rate of change less than 100 per cent. The above statistics shows a high level of tractorization or in other words high level of agricultural development in this mountainous region.

Trend of Sugarcane Crushers Development

Though the topography of the region is not favourable for sugarcane but it is grown on some levelled lands in the

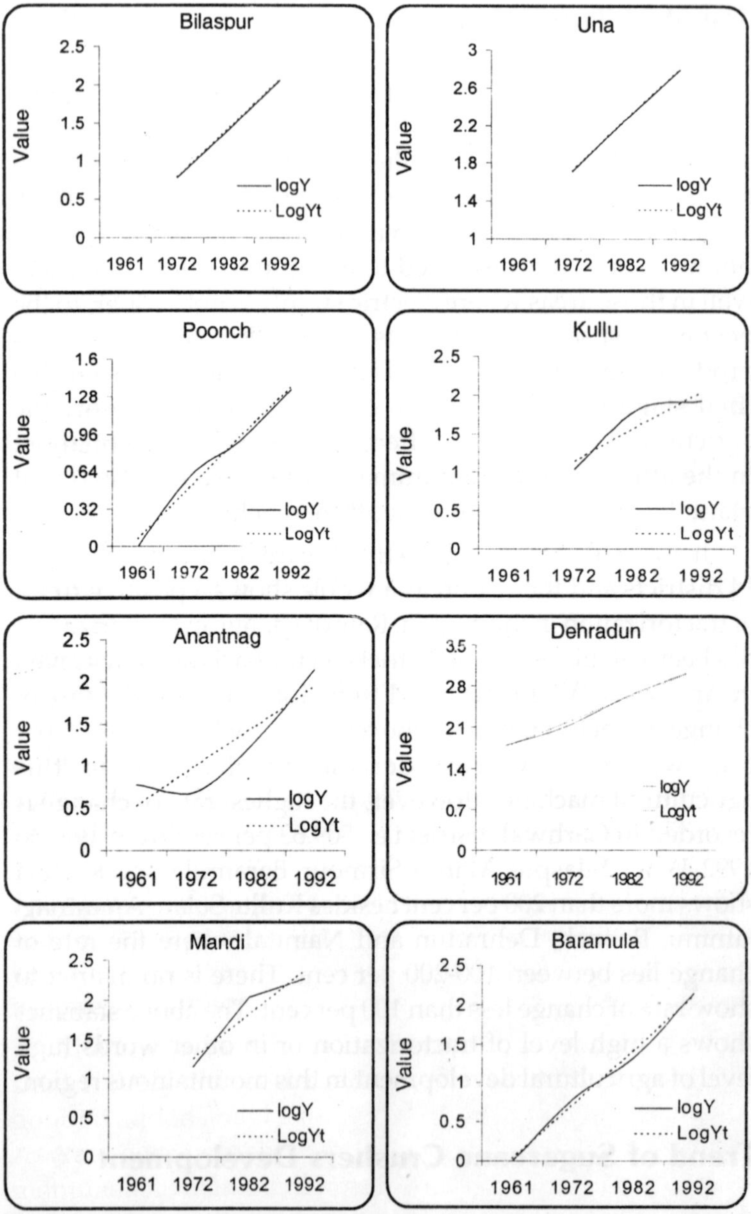

Fig. 4.4 : Trend line of Tractors Development in Selected Districts of North-Western Himalaya

Fig. 4.4 : *(Contd.)*

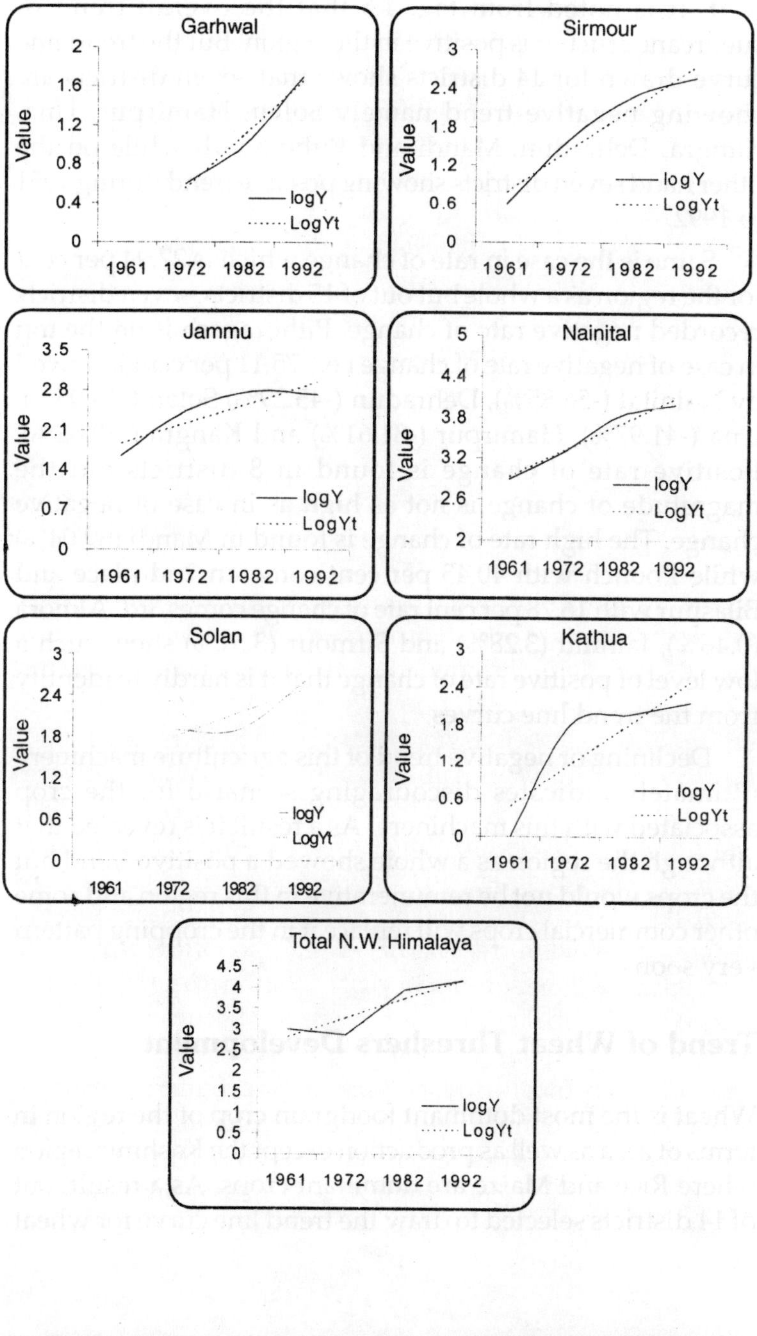
Garhwal
Value
2
1.6
1.2
0.8
0.4
0
1961 1972 1982 1992
logY
LogYt
Sirmour
Value
3
2.4
1.8
1.2
0.6
0
1961 1972 1982 1992
logY
LogYt
Jammu
Value
3.5
2.8
2.1
1.4
0.7
0
1961 1972 1982 1992
logY
LogYt
Nainital
Value
5
4.4
3.8
3.2
2.6
2
1961 1972 1982 1992
logY
LogYt
Solan
Value
3
2.4
1.8
1.2
0.6
0
1961 1972 1982 1992
logY
LogYt
Kathua
Value
3
2.4
1.8
1.2
0.6
0
1961 1972 1982 1992
logY
LogYt
Total N.W. Himalaya
Value
4.5
4
3.5
3
2.5
2
1.5
1
0.5
0
1961 1972 1982 1992
logY
LogYt

region, and needed sugarcane crusher for further processing of it. It is noted from Fig. 4.5 that the overall trend of sugarcane crusher is positive in the region. But the trend line curve drawn for 14 districts shows that seven districts are showing negative trend namely Solan, Hamirpur, Una, Kangra, Dehradun, Mandi and Pithoragarh while on the other hand seven districts showing positive trend during 1951 to 1992.

Same is the case in rate of change which is 27.94 per cent for the region as a whole but out of 15 districts, seven districts recorded negative rate of change. Pithoragarh is on the top in case of negative rate of change i.e. –75.11 per cent followed by Nainital (-56.85%), Dehradun (-45.29%) Solan (-45.05%), Una (-41.97%), Hamirpur (-31.61%) and Kangra (-30.82%). Positive rate of change is found in 8 districts but the magnitude of change is not as high as in case of negative change. The high rate of change is found in Mandi (69.04%) while Poonch with 40.45 per cent come on 2nd place and Bilaspur with 16.78 per cent rate of change comes 3rd. Almora (0.46%), Jammu (3.28%) and Sirmour (3.51%) show such a low level of positive rate of change that it is hardly to identify from the trend line curves.

Declining or negative trend of this agriculture machinery ultimately indicates discouraging scenario for the crop associated with this machinery. As a result it is revealed that although the region as a whole showed a positive trend but the crops would not be remunerative in this region and some other commercial crops will replace it in the cropping pattern very soon.

Trend of Wheat Threshers Development

Wheat is the most dominant foodgrain crop of the region in terms of area as well as production except the Kashmir region where Rice and Maize are dominant crops. As a result, out of 14 districts selected to draw the trend line curve for wheat

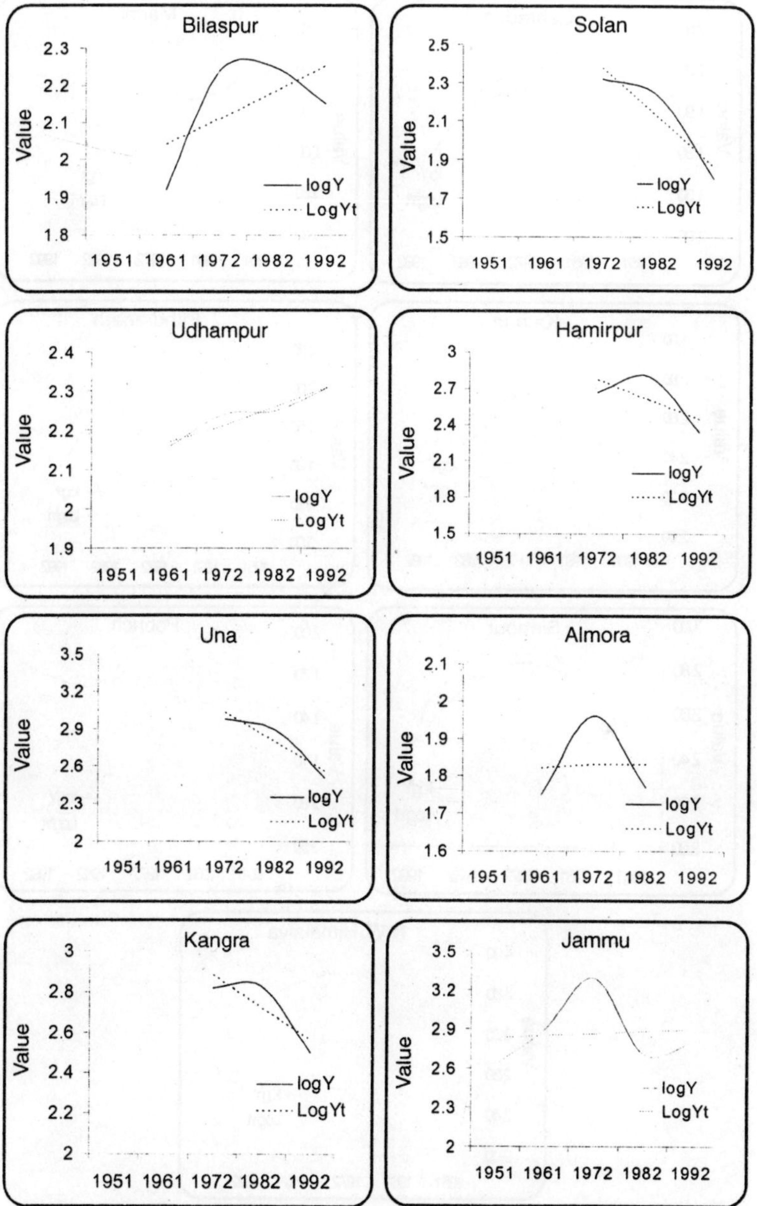

Fig. 4.5 : Trend line of Sugarcane Crushers Development in Selected Districts of North-Western Himalaya

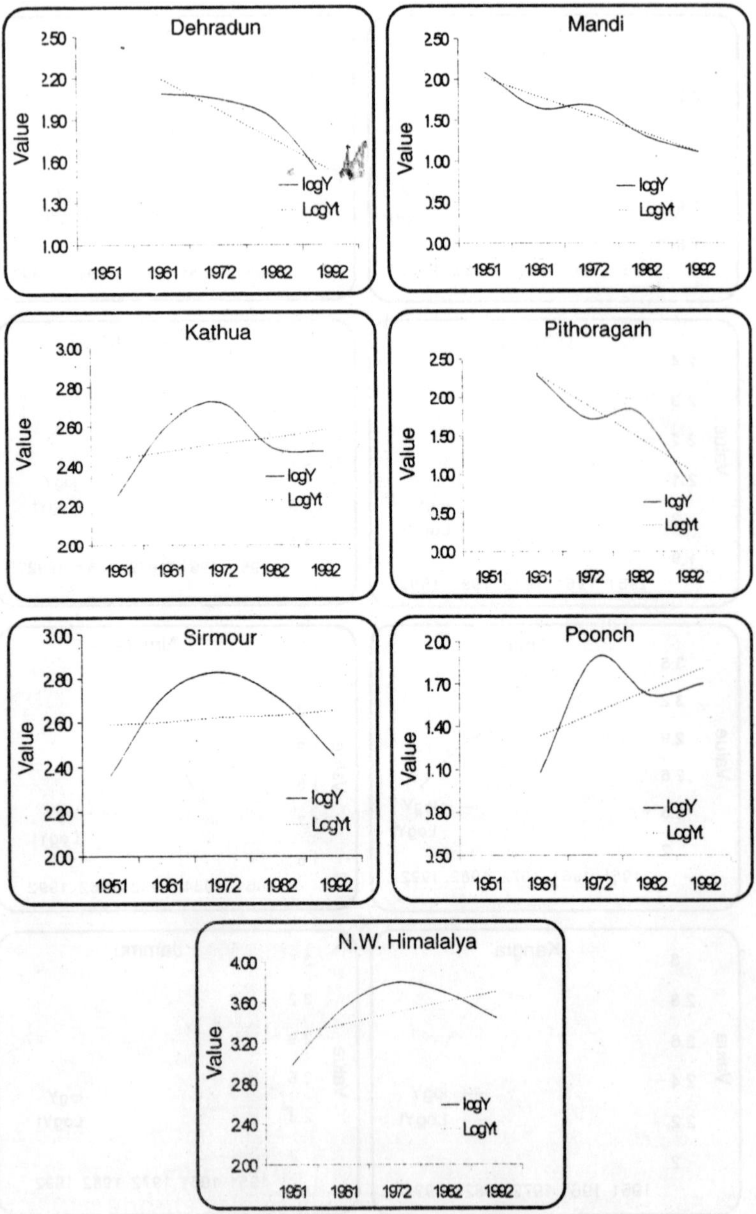
Dehradun
Mandi
Kathua
Pithoragarh
Sirmour
Poonch
N.W. Himalalya
Value
logY
LogYt
1951
1961
1972
1982
1992

threshers no district from Kashmir valley exists in it. It is evident from Fig. 4.6, that all the curve show convex appearance means that the actual growth of development is not at par with the earlier growth or in other words it is increased with reduced rate with the exception of Garhwal which shows concave appearance means the case is reverse with other districts.

The rate of change in development of wheat threshers for the region show very high magnitude i.e. 530.96 per cent from 1972 to 1992. The variations in rate of change also exists at district level. Chamba and Hamirpur show abnormally high positively rate of change i.e. 2530.12 per cent and 1142.51 per cent respectively followed by Sirmour (888.55%) Shimla (781.05%) and Kullu (751.14%). No district shows less than 100 per cent rate of change. The minimum rate of change found in Garhwal i.e. 145.47 per cent, Bilaspur (148.31%), and Kullu (154.10%). While rest of the districts found between these limits. It is revealed from the district level analysis that the Himachal region is on the top in terms of rate of change of wheat threshers. The wheat threshers is an input of post harvest technology, the efficiency of which is helpful in early processing of harvested crop and to make the field ready for next crop. It is also helpful to avoid damages of harvested crops by rain or other natural calamities by efficient methods of processing.

Trend of Rice Threshers Development

Rice is one of the dominant crops of the region in average and in production as well. In terms of productivity, it is only next to Maize. Irrigation is the prerequisite for rice. Its acreage and productivity in any region is directly related with the assured irrigation. Rice thresher is a post-harvest technology input. It helps in early processing of harvested paddy crop. So its production and distribution can be done on time and with efficiency as well. It is also an indicator of agricultural development.

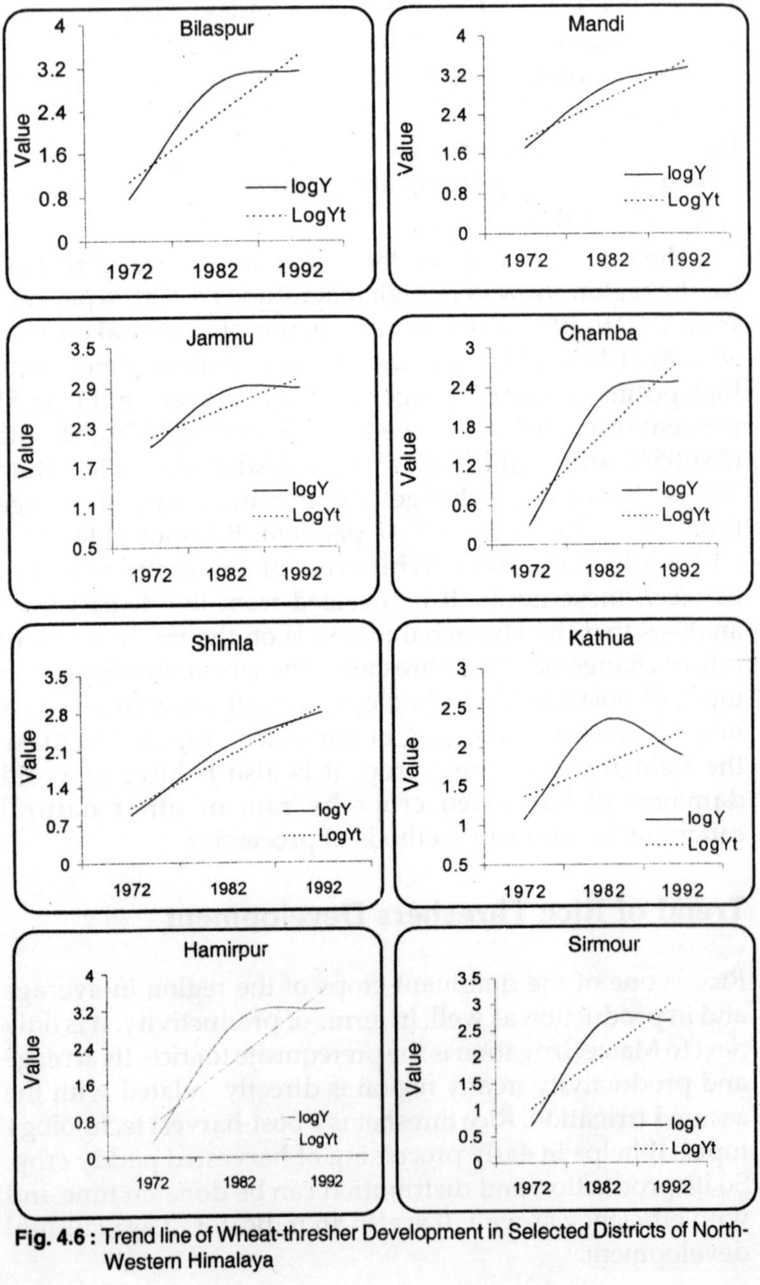

Fig. 4.6 : Trend line of Wheat-thresher Development in Selected Districts of North-Western Himalaya

Fig. 4.6 : *(Contd.)*

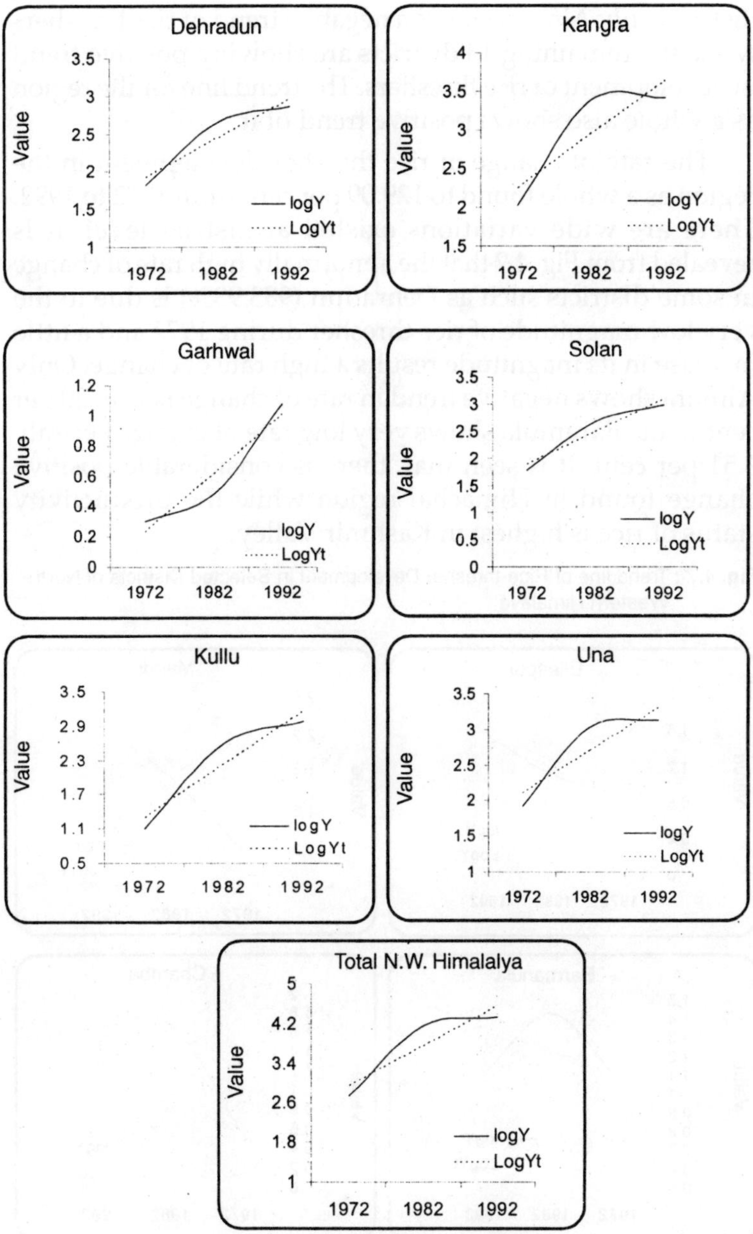
Dehradun
Value
3.5
3
2.5
2
1.5
1
1972
1982
1992
logY
LogYt
Kangra
Value
4
3.5
3
2.5
2
1.5
1972
1982
1992
logY
LogYt
Garhwal
Value
1.2
1
0.8
0.6
0.4
0.2
0
1972
1982
1992
logY
LogYt
Solan
Value
3.5
3
2.5
2
1.5
1
0.5
0
1972
1982
1992
logY
LogYt
Kullu
Value
3.5
2.9
2.3
1.7
1.1
0.5
1972
1982
1992
logY
LogYt
Una
Value
3.5
3
2.5
2
1.5
1
1972
1982
1992
logY
LogYt
Total N.W. Himalalya
Value
5
4.2
3.4
2.6
1.8
1
1972
1982
1992
logY
LogYt

It is evident from Fig. 4.7 that the trend line drawn for 14 districts only Almora shows a negative trend of rice threshers while the remaining 13 districts are showing positive trend in development of rice threshers. The trend line for the region as a whole also shows positive trend of it.

The rate of change of rice thresher development in the region as a whole found to 129.09 per cent from 1972 to 1992. There are wide variations existed at district level. It is revealed from Fig. 4.7 that the abnormally high rate of change in some districts such as Dehradun (983.93%) is due to the very low magnitude of rice thresher during 1972 and a little increase in its magnitude results a high rate of change. Only Almora shows negative trend in rate of change i.e.–46.30 per cent while Baramula shows very low rate of change i.e. only 3.51 per cent. It is seen that, there is considerable positive change found in Himachal region while the productivity status of rice is highest in Kashmir Valley.

Fig. 4.7 : Trend line of Rice-thresher Development in Selected Districts of North-Western Himalaya

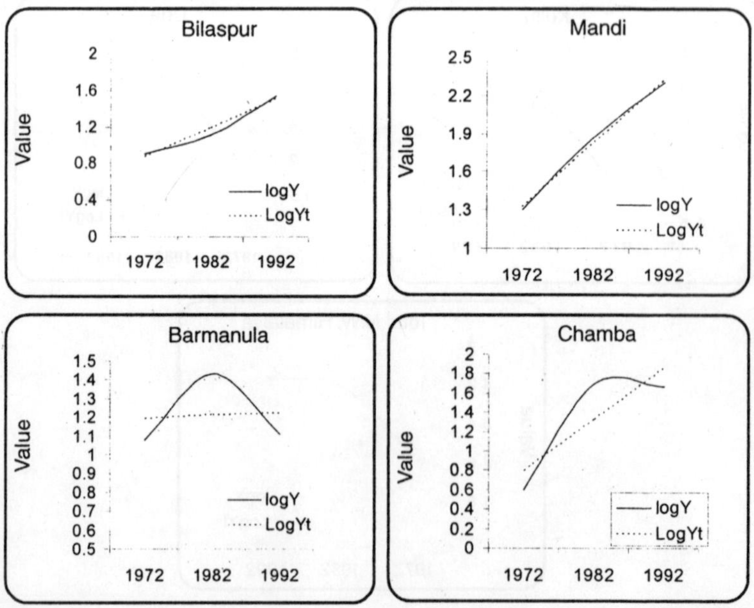

(Contd....)

Fig. 4.7 : *(Contd.)*

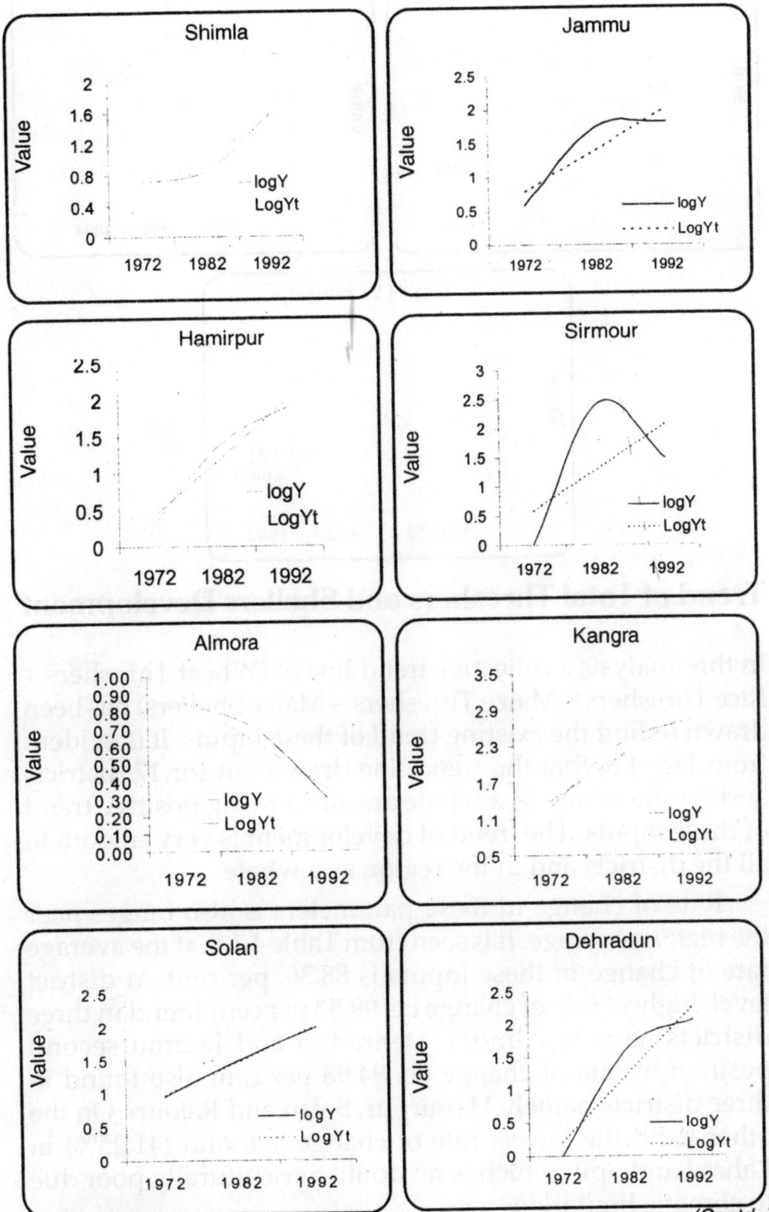

(Contd..)

Fig. 4.7: *(Contd.)*

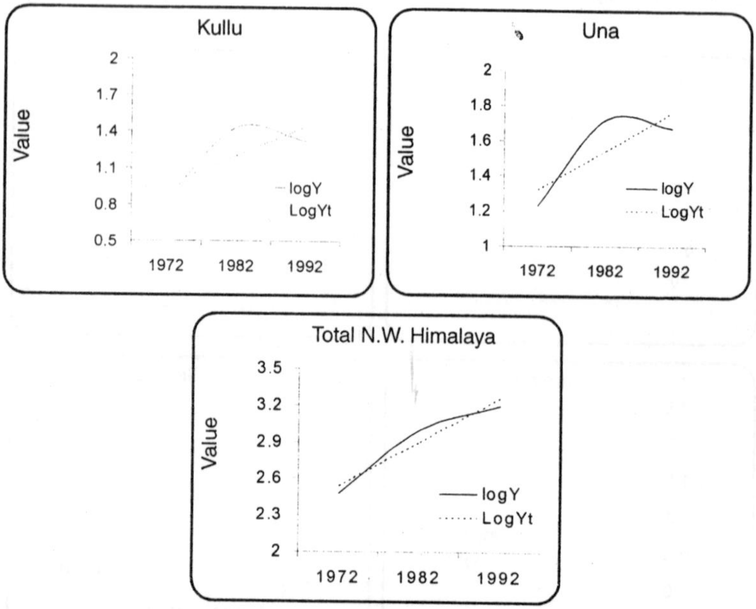

Trend of Total Threshers and Shellers Development

In this analysis a collective trend line of (Wheat Threshers + Rice Threshers + Maize Threshers + Maize Shellers) has been drawn to find the existing trend of these inputs. It is evident from Fig. 4.8, that the trend line drawn out for 17 districts and for the region as a whole are all showing positive trend of these inputs. The trend of development is very smooth in all the districts and in the region as a whole.

Rate of change in these parameters is also ranges near the regions average. It is seen from Table 4.1 that the average rate of change in these inputs is 88.36. per cent.At district level, highest rate of change i.e. 99.53 per cent found in three districts namely Almora, Dehradun and Jammu second position in rate of change i.e. 94.98 per cent also found in three districts namely Hamirpur, Solan and Rajouri. On the other hand the lowest rate of change is found (41.25%) in Lahaul and Spiti which is no doubt agriculturally poor due to climatic limitations.

Fig. 4.8 : Trend line of Total Threshers and Shelliers Development in Selected Districts of North-Western Himalaya

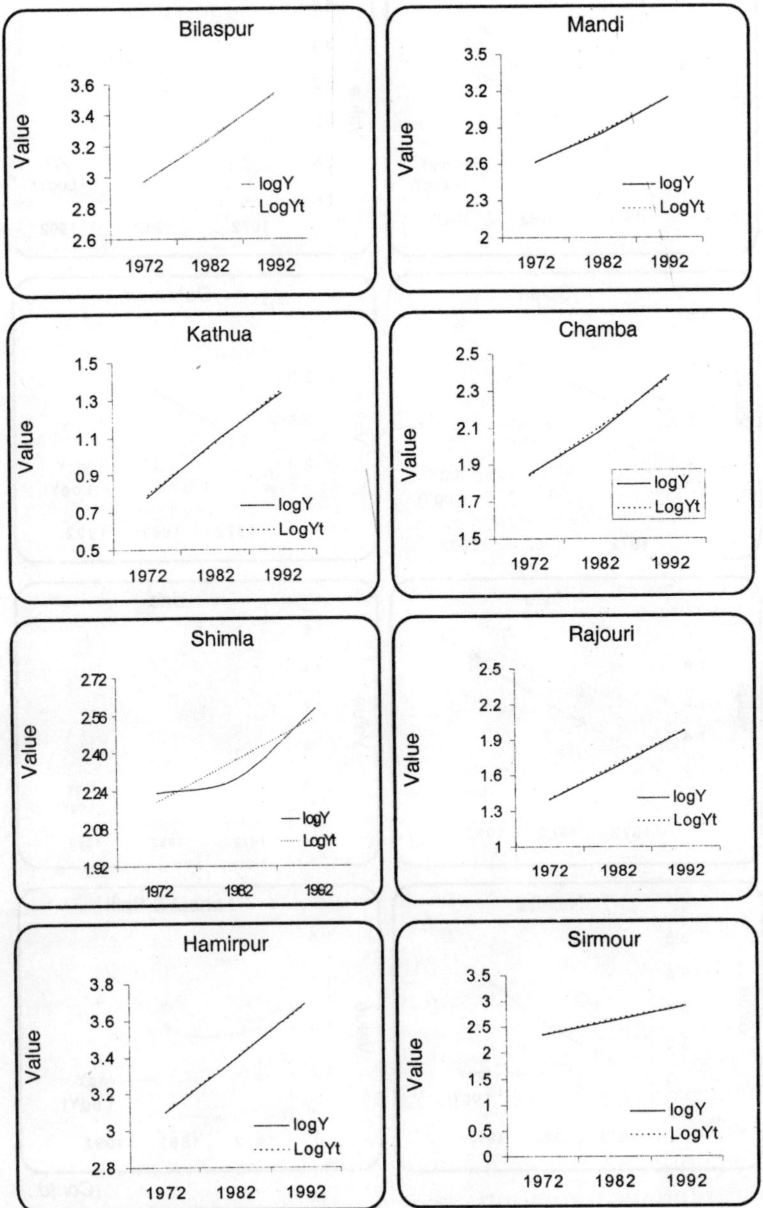

(Contd..)

Fig. 4.8 : *(Contd.)*

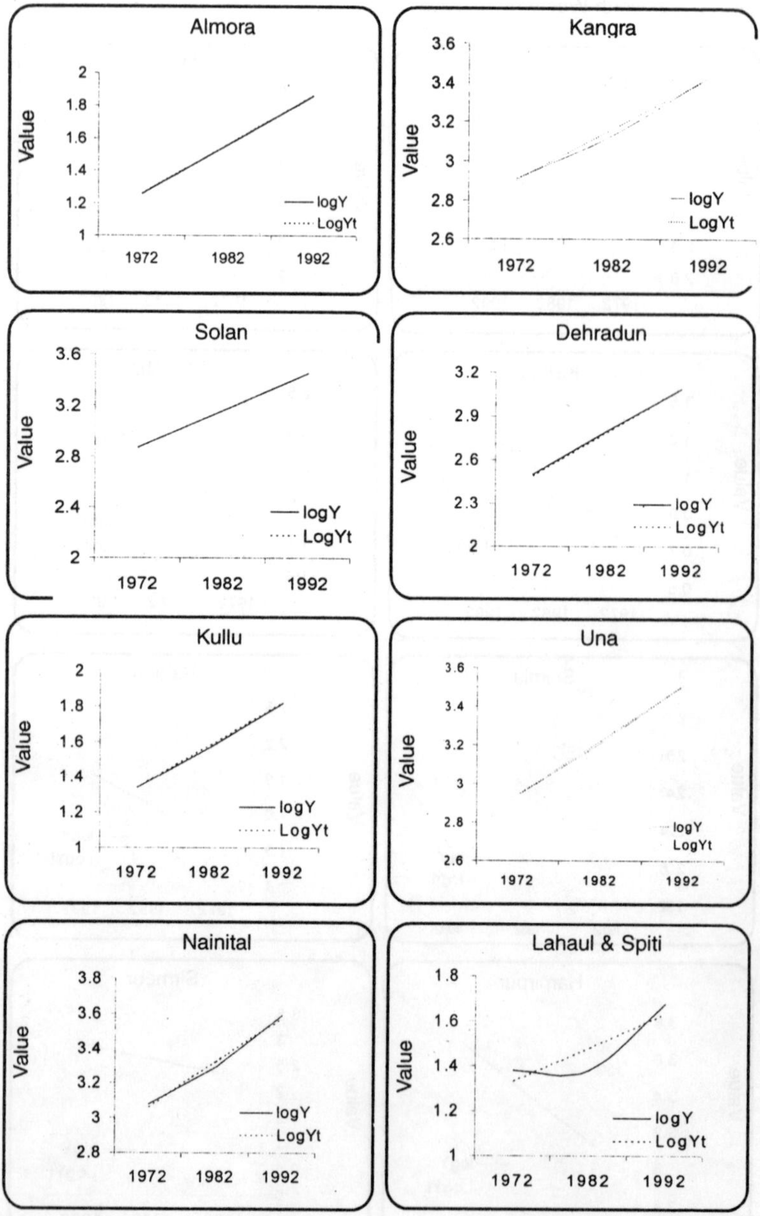

(Contd..)

Fig. 4.8 : *(Contd.)*

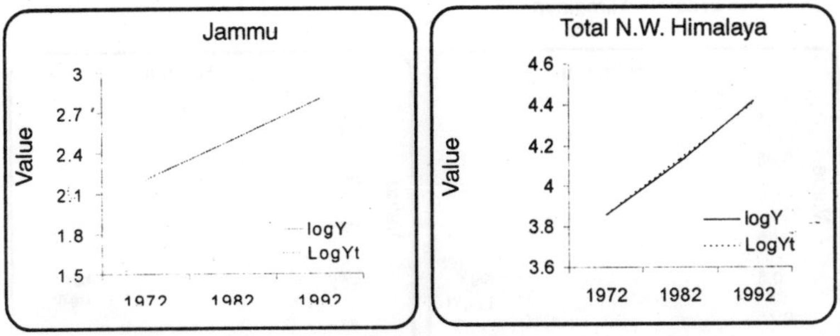

Trend of Irrigation Development

Irrigation is one of the basic necessities of agricultural development and all other inputs of agricultural development are useless and helpless without irrigation. To find out the trend of irrigation development in the region Trend line curve have been drawn for 17 districts out of 34 and also for the region as a whole.

It is revealed from Fig. 4.9 that the trend of irrigation in the region as a whole is positive. But on the other hand, out of 17 districts, six showing negative trends namely, Chamba, Kinnaur, Mandi, Shimla, Doda and Udhampur while eleven districts showing positive trend in the development of irrigation.

The rate of change in irrigation from 1961-63 to 1992-95 is recorded (25.18%) in the region as a whole (Table 4.1). Abnormally high rate of change is founded in Kullu district i.e. 338.31 per cent followed by Una 185.50 per cent, because the condition of irrigation in these districts was very poor during the pre-green revolution period and increased tremendously in later stages of post-green revolution scenes. On the other hand in some of the districts the rate of change is negative i.e. –9 per cent in Chamba, –7.01 per cent in Kinnaur, –2.97 per cent in Mandi, –12.70 per cent in Shimla, –7.90 per cent in Doda and –9.87 per cent in Udhampur. This negative trend is because of several reasons for example high growth of urbanization which encroached the fertile lands having irrigated as well besides agricultural policy and

Fig. 4.9 : Trend line of Irrigation Development in Selected Districts of North-Western Himalaya

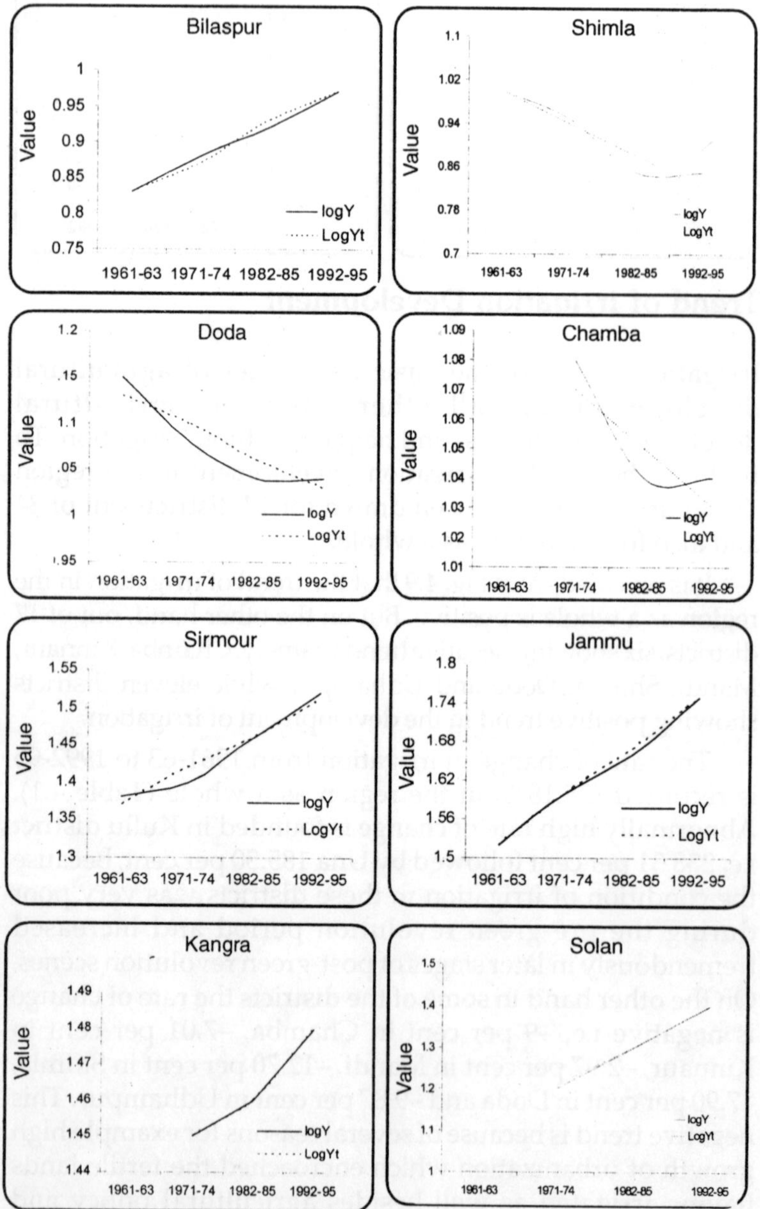

(Contd.)

Fig. 4.9 : *(Contd.)*

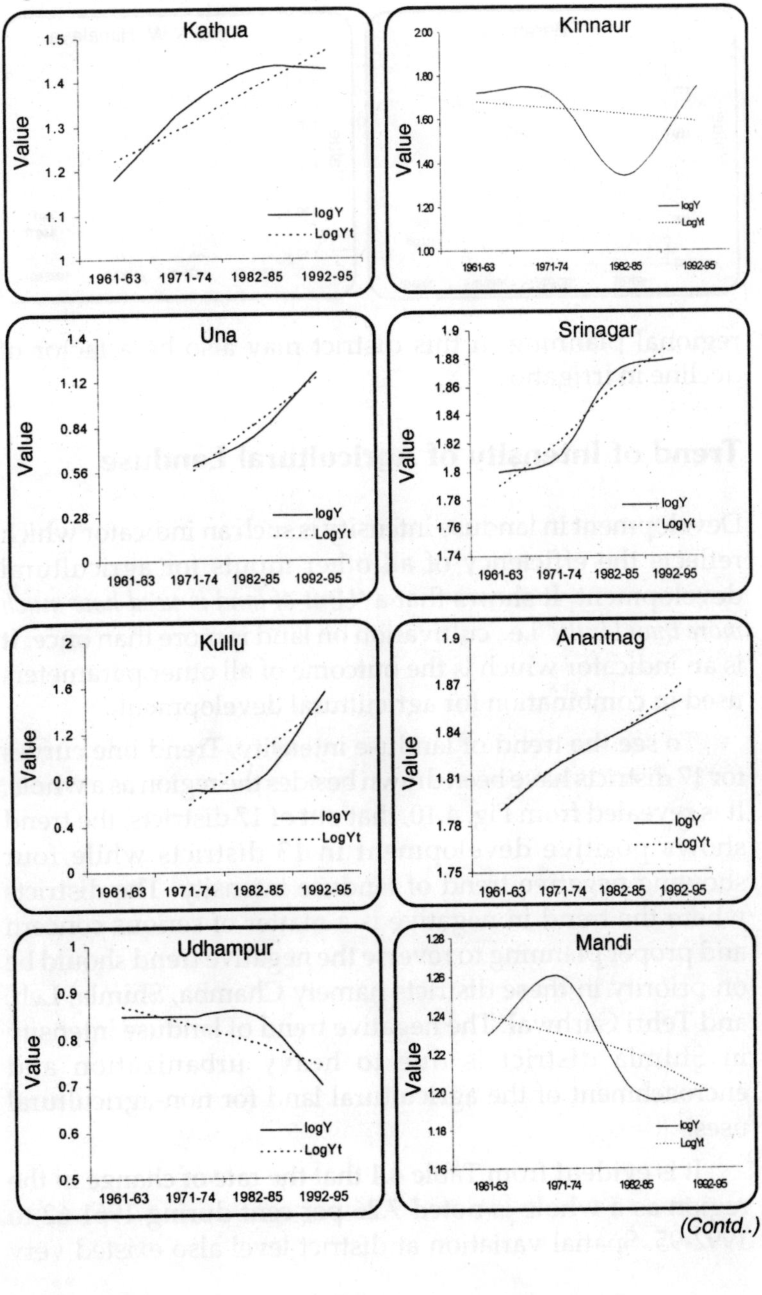

(Contd..)

Fig. 4.9 : *(Contd.)*

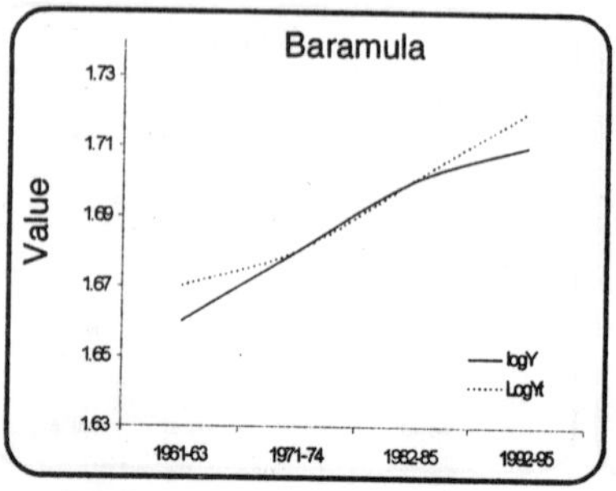

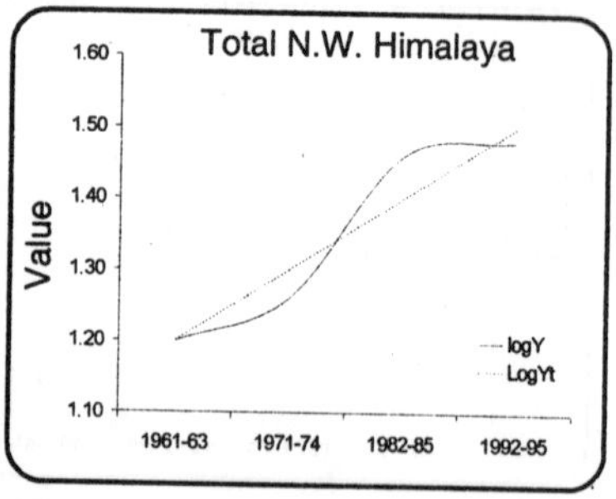

regional planning in this district may also be a factor of decline in irrigation.

Trend of Intensity of Agricultural Landuse

Development in landuse intensity is such an indicator which reflects the efficiency of all other inputs for agricultural development. It shows that a *'Unit of land is used how much more than Unity'* i.e., cultivation on land is more than once. It is an indicator which is the outcome of all other parameters used in combination for agricultural development.

To see the trend of landuse intensity, Trend line curves for 17 districts have been drawn besides the region as a whole. It is revealed from Fig. 4.10, that out of 17 districts, the trend shows positive development in 13 districts while four showing negative trend of landuse intensity. The districts where the trend in negative is a matter of serious concern and proper planning to reverse the negative trend should be on priority in these districts namely Chamba, Shimla, Leh, and Tehri Garhwal. The negative trend of landuse intensity in Shimla district is due to heavy urbanization and encroachment of the agricultural land for non-agricultural uses.

It is evident from Table 4.1 that the rate of change in the region as a whole is noted 7.26 per cent during 1961-62 to 1992-95. Spatial variation at district level also existed very

Fig. 4.10 : Trend line of Intensity of Landuse Development in Selected Districts of North Western Himalaya

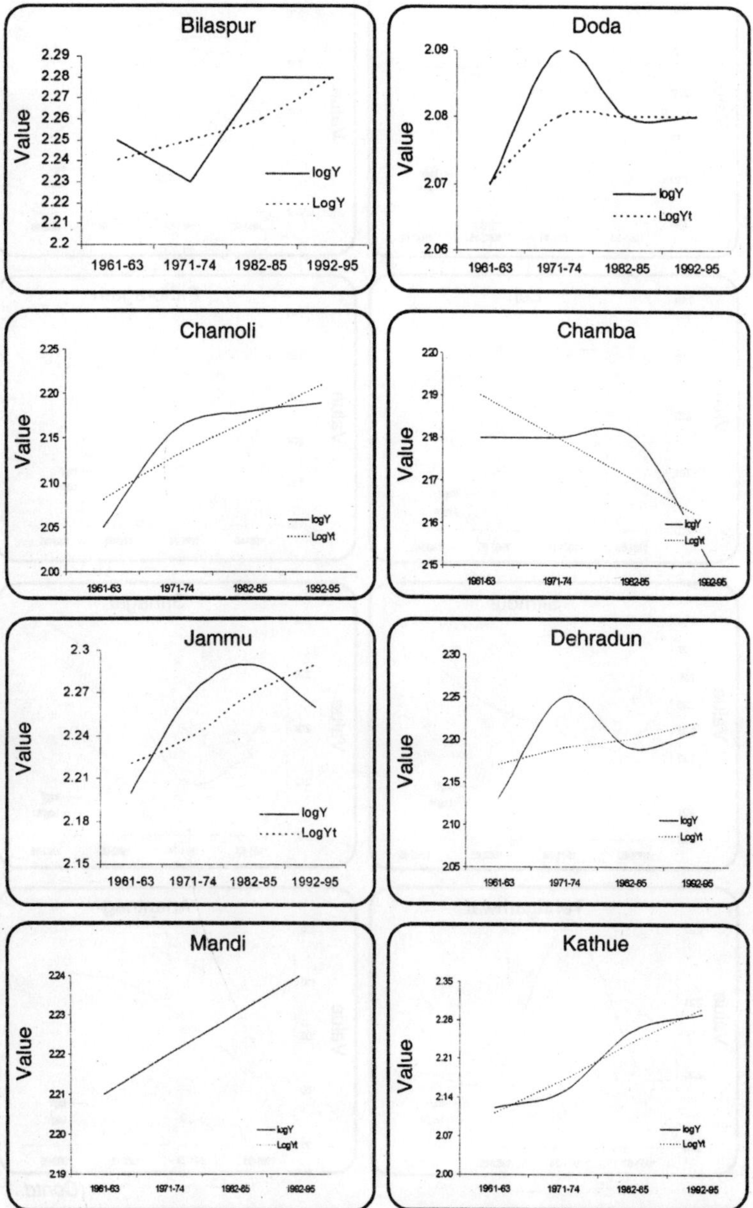

(Contd.)

Fig. 4.10 : *(Contd.)*

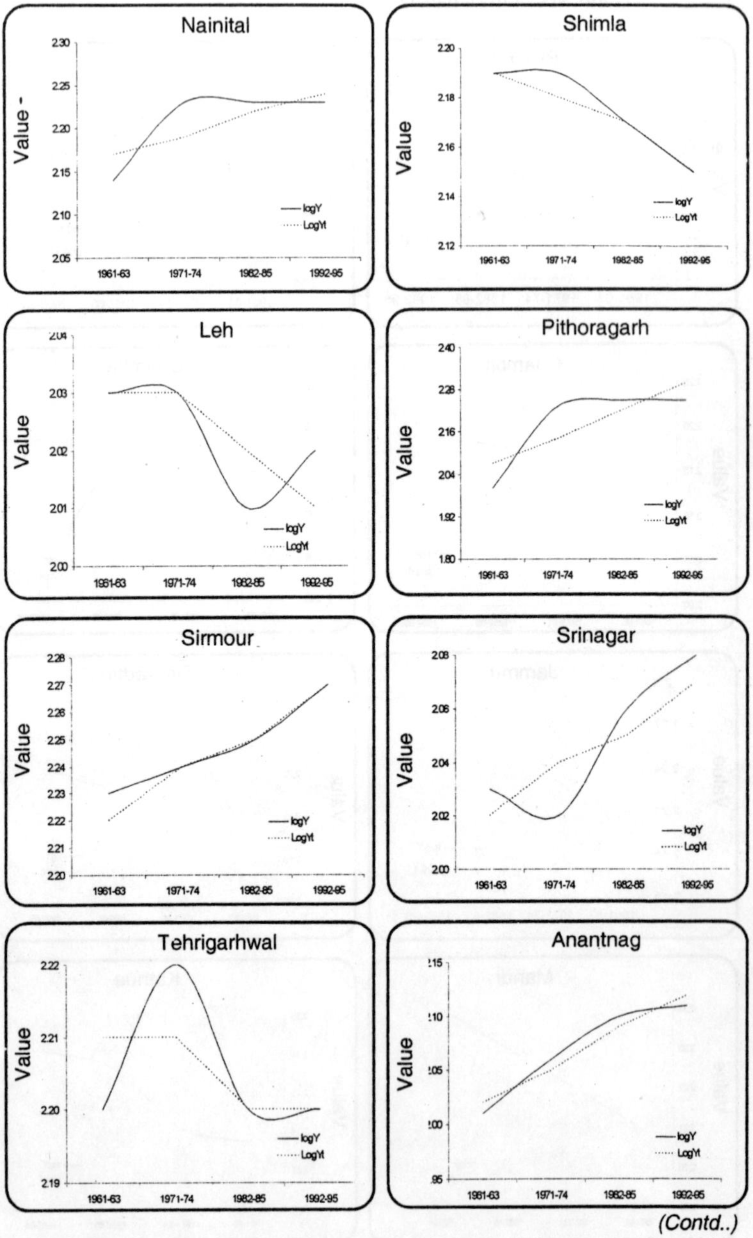

(Contd..)

Fig. 4.10 : *(Contd.)*

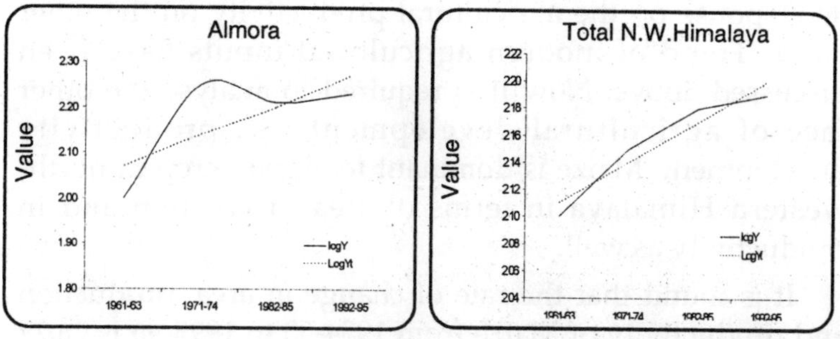

prominently. In terms of rate of change Pithoragarh is found to be the most developed where it is equal to 18.29 per cent followed by Almora and Kathua where it is noted 15.29 per cent and 14.55 per cent respectively. On the other hand, some district show negative rate of change in this indicator of agricultural development namely (–2.97%) in Chamba, –2.06 per cent in Shimla –.37 per cent, in Leh and –0.46 per cent in Tehri Garhwal. There are several reasons for this decline but climatic factor is the main in Leh while in other districts urbanization and sometimes government policies also play their role.

AGRICULTURAL PRODUCTIVITY

The rate of change and trend of area, production and productivity of cereals and pulses has been analyzed for six crops namely Maize, Rice, Wheat, Barley, Total Pulses and Total Foodgrains. The rate of change has been calculated and trend line curve has been drawn for 14 districts to analyze the spatial variations and also for the region as a whole, to determine the average trend of agricultural development in the region.

Trend of Agricultural Development in the Region

Trend of agricultural development can be assessed by the

adoption of modern agricultural inputs on the one hand and its response on the agricultural productivity on the other hand. Trend of modern agricultural inputs have been discussed above. Now it is required to analyse the other face of agricultural development i.e., productivity development. Maize is dominant foodgrain crop in north-western Himalaya in terms of area production and in productivity as well.

It is found that the rate of change in area, production and productivity of Maize from 1954-55 to 1995-96 is 26.77 per cent, 66.34 per cent and 31.22 per cent respectively. Rice is also a dominant crop specially in Kashmir Valley which shows a lower rate of change than maize in the area, production and productivity in the region as a whole. It is evident from Tables 4.2, 4.3 and 4.4 that the rate of change in area production and productivity is noted as 23.59 per cent, 49.28 per cent and 29.13 per cent respectively.

Table 4.2 : Rate of Change (%) in Area Under Cereals and Pulses

Districts	*Maize*	*Rice*	*Wheat*	*Barley*	*Pulses*	*Foodgrain*
Bilaspur	9.40	-15.67	13.24	26.47	-30.02	6.41
Chamba	3.75	-3.39	7.40	-11.08	448.28	3.75
Kinnaur	N.A.	N.A.	-19.46	N.A.	N.A.	29.12
Mandi	16.41	-1.14	13.51	-4.72	35.21	8.39
Shimla	-4.72	-21.48	-8.59	-22.02	26.77	-11.28
Sirmour	8.14	-1.60	7.15	-4.94	2.33	5.20
Jammu	22.74	36.46	28.82	-52.79	29.12	23.59
Udhampur	52.76	48.94	29.12	-25.87	90.11	35.52
Kathua	12.98	21.06	31.22	-13.10	37.09	21.90
Doda	45.21	4.23	-2.95	-8.80	N.A.	25.60
Poonch	-11.08	-15.28	-3.84	-54.40	18.58	-9.64
Srinagar	-22.02	-25.53	-82.34	-79.20	-53.43	-26.21
Anantnag	16.14	-13.70	-87.09	-81.97	-36.87	-11.49
Baramula	-0.23	-7.10	-56.85	-76.29	-19.28	-7.32
Leh	N.A.	N.A.	N.A.	-30.18	-30.50	-4.28
Dehradun	8.14	5.20	3.51	-2.58	-18.15	0.24
Nainital	-28.55	21.90	31.52	-27.89	-34.69	12.46
Total	26.77	23.59	28.53	0.46	15.08	29.12

Table 4.3 : Rate of Change (%) in Production of Cereals and Pulses (1954-55 to 1995-96)

Districts	*Maize*	*Rice*	*Wheat*	*Barley*	*Pulses*	*Foodgrain*
Bilaspur	42.56	-3.17	29.12	42.23	-47.76	28.53
Chamba	53.46	-2.73	27.94	15.35	20.23	37.72
Kinnaur	N.A.	N.A.	-23.62	N.A.	N.A.	23.31
Mandi	73.78	15.88	39.64	11.43	34.59	41.91
Shimla	21.06	-16.63	8.39	-4.72	8.14	5.93
Sirmour	50.66	8.64	33.05	9.40	-4.72	35.21
Jammu	59.22	70.61	78.24	-27.89	-11.28	62.55
Udhampur	112.32	78.65	63.68	-9.22	77.01	74.18
Kathua	54.17	49.28	66.34	-6.67	5.68	51.36
Doda	90.11	20.50	13.24	-20.20	3.28	48.94
Poonch	30.02	25.03	41.91	-56.95	3.04	24.74
Srinagar	-11.49	17.32	-83.02	-78.42	-71.02	-9.84
Anantnag	28.53	2.09	-86.26	-73.27	-27.01	6.66
Baramula	5.20	10.66	-55.02	-73.27	283.64	N.A.
Leh	N.A.	N.A.	N.A.	-27.39	151.77	-6.24
Dehradun	35.83	31.83	45.55	14.02	-10.87	30.62
Nainital	-19.28	79.47	82.81	-13.30	-25.01	60.32
Total	66.34	49.28	64.06	19.12	12.72	60.32

Table 4.4 : Rate of Change (%) in productivity of Cereals and Pulses (1954-55 to 1995-96)

Districts	*Maize*	*Rice*	*Wheat*	*Barley*	*Pulses*	*Foodgrain*
Bilaspur	30.32	14.82	14.02	12.46	-25.53	19.95
Chamba	47.91	0.69	15.88	29.42	17.49	32.74
Kinnaur	7.65	N.A.	-5.16	N.A.	N.A.	-4.50
Mandi	49.28	17.22	23.03	17.22	-0.46	30.92
Shimla	27.06	5.93	18.58	22.46	-14.69	19.40
Sirmour	39.32	10.15	24.17	14.82	-6.89	28.53
Jammu	29.72	25.31	38.68	52.41	-31.29	31.52
Udhampur	38.68	19.95	26.77	22.18	-6.89	28.53
Kathua	36.14	23.31	26.47	7.40	-22.73	23.88
Doda	30.92	15.61	16.68	-12.50	6.55	18.58
Poonch	46.55	47.57	47.91	-5.38	-13.10	38.36
Srinagar	13.50	24.45	-4.06	-73.94	-23.99	21.90
Anantnag	10.41	18.30	6.41	-61.98	-12.09	12.98
Baramula	4.95	19.12	4.23	12.72	28.05	18.85
Leh	N.A.	N.A.	N.A.	3.99	16.95	-2.05
Dehradun	25.31	25.31	40.60	16.68	8.89	30.61
Nainital	12.72	47.23	39.00	20.23	14.82	42.56

Wheat is the most dominant crop of the region and its rate of change shows that its production is increased tremendously in the region. The rate of change in area, production and productivity is found to be 28.53 per cent, 64.06 per cent and 27.64 per cent respectively.

Barley is not a significant crop in the region showing very little positive rate of change in area i.e. only 0.46 per cent while it is 19.12 per cent in production and 43.55 per cent in productivity.

Pulses are in crisis, which shows positive rate of change in area and production i.e. 15.08 per cent and 12.72 per cent but it is negative in case of productivity i.e. -2.05 per cent which is a matter of great concern because the green revolution could not improve the conditions of pulses in this mountainous region.

On the other hand, condition of total foodgrain is encouraging as the rate of change in area, production and productivity is as such 29.12 per cent, 60.32 per cent and 24.45 per cent respectively. High rate of change in production is obviously a reflection of green revolution in this region.

Trend of agricultural development in the region is more clear from fig. 4.11 which shows the trend line curve of area, production and productivity of maize, rice, wheat, barley, pulses and foodgrains and it is seen that besides the ups and downs the overall trend of area, production and productivity is positive for maize, rice, wheat, barley and foodgrains but in case of pulses the trend of productivity is negative which shows that the trend of agricultural development in the region is positive but pulses need special attention.

Trend of Agricultural Development in Nainital

Nainital is one of the most important districts of the region in terms of agricultural development. A considerable portion of Nainital is a plain area included in this study where the productivity is very high besides the technological inputs the soil fertility.

It is evident from Tables 4.2, 4.3 and 4.4 that, Maize is

Fig. 4.11 : Trend line of Area, Production and Productivity of Cereals and Pulses (North-Western Himalaya-1954-55 to 1995-96)

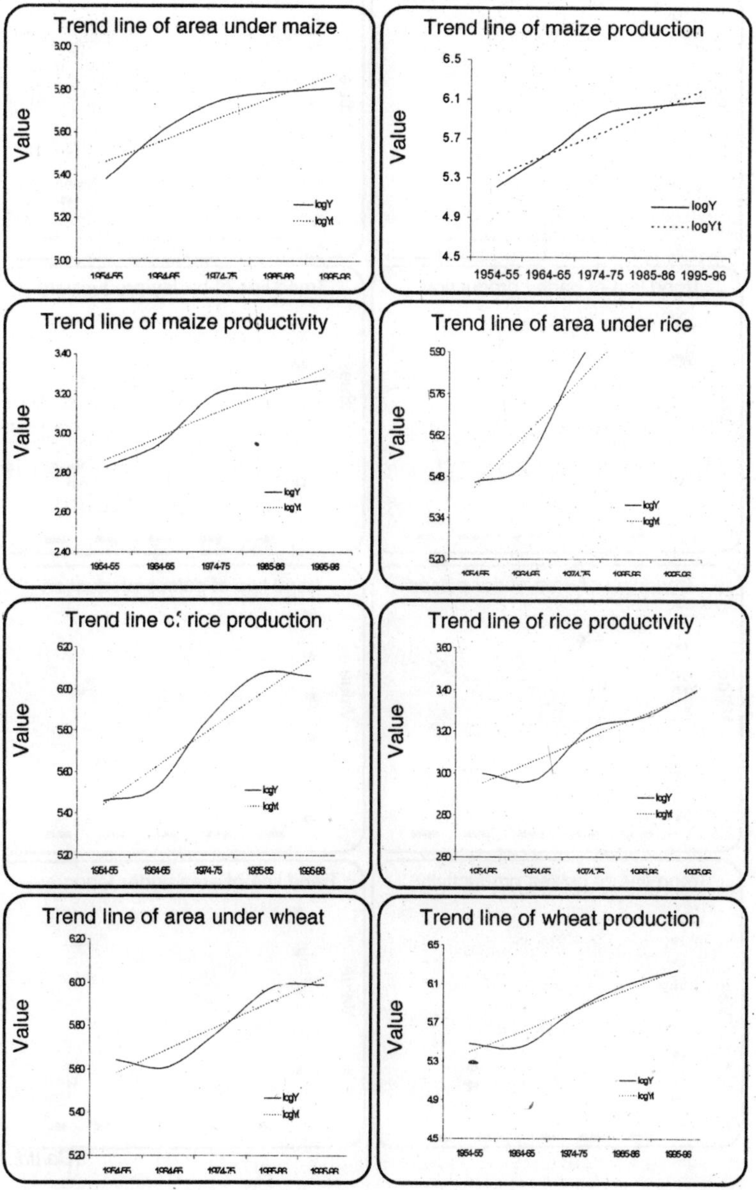

(Contd..)

Fig. 4.11 : *(Contd.)*

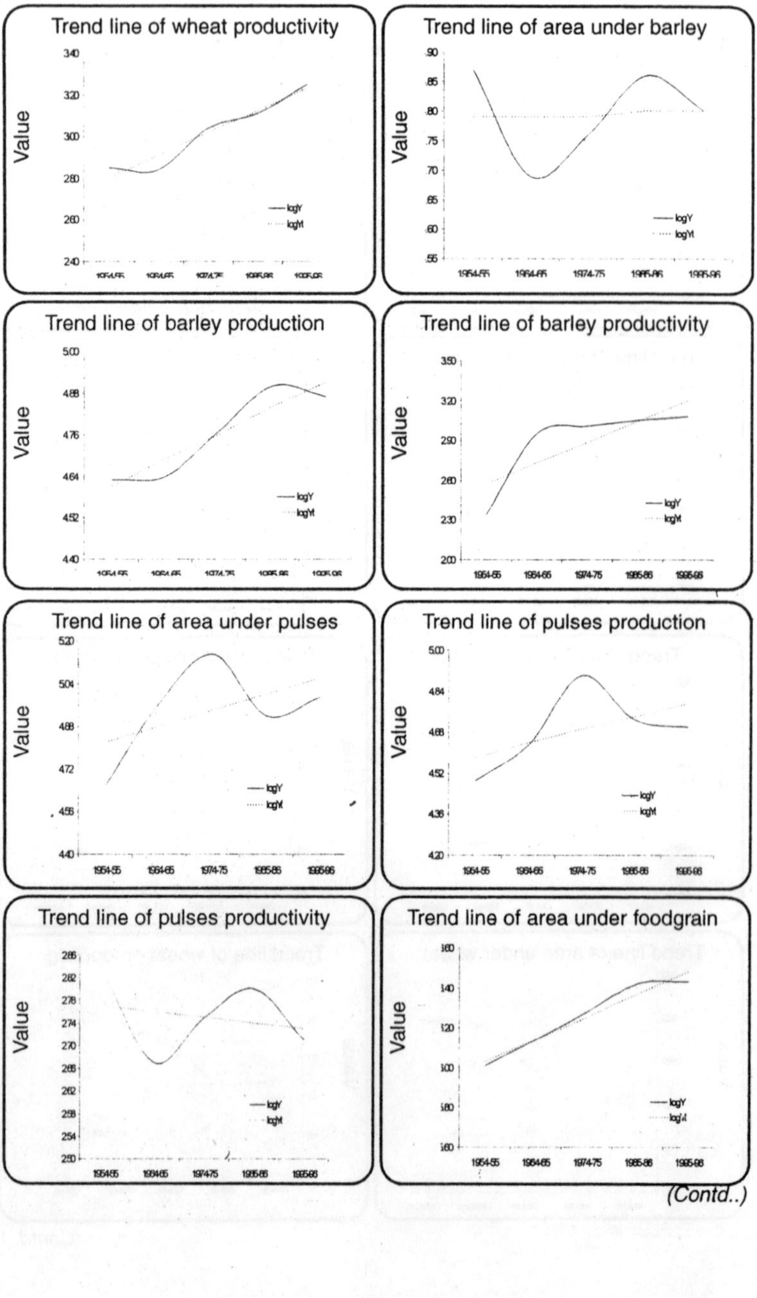

(Contd..)

Fig. 4.11 : *(Contd.)*

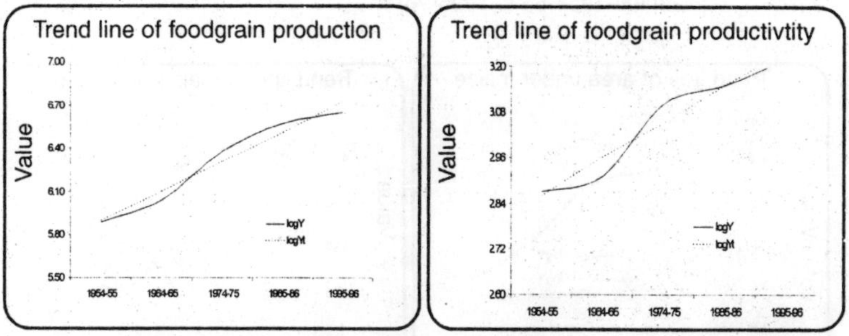

not a significant crop of this district. Where area under maize shows negative rate of change i.e. –28.55 per cent, while the production decline with –19.28 per cent of rate of change. But the productivity shows positive rate of change i.e. 12.72 per cent. Besides the negative rate of change in area and production, the productivity shows positive rate of change which is an impact of green revolution especially the high yielding variety seeds.

In case of rice, it is seen that the rate of change in wheat expansion is encouraging where it is found 21.90 per cent from 1954-55 to 1995-96. But production shows tremendous increase with rate of change equal to 79.47 per cent. As a result the productivity ultimately shows positive rate of change which is found to be 29.13 per cent during this period.

Wheat is the most dominant crop of the district where the rate of change in area, production and productivity is noted as 31.52 per cent, 82.81 per cent and 39.00 per cent respectively. The high rate of change in production and productivity is obviously due to green revolutionary measures.

The condition of barley is poor in this district too, where the negative rate of change existed in area as well as in production i.e. –27.09 per cent and –13.30 per cent respectively but because of green revolution, productivity shows positive rate of change which is equal to 20.23 per cent (Tables 4.2, 4.3 and 4.4).

Pulses are also facing crisis in this district with negative

Fig. 4.12 : Trend line of Area, Production and Productivity of Cereals and Pulses (Naimital-1954-55 to 1995-96)

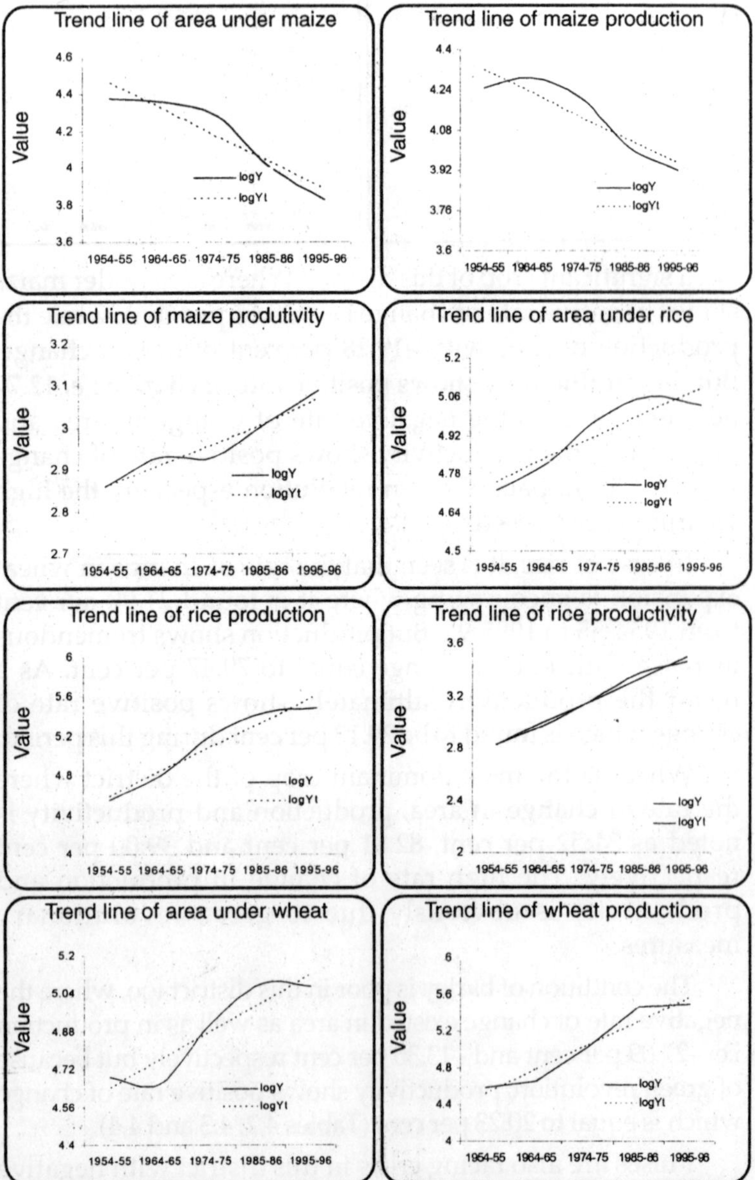

(Contd..)

Fig. 4.12 : *(Contd..)*

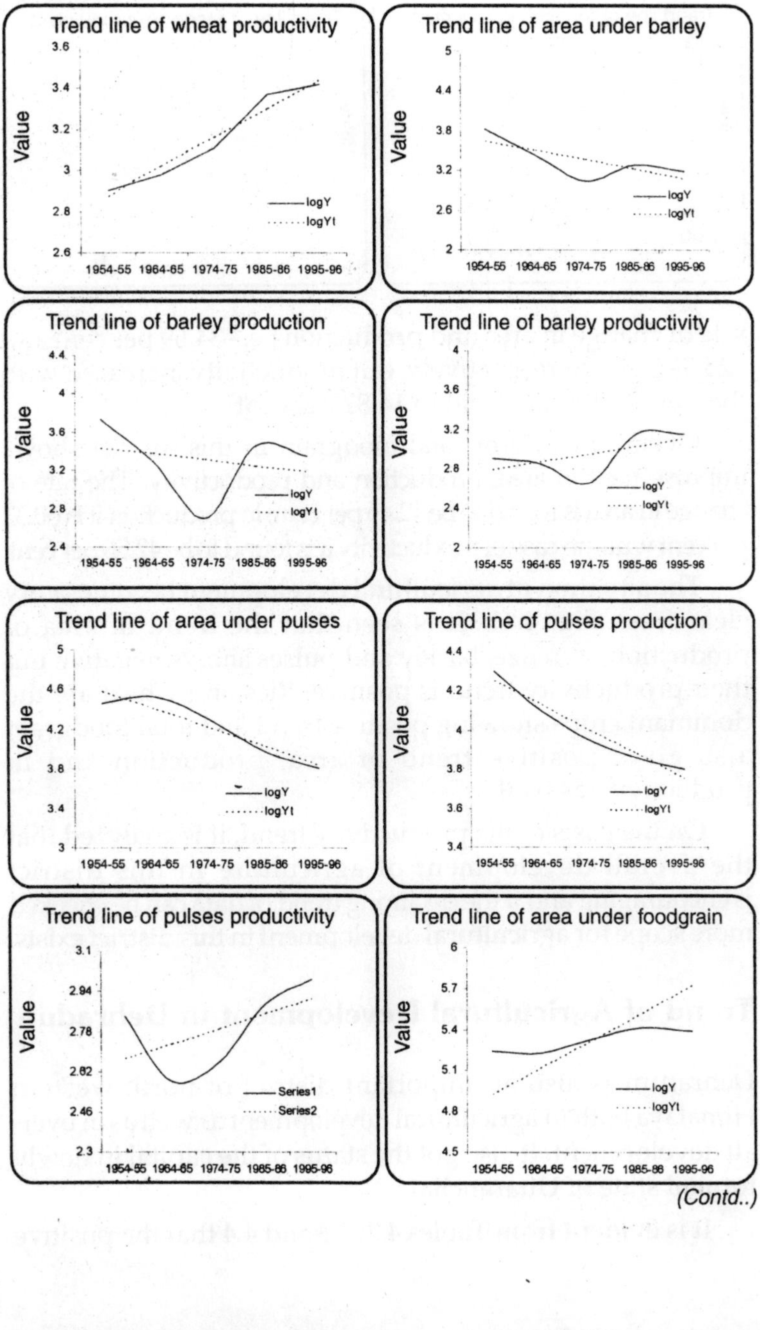

(Contd..)

Fig. 4.12 : *(Contd..)*

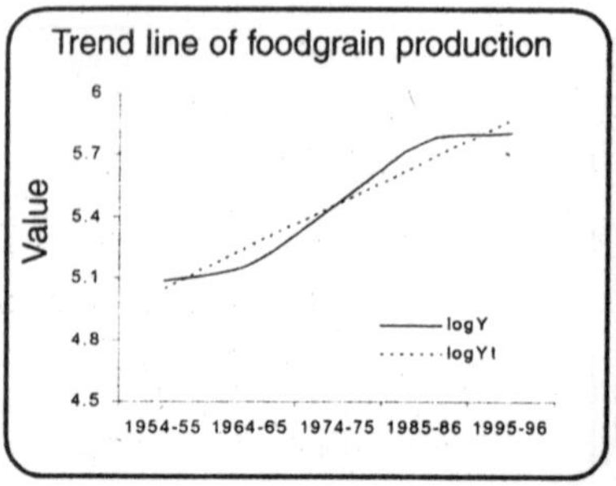

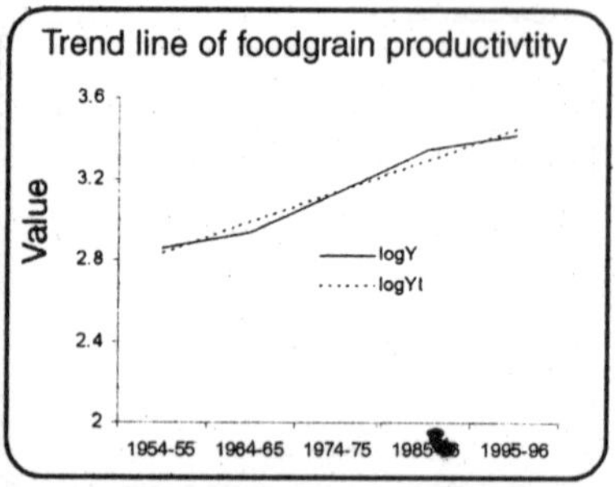

rate of change in area and production i.e. –34.69 per cent and –25.01 per cent respectively, but productivity increased with the rate of change equal to 14.82 per cent.

Overall picture of total foodgrain in this district shows improvement in area, production and productivity. The rate of change in area is found to be 12.46 per cent, in production it is 60.32 per cent while in case of productivity it is found to be 42.56 per cent.

The situation of agricultural development becomes very clear from Fig. 4.12. It is seen that the trend of area of production of maize, barley and pulses shows negative but their productivity trend is positive. Rice and wheat are the dominant crops showing positive trend and total foodgrain also gives positive trend of area, production and in productivity as well.

On the basis of the productivity trend, it is analyzed that the overall development of agriculture in this district is encouraging and if the declining trend of data can be checked more scope for agricultural development in this district exists.

Trend of Agricultural Development in Dehradun

Dehradun is also an important district of north-western Himalaya both in agricultural development as well as in overall development. It also got the status of the capital in newly formed state of Uttaranchal.

It is evident from Tables 4.2, 4.3 and 4.4 that the positive

Fig. 4.13 : Trend line of Area, Production and Productivity of Cerelas and Pulses (Dehradun—1954-55 to 1995-96)

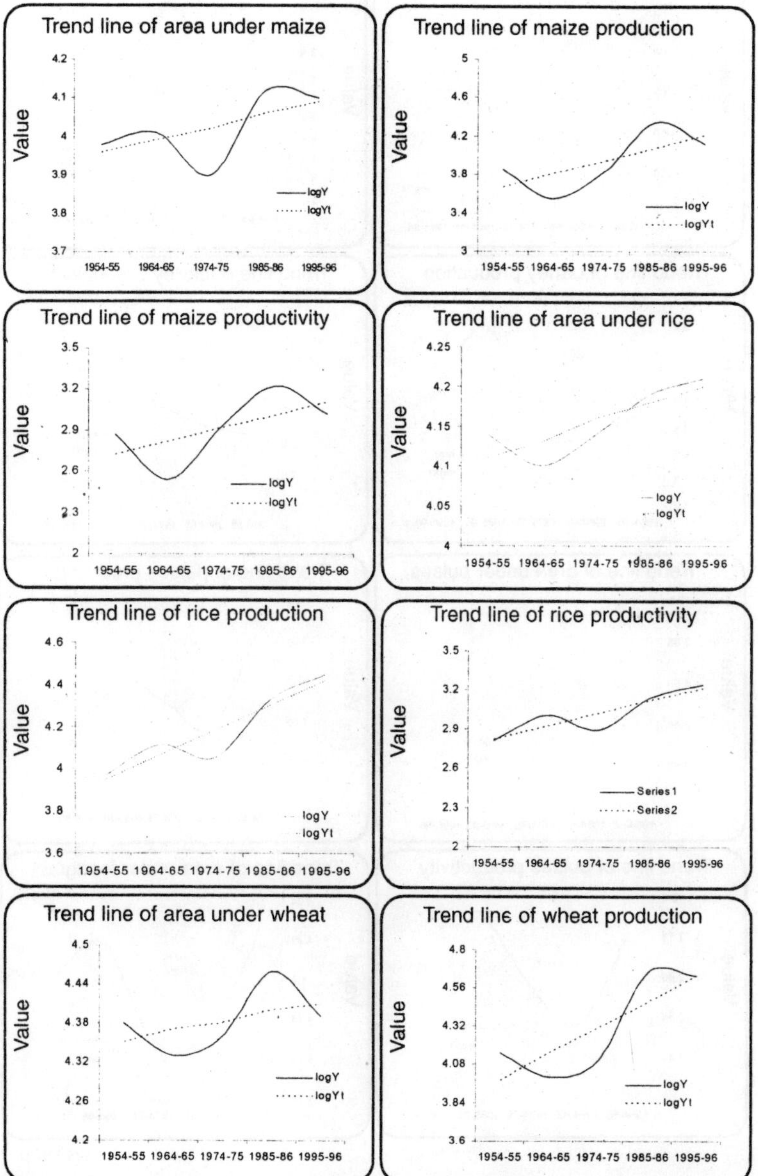

(Contd..)

Fig. 4.13 : *(Contd.)*

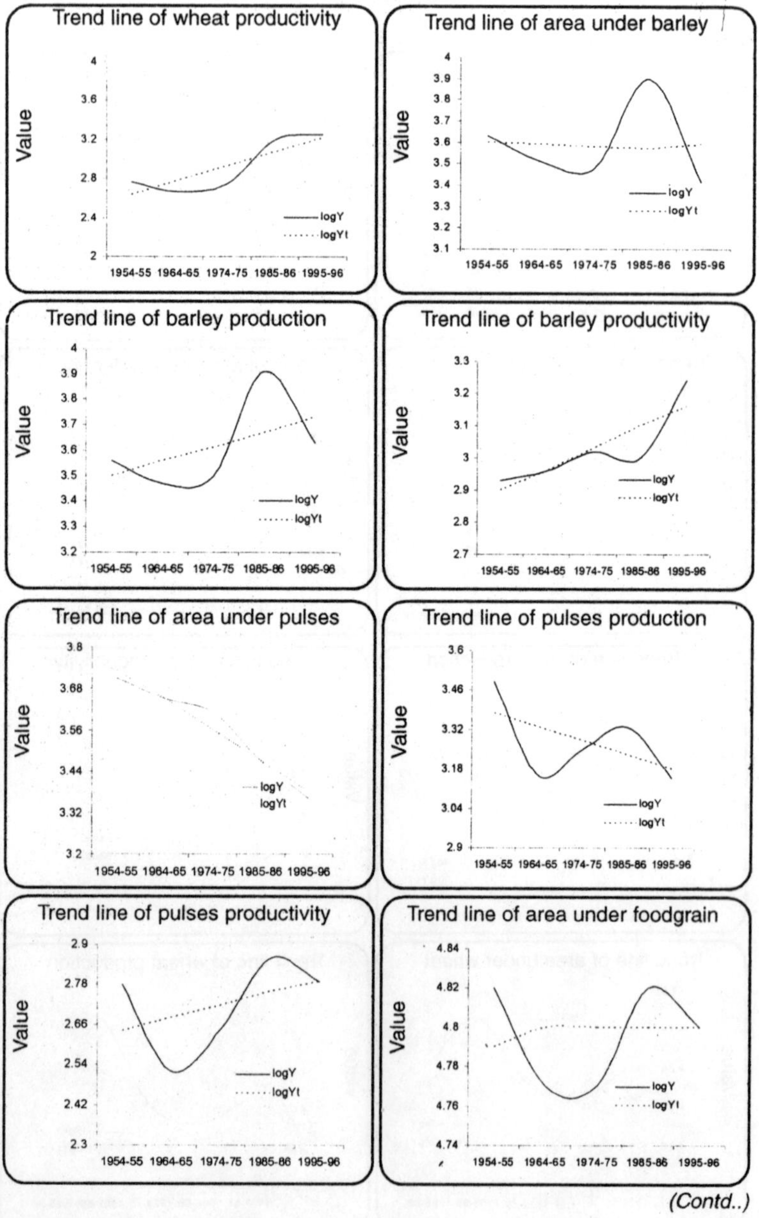

(Contd..)

Fig. 4.13 : *(Contd.)*

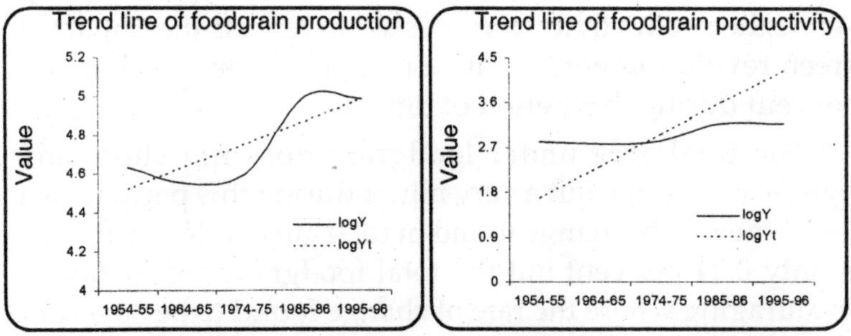

rate of change found in area, production and productivity of maize is 8.14 per cent, 35.83 per cent and 25.3 per cent respectively. It shows that the condition of maize in this district is far better than Nainital.

The condition of rice is also improving in the district with the rate of change in area is 5.20 per cent in production it is 31.83 per cent and in productivity it is 25.31 per cent. Agricultural technology used also show positive trend except in wooden plough and sugarcane crushers where it is negative. The rate of change in rice thresher is found extremely high i.e. 983.93 per cent.

Wheat shows tremendous development in this district. It is again evident from Tables 4.2, 4.3 and 4.4 that the rate of the change in area under wheat from 1954-55 to 1995-96 is 3.51 per cent which is very little as compared to rate of the change in production i.e. 45.55 per cent. The productivity of wheat is also showing very encouraging trend in this district with the rate of change noted as 40.60 per cent. As a whole it can be said that the wheat is gaining its position and importance here.

Condition of barley shows decline in areal extent, the rate of change found to be –2.58 per cent. But the rate of change in production and productivity is positive i.e. 14.02 per cent and 16.68 per cent respectively. It is analyzed that barley losing its space but due to green revolutionary measures it improved in production and in productivity as well.

The condition of pulses showing poor development in area as well as in production, the rate of change found negative i.e. –18.15 per cent and –10.87 per cent respectively. While the productivity of pulses again reflects the impact of green revolution with positive rate of change equal to 8.89 per cent during this period of time.

The total area under foodgrain does not show any significant change but a very minor during this period. As a result the rate of change found in total area under foodgrain is only 0.24 per cent but the total foodgrain production is encouraging where the rate of change found to be 30.63 per cent. Because of this significant increase in foodgrain production, its productivity also changes positively with the rate of change equal to 30.61 per cent.

The scenario of agricultural development becomes very clear from Fig. 4.13. Where the solid line shows actual fluctuation in area, production and productivity and the dotted line shows the overall trend of development in area, production and productivity. Developmental trend of area, production and productivity of maize, rice, wheat and total foodgrain is positive but area under barley and area with production under pulses shows negative trend of agricultural development. But overall trend of productivity of these crops is found positive.

Trend of Agricultural Development in Shimla

Shimla is an important urban centre as well as a tourist spot in this mountainous region. In the core zone of Shimla there is no place for agriculture as every inch of land is extensively covered by urban infrastructure.

The overall development in this district plays a negative role in agricultural development using modern agricultural inputs. A significant increase found in number of iron ploughs, wheat and rice threshers utilization in this district. Agricultural development on the basis of few dominant food crops is given below.

Fig. 4.14 : Trend line of Area, Production and Productivity of Cereals and Pulses (Shimla-1954-55 to 1995-96)

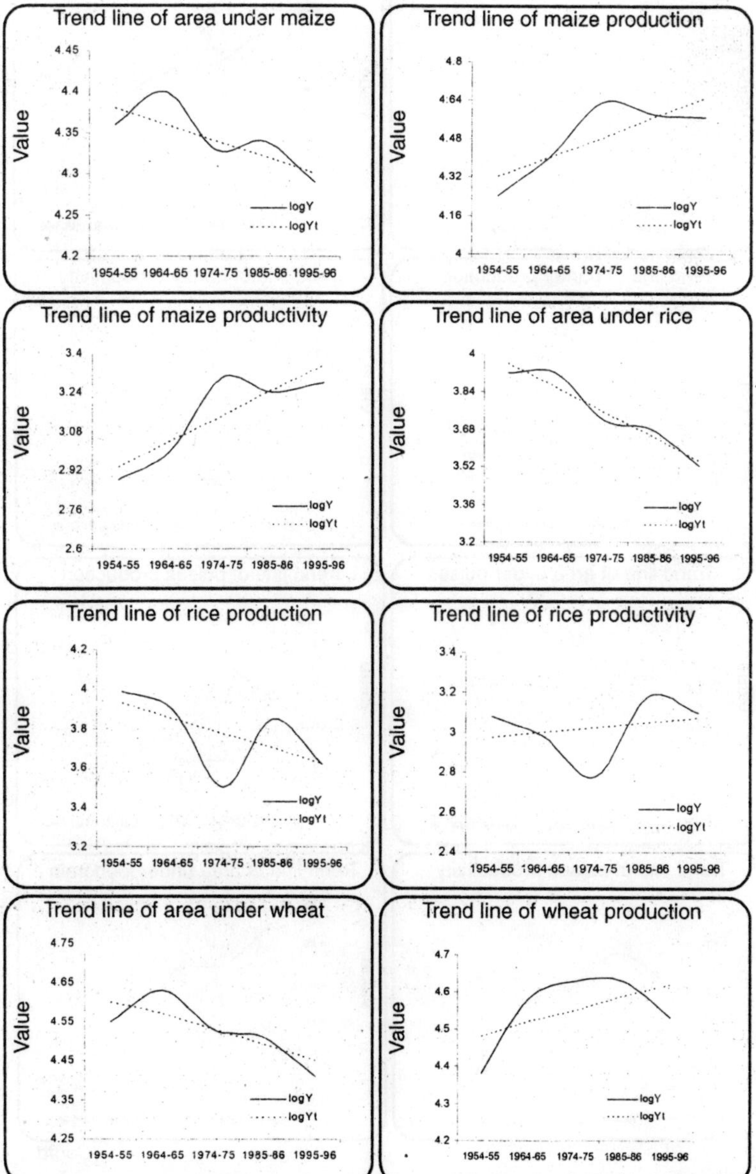

(Contd..)

Fig. 4.14 : *(Contd..)*

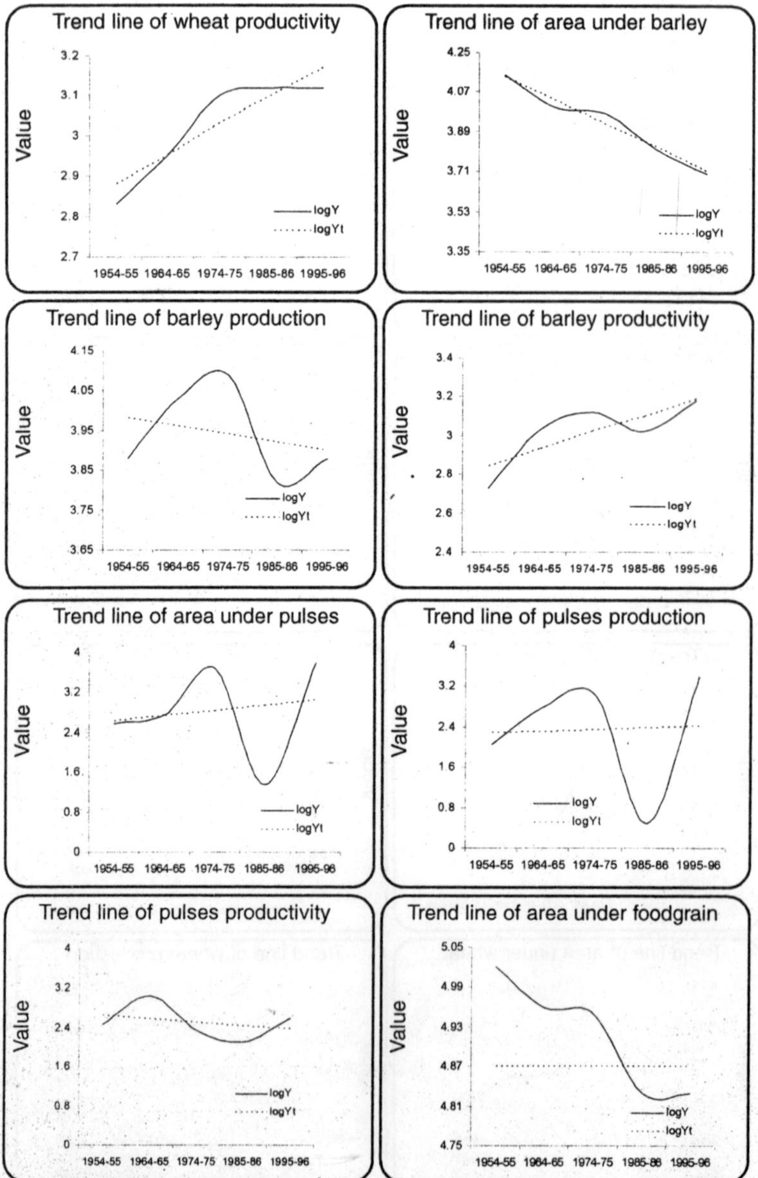

(Contd..)

Fig. 4.14 : *(Contd..)*

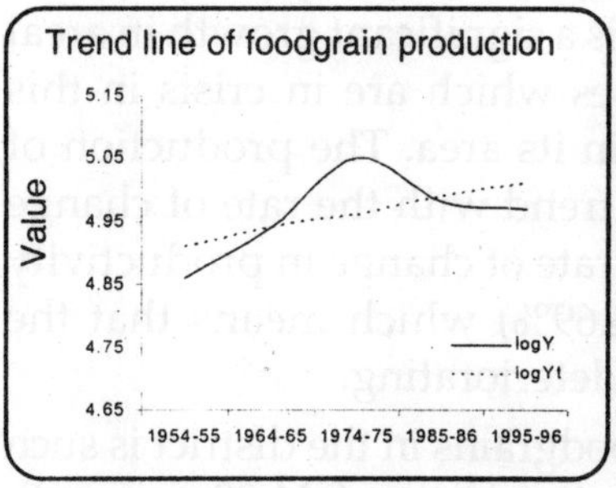

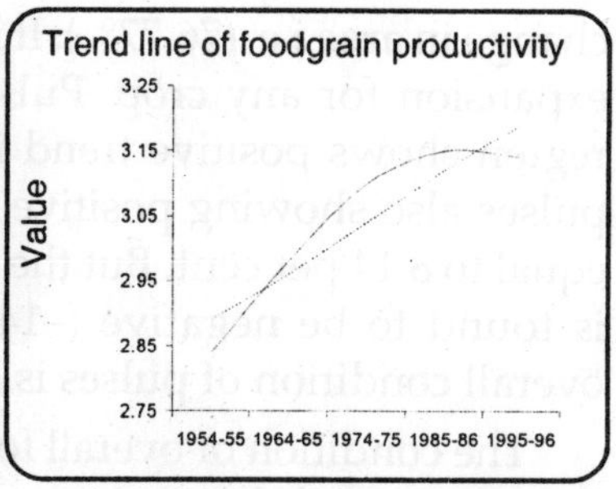

Maize is an important crop but negative trend is found in the rate of change in area under maize i.e. –4.72 per cent (Table 4.2). Due to modernization of agriculture its production does not decrease. As a result, the positive rate of change is noted as 21.06 per cent from 1954-55 to 1995-96 (Table 4.3). The condition of productivity also improved during this period and its rate of change found as 27.06 per cent.

The condition of rice in terms of area and production is shown discouraging, the area and production of rice shows negative trend with rate of change –21.48 per cent and –16.63 per cent respectively. Besides the declining area and production the productivity of rice shows positive rate of change equal to 5.93 per cent.

Wheat is in better condition than rice because the area under wheat is shrinking but its production is increased. The rate of change in area and production during this period is 8.59 per cent and 8.39 per cent respectively. The productivity of wheat is also improving significantly and the rate of change during this period is found to be 18.58 per cent.

The condition of Barley is poor in this district too. Its area as well as production shows negative trend in the rate of change which is equal to –22.02 per cent and –4.72 per cent. But the reflection of green revolution is seen in case of productivity which is increased during this period with the rate of change equal to 22.46 per cent.

Pulses are expanding in this district with positive rate of change in area i.e. (76.77%). It is a significant growth in areal expansion for any crop. Pulses which are in crisis in this region shows positive trend in its area. The production of pulses also showing positive trend with the rate of change equal to 8.14 per cent. But the rate of change in productivity is found to be negative (–14.69%) which means that the overall condition of pulses is deteriorating.

The condition of overall foodgrains in the district is such that the rate of change in area is negative (–11.28 per cent), while it is 5.93 per cent and 19.40 per cent in production and productivity respectively. Though the intensity of landuse is decreasing, the improvement in production and productivity is due to the adoption of modern agricultural technology.

A very clear picture of agricultural development can be revealed from Fig. 4.14, where the solid line shows the fluctuation while dotted line shows the exact trend of development in area, production and productivity. It is evident from the given figure that trend of area under maize, rice, wheat, barley and total foodgrains is declining while pulses shows positive trend in areal expansion. In case of production positive trend is found in maize, wheat, pulses and total foodgrain while rice and barley shows negative trend and the productivity of maize, wheat, rice, barley and total foodgrain shows positive trend. As per the above discussion it is said that the agricultural development is going on in Shimla district besides the sprawl of urbanization.

Trend of Agricultural Development in Sirmour

All the modern inputs used for agricultural development in Sirmour showing positive trends. Now it is evident to analyze the impact of these inputs on area, production and productivity of food crops in Sirmour.

In case of maize, it is seen that there is positive rate of

Fig. 4.15 : Trend line of Area, Production and Productivity of Cereals and Pulses (Sirmour—1954-55 to 1995-96)

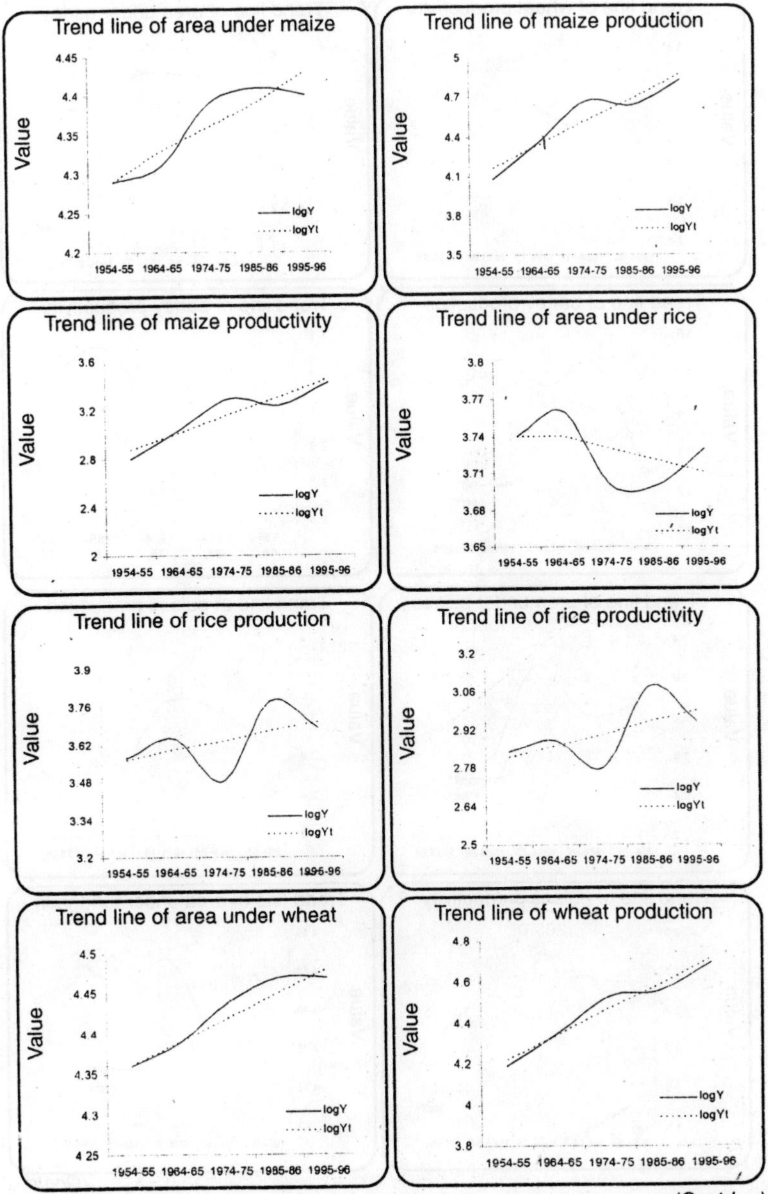

(Contd....)

Fig. 4.15 : *(Contd..)*

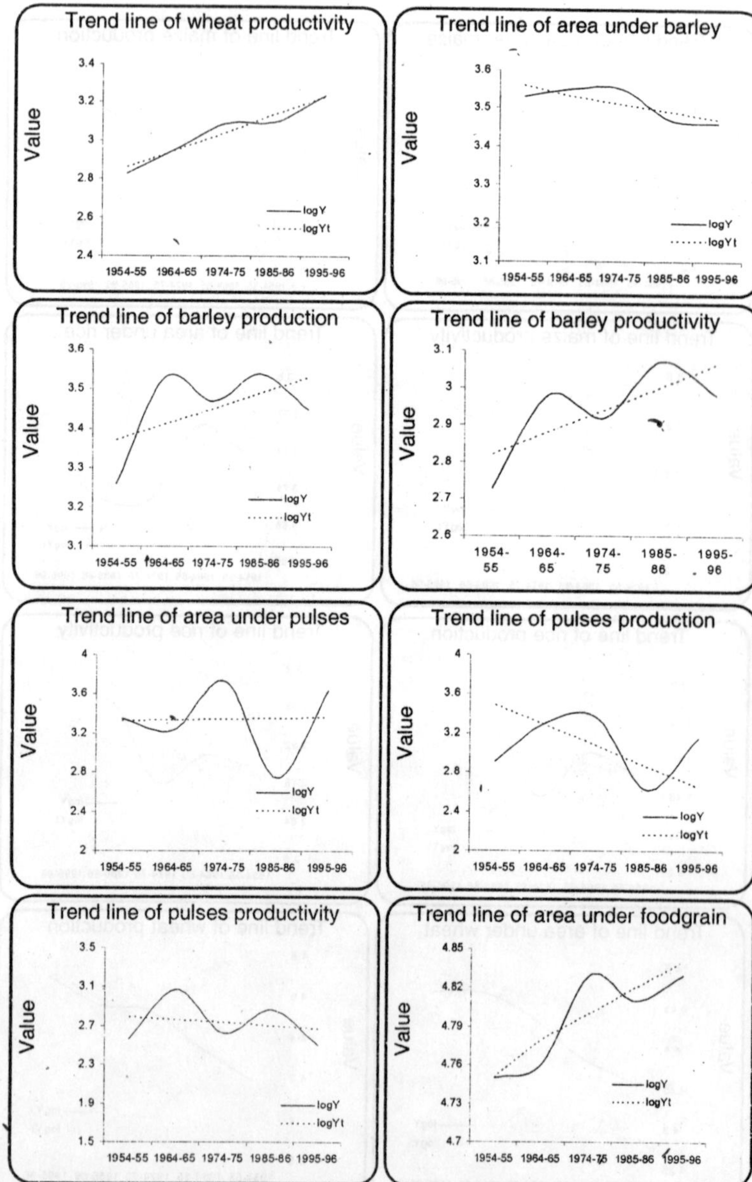

(Contd..)

Fig. 4.15 : *(Contd....)*

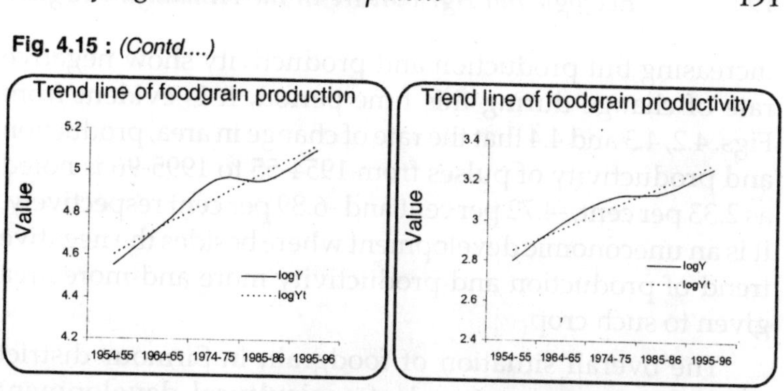

change in area, production and productivity from 1954-55 to 1995-96. But the rate of change in area is 8.14 per cent which is far below the production i.e. 50.66 per cent and productivity i.e. 39.32 per cent. High rate of change in production and productivity is due to the adoption of green revolutionary input in this district.

Rice shows a decline in its expansion with the rate of change equal to –01.60 per cent. Besides the negative magnitude of areal expansion its production increased with the rate of change equal to 8.64 per cent. Decline in area and increase in production ultimately increased the productivity with the rate of change equal to 10.15 per cent in the same period of time.

Wheat has also shown development in area, production and productivity but the rates of change are different in each case. It is evident from Tables 4.2, 4.3 and 4.4 that the rate of change in area, production and productivity of wheat during the said time period is recorded as 7.15 per cent, 33.05 per cent and 24.17 per cent respectively.

The condition of barley shows declining trend in acreage while it is improving in production and productivity with the rate of change equal to –4.94 per cent, 9.40 per cent and 14.82 per cent respectively. Area under barley shows shrinking in a number of districts which indicates its declining importance either because of uneconomic or inecological for these mountainous landscapes.

Pulses are again showing economically negative trend of agricultural development because its areal expansion is increasing but production and productivity show negative rate of change during this time period. It is evident from Figs. 4.2, 4.3 and 4.4 that the rate of change in area, production and productivity of pulses from 1954-55 to 1995-96 is noted as 2.33 per cent, -4.72 per cent and -6.89 per cent respectively. It is an uneconomic development where besides the negative trend of production and productivity more and more area given to such crop.

The overall situation of foodgrain in Sirmour district shows encouraging trend of agricultural development because area, production and productivity of foodgrains show positive trend and rate of change from 1954-55 to 1995-96. It is seen that the rate of change is found to be 5.20 per cent, 35.21 per cent and 28.53 per cent respectively for area, production and productivity of foodgrain.

The actual trend of area, production and productivity of the above discussed crops has been very clear as shown in Fig. 4.15. The trend line curve of area under maize, wheat, pulses and foodgrains show positive while rice and barley show negative trend of areal expansion. In case of production only pulses shows negative trend while the remaining crops showing positive trend, and the same condition existed in productivity too, where all crops show positive trend except pulses.

It is revealed from the above analysis that the agricultural development is on its improvement but the pulses are in crisis which needs special attention, priority should also be given for barley and rice to increase its share in cropping pattern in this district.

Trend of Agricultural Development in Mandi

Mandi is an important district for agricultural practices in this mountainous track where all the modern inputs of agricultural development shows positive rate of change except irrigation. The impact of these modern technological

Fig. 4.16 : Trend line of Area, Production and Productivity of Cereals and Pulses (Mandi—1954-55 to 1995-96)

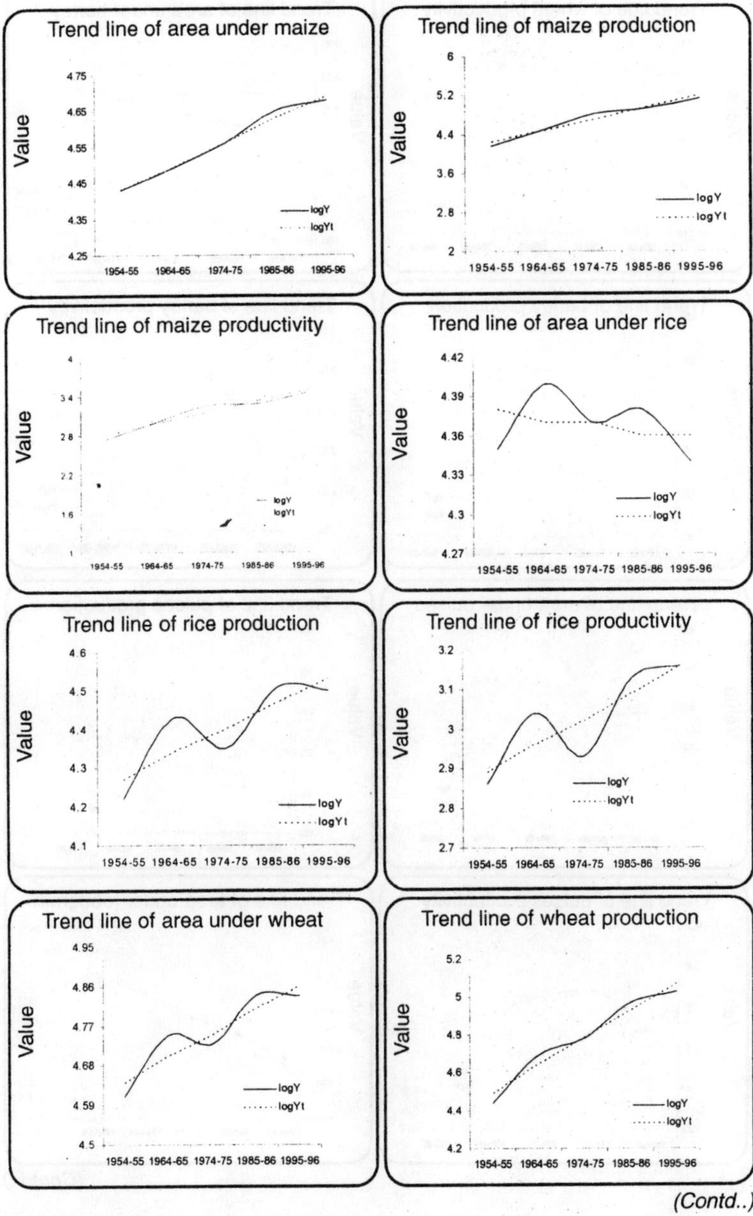

(Contd..)

Fig. 4.16 : *(Contd..)*

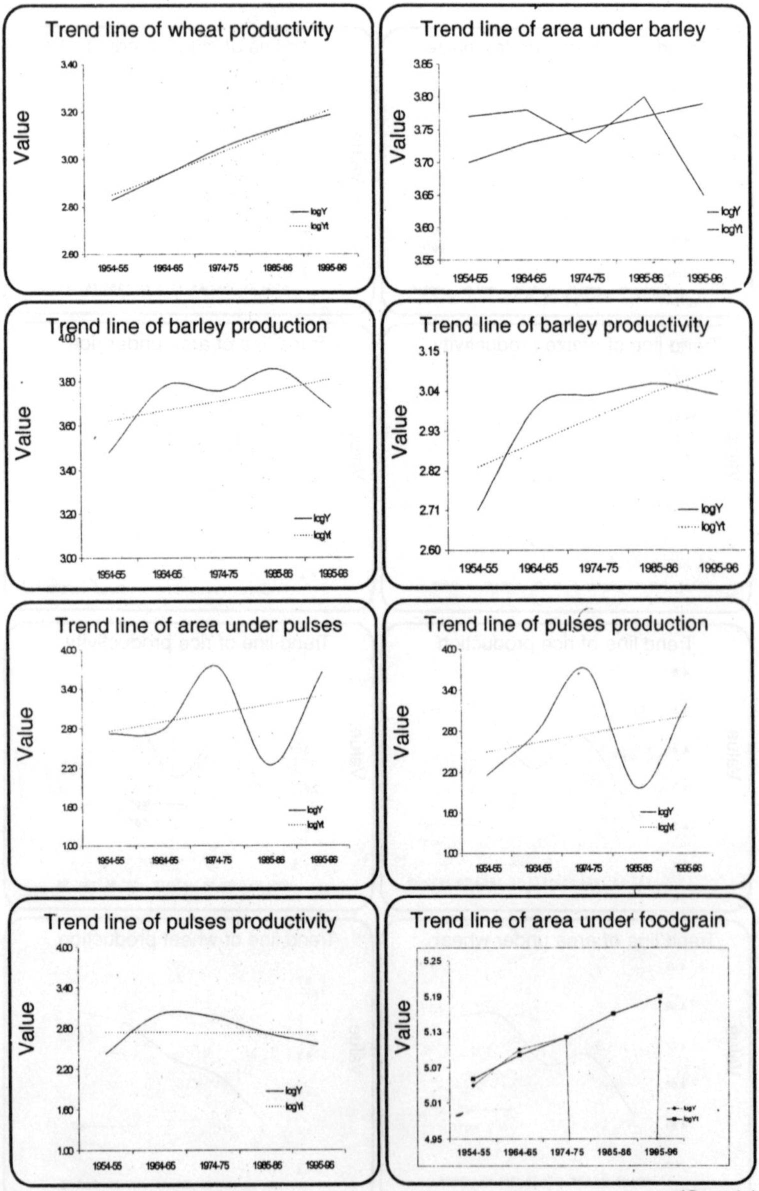

(Contd..)

Fig. 4.16 : *(Contd..)*

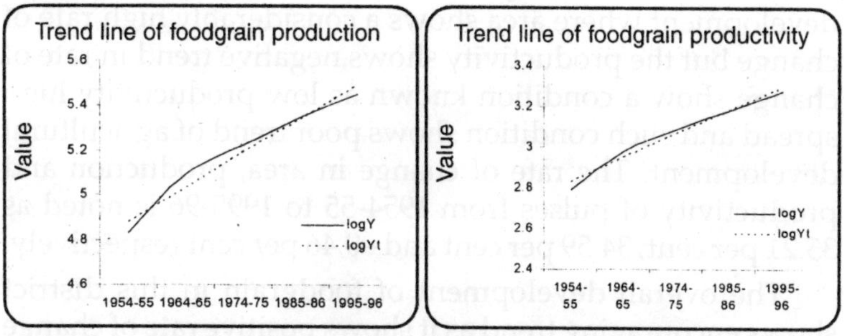

inputs on development of area, production and productivity of crops is discussed below.

Maize shows its superiority in this district too, as area production and productivity changes positively with the rate of change equal to 16.41 per cent, 73.78 per cent and 49.28 per cent respectively. It is revealed that increase in productivity and area of maize gives tremendous increase in production of maize crop in this district.

Rice is not as remunerative as maize and wheat in the agro-ecology of Himachal region as a result its area shows negative rate of change from 1954-55 to 1995-96 but adoption of modern inputs has increased its production and productivity as well. The rate of change in area, production and productivity of rice is noted as –1.14 per cent, 15.88 per cent and 17.22 per cent respectively.

The condition of wheat is encouraging in this district with expansion of area besides increase in productivity and production as well. It is evident from Tables 4.2, 4.3 and 4.4 that the rate of change in area production and productivity of wheat are all positive i.e. 13.51 per cent, 39.64 per cent and 23.03 per cent respectively.

The condition of barley shows shrinking in areal expansion but increasing trend of production and productivity. The rate of change in its area from 1954-55 to 1995-96 is recorded negative i.e. –4.72 per cent but the production and productivity noted with positive rate of change which is equal to 11.43 per cent and 17.22 per cent respectively.

Pulses are showing uneconomic trend of agricultural development where area shows a considerably high rate of change but the productivity shows negative trend in rate of change show a condition known as low productivity high spread and such condition shows poor trend of agricultural development. The rate of change in area, production and productivity of pulses from 1954-55 to 1995-96 is noted as 35.21 per cent, 34.59 per cent and –0.46 per cent respectively.

The overall development of foodgrain in this district shows encouraging trend as it shows positive rate of change in all the three fronts of area, production and productivity with the magnitude of 8.39 per cent, 41.91 per cent and 30.92 per cent respectively.

Figure 4.16 shows the trend of agricultural development in terms of area, production and productivity. Maize shows positive trend in area, production and producitivity, while rice shows negative trend in area and positive in production and productivity. The trend of wheat is positive in area, production and productivity. Besides, the fluctuations the overall development of barley also shows positive trend. With high ups and downs the area and production of pulses shows positive trend but productivity shows negative trend. Overall foodgrains are showing positive trend in area, production and productivity as well. From the above discussion it is revealed that the overall development of agriculture in Mandi is satisfactory but pulses are in crisis which need special attention not only in expansion but their productivity is also a matter of great concern.

Trend of Agricultural Development in Bilaspur

The status of agriculture in Bilaspur shows positive trend of development where adoption of all modern inputs of agricultural development i.e. green revolution technology is very high. The impact of green revolution on the agricultural scenario is very prominent specially on agricultural productivity of different crops given below.

Fig. 4.17 : Trend line of Area, Production and Productivity of Cereals and Pulses (Bilaspur—1954-55 to 1995-96)

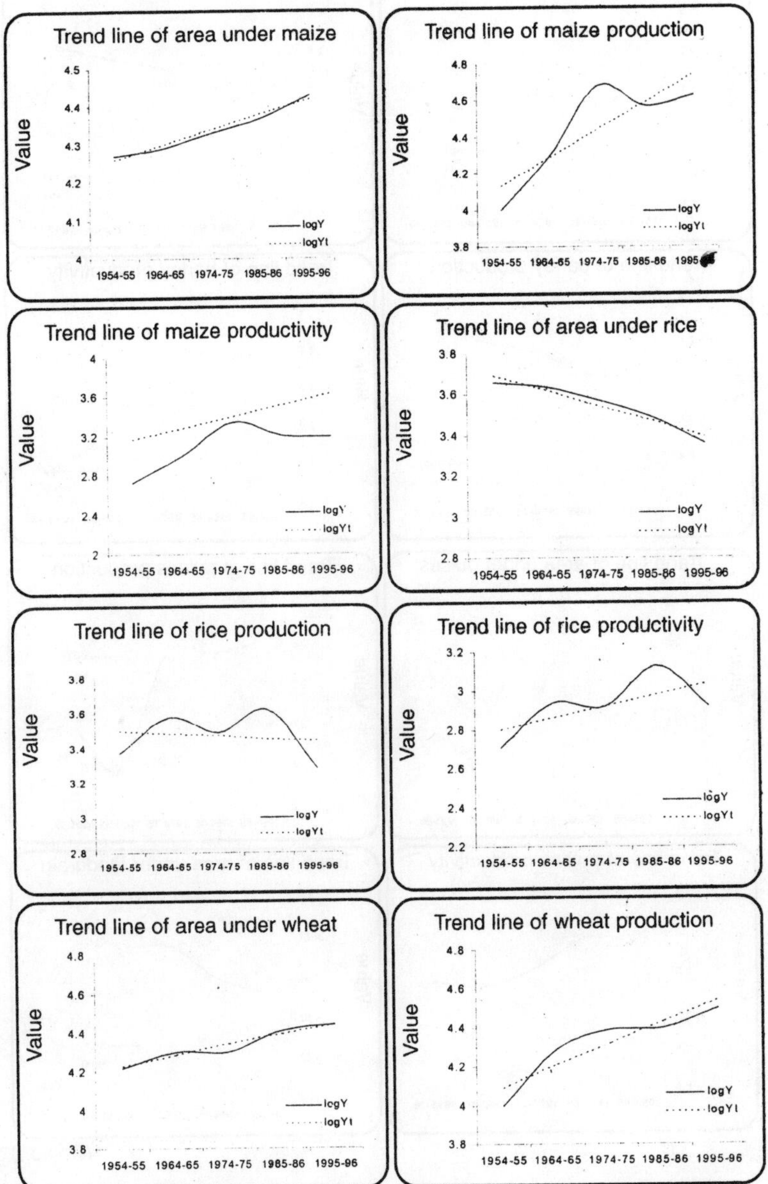

(Contd..)

Fig. 4.17 : *(Contd..)*

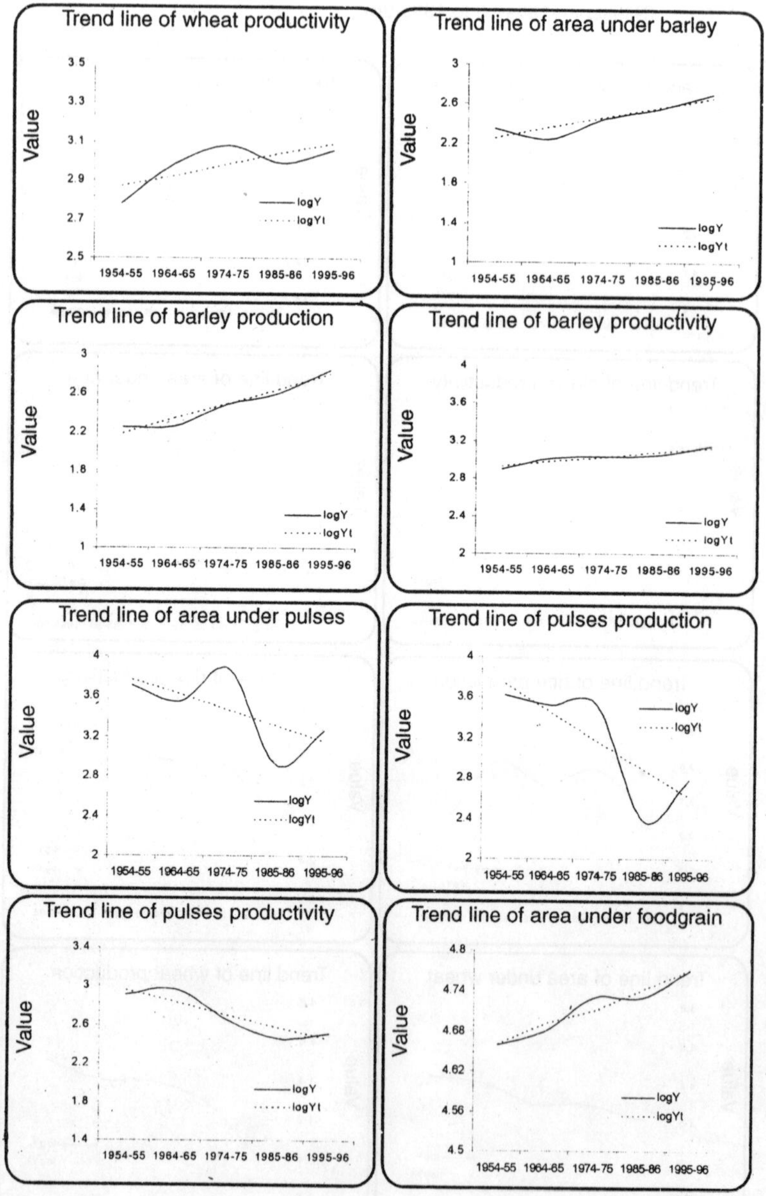

(Contd..)

Fig. 4.17 : *(Contd..)*

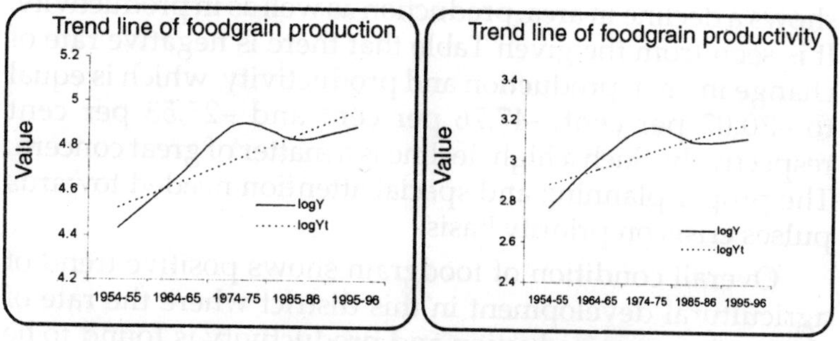

Maize is a dominant crop of this district which shows positive trend in area, production and productivity. It is noted from Tables 4.2, 4.3 and 4.4 that the rate of change in area, production and productivity of maize during 1954-55 to 1995-96 is 9.40 per cent, 42.56 per cent and 30.32 per cent respectively. Significant increase in productivity is an important indication of agricultural development.

The condition of rice shows a decline in areal expansion and in production as well but the productivity is increasing which is a positive indicator of its development. It is evident from Tables 4.2, 4.3 and 4.4 that the rate of change in area, production and productivity is –15.67 per cent, -3.17 per cent and 14.82 per cent respectively. It is said that rice is not as significant as maize and wheat in this district.

Wheat is showing positive trend of development in areal expansion, and improvement of production as well as productivity. The overall rate of change in area, production and productivity of wheat during 1954-55 to 1995-96 is found to be 13.24 per cent, 29.12 per cent and 14.02 per cent respectively, which shows that wheat improves its position in the cropping pattern.

Barley is showing very encouraging results of development. As its area shows high rate of change besides the positive rate of change in productivity. It is revealed from Tables. 4.2, 4.3 and 4.4 that the rate of change in area, production and productivity is found to be 26.47 per cent, 42.23 per cent and 12.46 per cent respectively. Berley does not show such an encouraging condition in any dastrict as it shows in Bilaspur.

Pulses are passing through the worst phase of crisis and shows a decline in area, production as well as in productivity. It is seen from the given Table that there is negative rate of change in area, production and productivity which is equal to –30.02 per cent, -47.76 per cent and –25.53 per cent respectively. Such a high decline is a matter of great concern. The proper planning and special attention needed towards pulses crisis on priority basis.

Overall condition of foodgrain shows positive trend of agricultural development in this district where the rate of change in area, production and productivity is found to be 6.41 per cent, 28.53 per cent and 19.95 per cent respectively.

Figure 4.17 clearly shows the trend of agricultural development in Bilaspur. It is seen that the trend is positive in case of area, production and productivity of maize, wheat, barley and total foodgrain in case of rice, area and production shows negative trend of agriculture development where productivity is showing positive trend. Pulses showing negative trend of agricultural development and needs special attention.

Trend of Agricultural Development in Chamba

Chamba is the northern most district of Himachal region adjacent to Kathua of Jammu and Kashmir. The landscape of Chamba is very undulating and tough for agricultural practices. As a result negative trend is found in irrigation and intensity of landuse. While other inputs show positive trend of agricultural development. The reflection of these inputs on productivity of crops is summarized below.

Maize is the most dominant crop of the district in terms of rate of change in productivity i.e. 47.19 per cent while rate of change in area and production is found to be 3.75 per cent and 53.46 per cent. Such an ecouraging situation of maize is the result of green revolutionary measures adopted to increase the agricultural productivity.

The condition of rice is disouraging and the rate of change in its areal expansion and production is found to be negative while productivity shows a slight increase. The magnitude

Fig. 4.18 : Trend line of Area, Production and Productivity of Cereals and Pulses (Chamba—1954-55 to 1995-96)

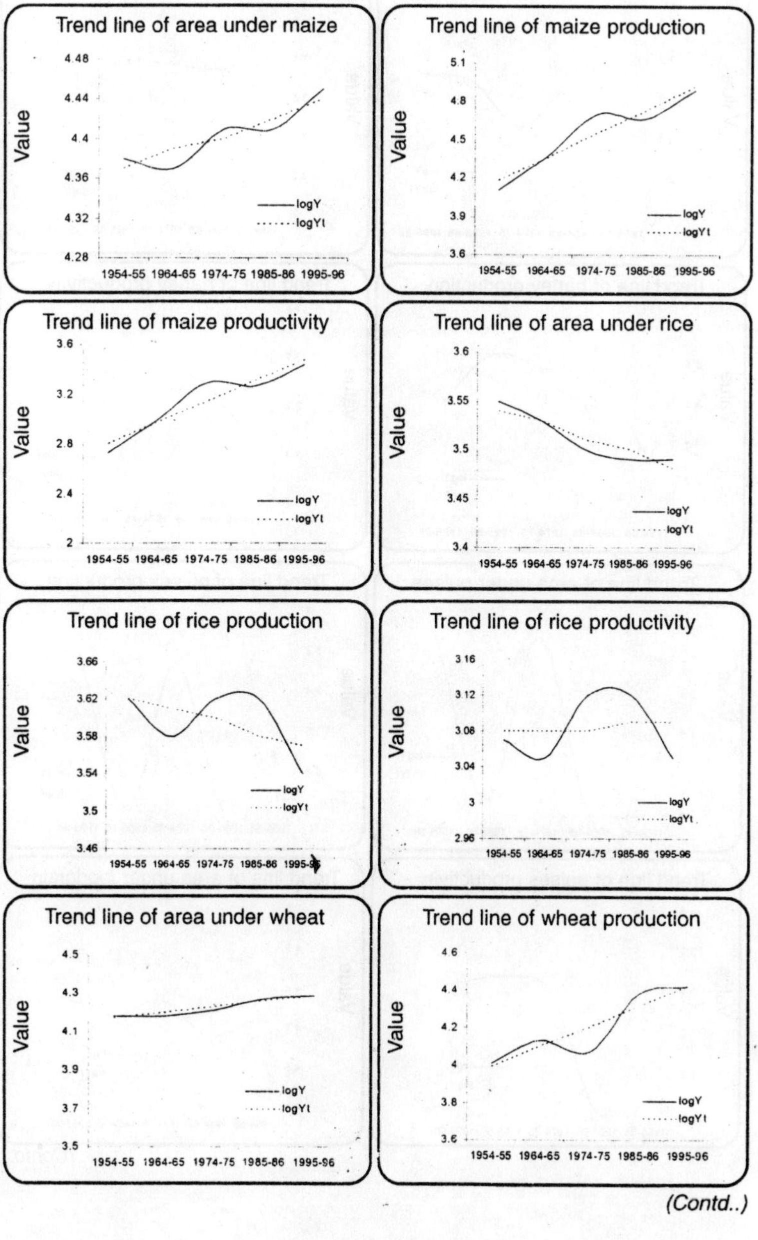

(Contd..)

Fig. 4.18 : *(Contd....)*

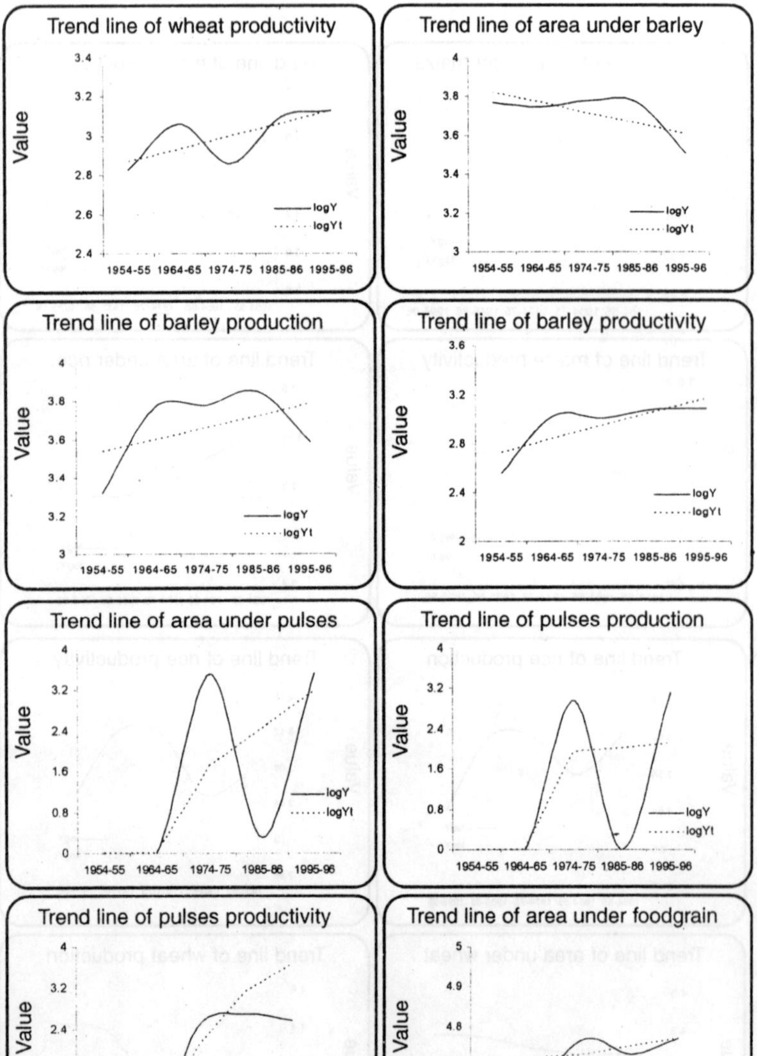

(Contd..)

Fig. 4.18 : *(Contd....)*

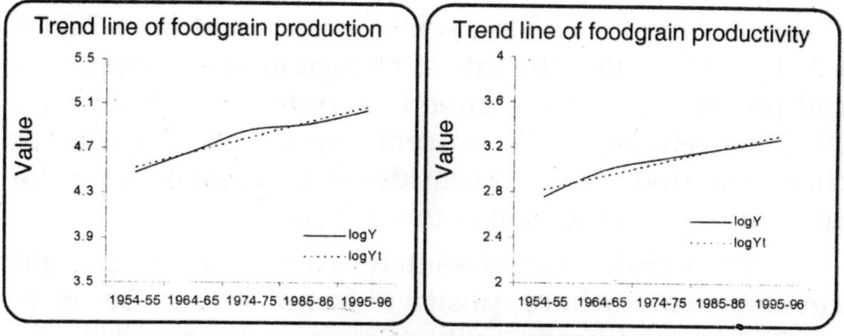

of rate of change in area, production and productivity during 1954-55 and 1995-96 recorded as –3.39 per cent, –2.73 per cent and 0.69 per cent respectively. Such an agricultural condition cannot be considered as satisfactory (Tables 4.2, 4.3 and 4.4).

Wheat is next to maize in terms of trend of development, where the rate of change in areal expansion is noted as 7.40 per cent while in production and productivity it is 27.94 per cent and 15.00 per cent respectively. Wheat occupied a considerable area and seen satisfactory in development of production and productivity due to adoption of modern agricultural technology.

Barley gives positive results in its productivity development. It is evident from Tables 4.2, 4.3 and 4.4, that its area is shrinking with the rate of change equal to –11.08 per cent during 1954-55 to 1995-96, while its production and productivity increased with the rate of change equal to 15.35 per cent and 29.42 per cent respectively. The condition seems to be satisfactory as in a number of districts mentioned above observation found negative.

Pulses are also giving good response in Chamba district where the rate of change in area, production and productivity is found to be 2.33 per cent, 20.23 per cent and 17.49 per cent respectively. It is seen satisfactory that the productivity of pulses is increasing, while a number of districts show negative rate of change in pulses productivity because it is grown on agriculturally uneconomic lands.

Overall condition of total foodgrain is such that its area shows positive trend of agricultural development. The increase in area shows a meagre rate of change while its production and producitivity show a considerable rate of change during 1954-55 to 1995-96. It is evident from Tables 4.2, 4.3 and 4.4 that the rate of change in area, production and productivity of foodgrain is recorded as 3.75 per cent, 37.72 per cent and 32.74 per cent respectively. This rate of change in productivity is considered as satisfactory for the agricultural development in this district.

Figure 4.18 shows an absolutely clear trend of agricultural development, where positive trend found in area, production, and productivity of maize, wheat, pulses and total foodgrain while in case of rice, area and production shows declining trend. Barley shows negative trend in area only. Overall it can be said that the agricultural development in the region is satisfactory but special attention should be needed in case of rice where the development of productivity is not at par with other crops.

Trend of Agricultural Development in Kathua

Kathua is the southern most district of Jammu and Kashmir state. It is found that trend of adoption of agriculture inputs in this district is positive. All the technological measures of agricultural development show a positive trend and rate of change as well, from 1961 to 1995. But the actual development of agriculture depends on the development of agricultural productivity or agricultural output for which all the technological measures applied. An analysis for trend of agricultural productivity of some crops is given below.

The rate of change in area, production and productivity of maize gives the positive trend of agricultural development. It is evident from Table 4.2, that the area under maize increased during 1954-55 to 1995-96 with the rate of change equal to 12.98 per cent in case of production (Table 4.3) which is an obvious impact of green revolution technology as a result the productivity of maize also shows positive rate of change equal to 36.14 per cent in same period of time (Table 4.4)

Fig. 4.19 : Trend line of Area, Production and Productivity of Cereals and Pulses (Kathua—1954-55 to 1995-96)

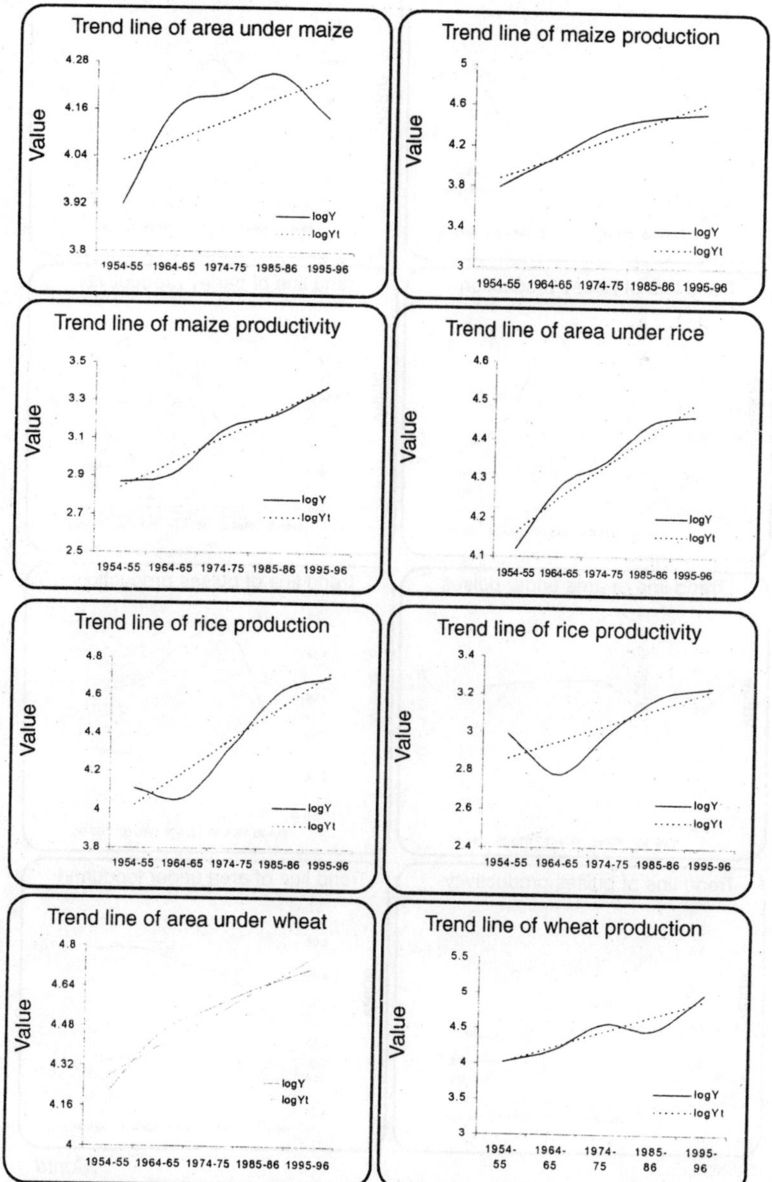

(Contd..)

Fig. 4.19 : *(Contd..)*

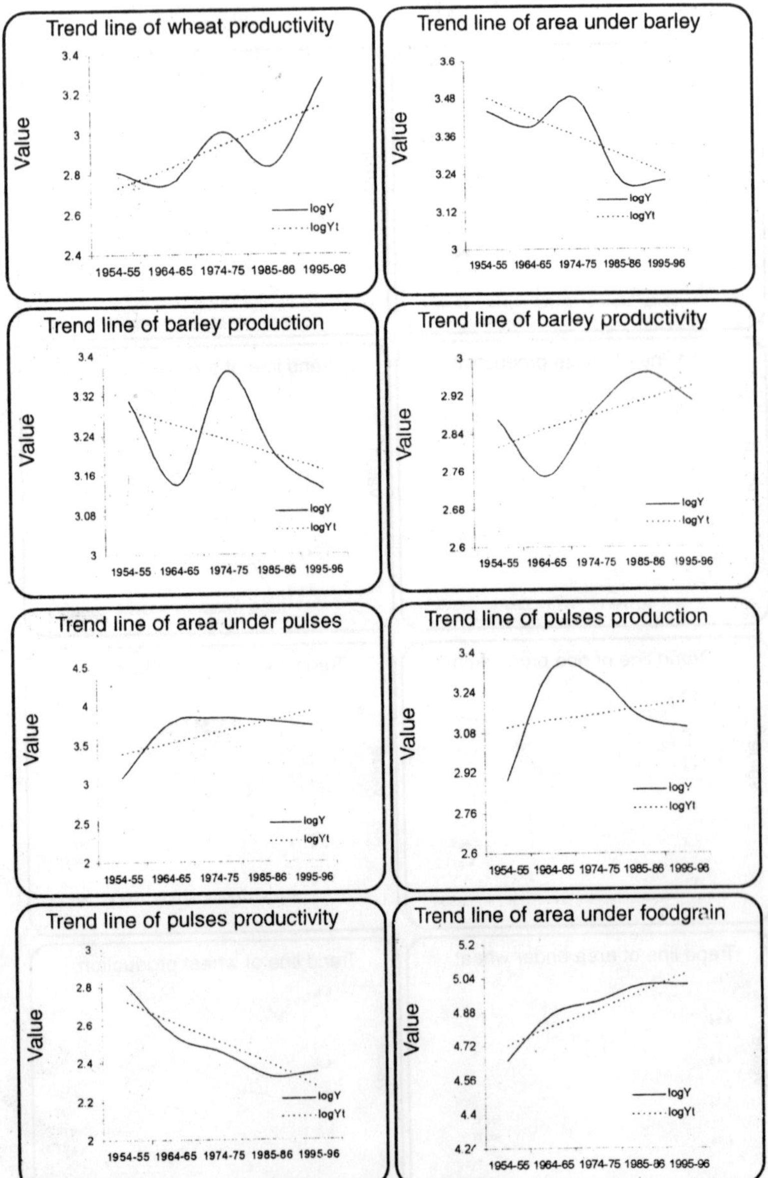

(Contd..)

Fig. 4.19 : *(Contd..)*

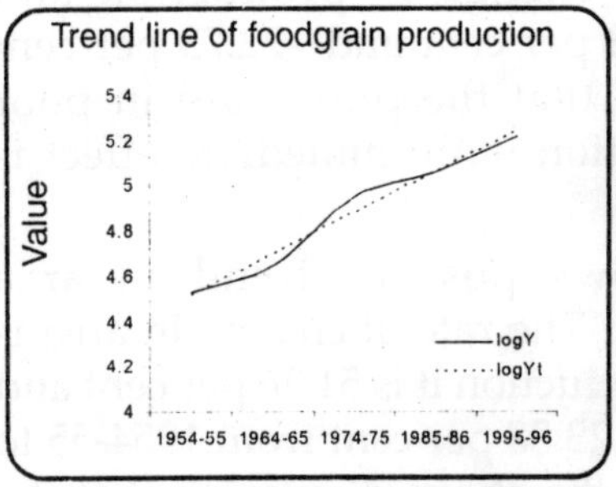

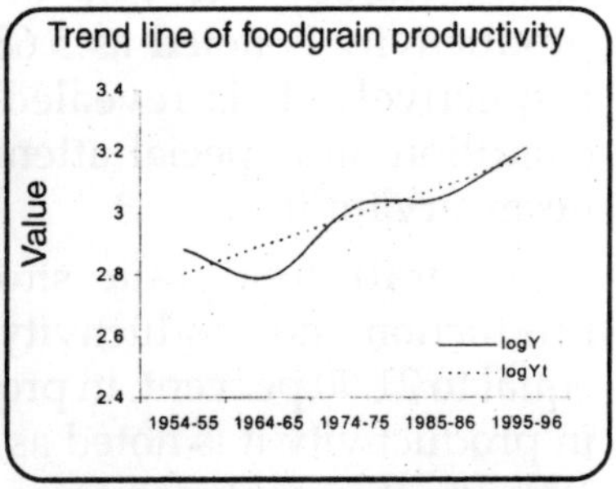

Rice is also showing positive trend of agricultural development. The area under rice shows positive change in areal expansion from 1954-55 to 1995-96. The magnitude of rate of change is found to be 21.06 per cent which is an ultimate effect of irrigation development in this district. The production shows positive trend with the rate of change equal to 49.28 per cent. As a result of significant increase in production its productivity has also increased significantly with the rate of change equal to 23.31 per cent in the same period of time.

Wheat shows significant development in its area production and productivity. Considerably improvement noted in the wheat condition from 1954-55 to 1995-96. It is evident from Tables 4.2, 4.3 and 4.4 that the rate of change in area, production and productivity during the said time period is noted as 31.22 per cent, 66.34 per cent and 26.47 per cent respectively. This trend of wheat development is said to be satisfactory.

The condition of barley is such that its area and production shows negative rate of change while productivity shows positive. As a result its area is shrinking and production is reducing with the rate of change equal to –13.10 per cent and –6.67 per cent but the productivity of barley shows positive rate of change which is equal to 7.40 per cent.

The area under pulses has increased by the rate of change equal to 37.09 per cent from 1954-55 to 1995-96. It is a significant increase in area but the production does not increase at par with it. As a result the productivity of pulses has declined. The rate of change in production and productivity is noted as 5.68 per cent and –22.73 per cent respectively. It is revealed that the pulses are in poor condition and special attention with immediate effect is necessary for it.

Overall foodgrain shows positive trend in area production and productivity. The rate of change in area is equal to 21.90 per cent, in production it is 51.36 per cent and in productivity it is noted as 23.88 per cent from 1954-55 to 1995-96. It is analyzed from the above discussion that the agricultural development in Kathua district is satisfactory but pulses are in crisis which need to be developed on priority basis.

The trend of agricultural development in Kathua district can be seen very clear from Fig. 4.19, where the trend line curve of area production and productivity of some dominant crops have been drawn. It is evident from Fig. 4.19 that the trend is positive in area and productivity of maize, rice, wheat and total foodgrain. In case of barely area, and production shows negative trend while productivity shows positive. In case of pulses productivity shows negative while area increases which is not a remunerative condition for any crop.

Trend of Agricultural Development in Jammu

Jammu is the winter capital of Jammu and Kashmir state with a considerable plain area. As a result it is the most developed district of Jammu region. Ultimately, the agricultural input adopted here must be on top priority. It is evident from Table 4.1 that the agricultural implements used in the district are on increasing trend between 1961 to 1995, and all the agricultural machinery taking into account showing positive rate of change. Adoption of agricultural machinery results in development of agricultural productivity. A brief account of trend of agricultural productivity of some crops is given below.

Fig. 4.20 : Trend line of Area, Production and Productivity of Cereals and Pulses (Jammu—1954-55 to 1995-96)

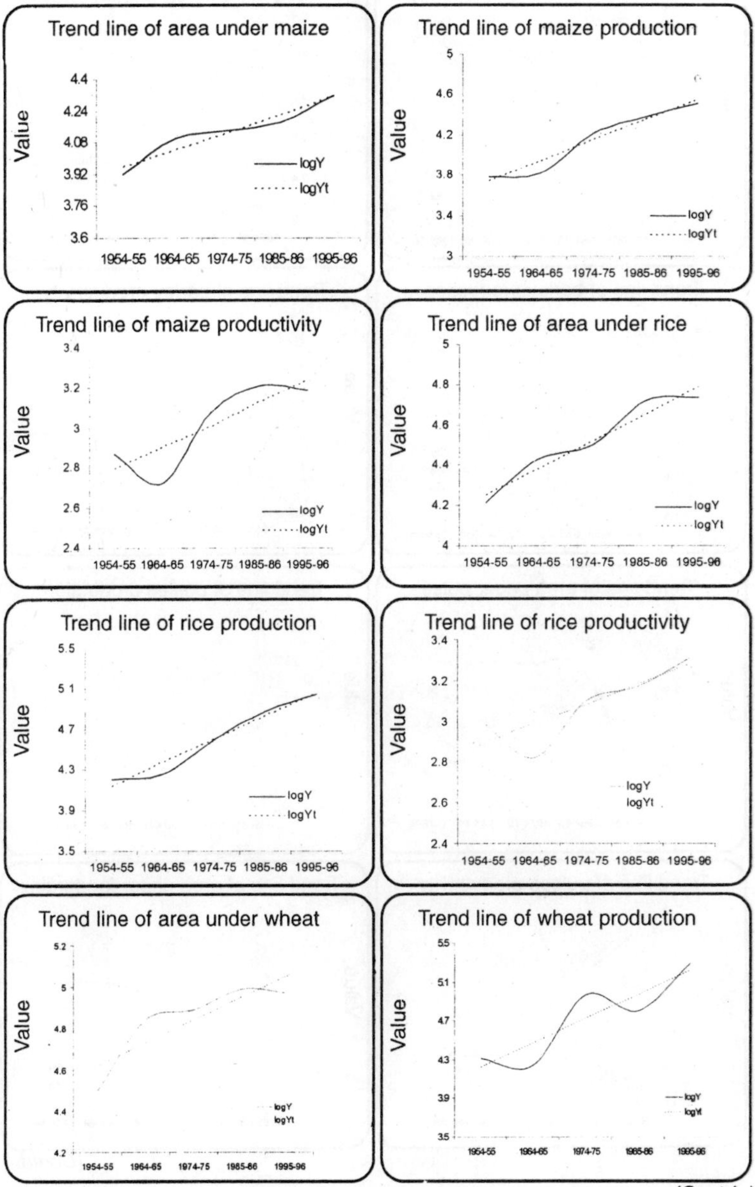

(Contd..)

Fig. 4.20 : *(Contd..)*

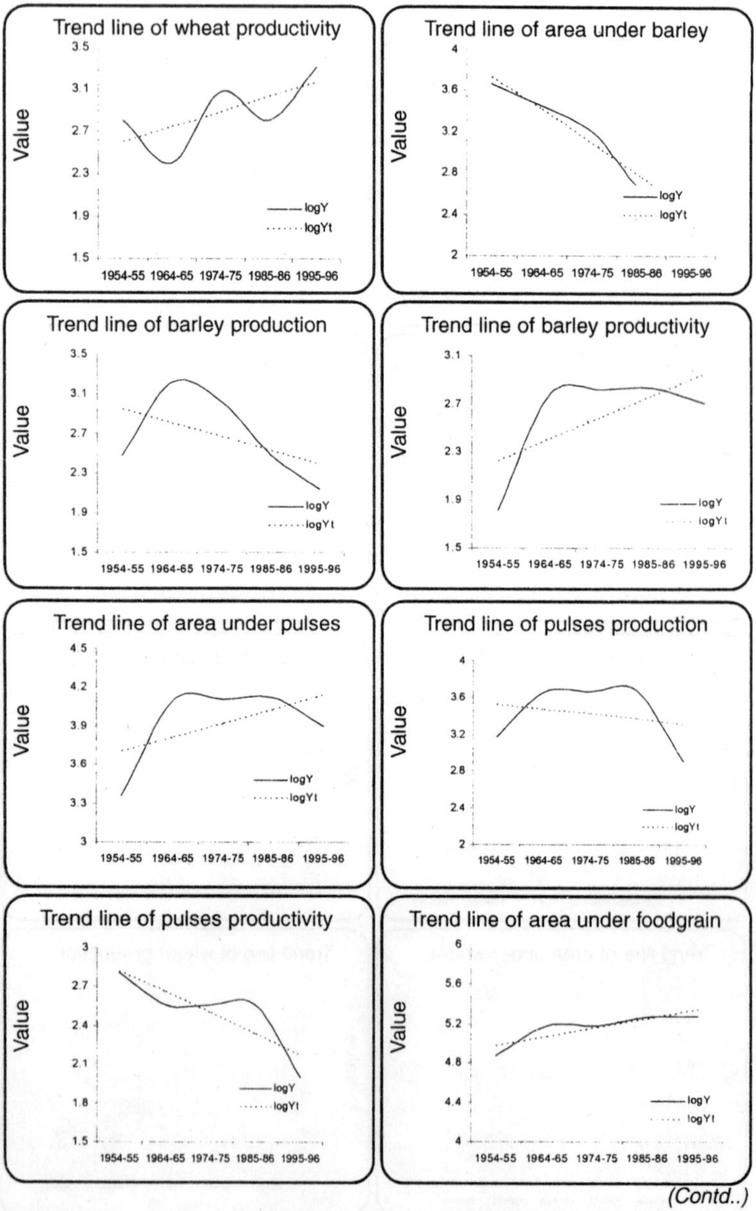

(Contd..)

Fig. 4.20 : *(Contd..)*

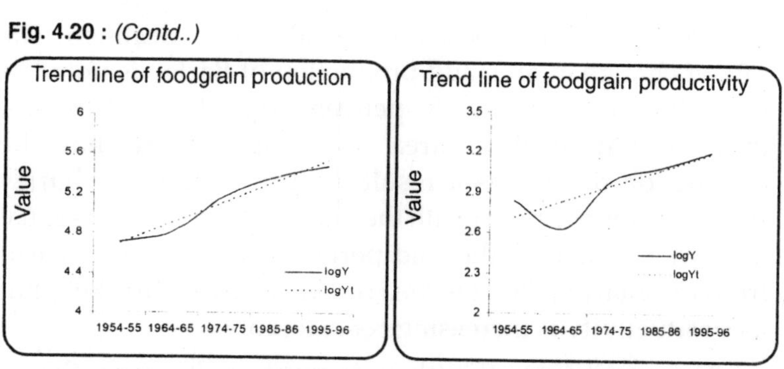

Maize is a dominant crop having significant share in cropping pattern. It is seen from Tables 4.2, 4.3, and 4.4 that the rate of change in area, production and productivity of maize between 1954-55 and 1995-96 is recorded as 22.74 per cent, 59.22 per cent and 29.72 per cent respectively. The change in area and production is prominent which is the direct impact of irrigation and other agricultural inputs.

Rice which is the most dominant crop of Kashmir valley shows significant improvement in this district too. It is seen from the given table that the positive trend of change is found in rice condition. There is a high positive rate of change found in area under rice i.e. 36.46 per cent due to the development of irrigation in the district. Besides the area, production and productivity of rice also shows significant increase with the rate of change equal to 70.61 per cent and 25.31 per cent respectively.

Wheat is the most important crop not only in the cropping pattern but also in its developmental change from 1954-55 to 1995-96. The level of development can be assessed from the rate of change in area, production and productivity of wheat during 1954-55 to 1995-96 which is found to be 28.82 per cent, 78.24 per cent and 38.60 per cent respectively. Such a high level development in area, production and productivity is the result of green revolution technology adopted in the district.

The condition of Barley is such that its area has been declined to a considerable extent from 1954-55 to 1995-96.

Not only area but production has also been declined. The rate of change in area and production of Barley is noted as – 52.79 per cent and –27.89 per cent respectively. It is seen that the magnitude of area shrinking is more than the production decline as a result the productivity of barley shows an improvement with the rate of change which is equal to 52.41 per cent in the said period of time. It is revealed from the analysis that the area under barley is shrinking but its productivity is increasing considerably.

The condition of pulses is poor as its area shows increasing trend but with decline productivity which is the most uneconomic condition i.e. low yield with high spread. The rate of change in area, production and productivity of pulses during 1954-55 to 1995-96 is 29.12 per cent, –11.28 and –31.29 per cent respectively.

Overall trend of foodgrain is showing positive trend of agricultural development in area, production and productivity. It is evident from Tables 4.2, 4.3 and 4.4 that the rate of change in area, production and productivity of foodgrain from 1954-55 to 1995-96 is found to be 23.59 per cent, 62.55 per cent and 31.52 per cent respectively. Development of irrigation status and adoption of modern agricultural technology results in such a high level of improvement in agricultural development besides the political importance of Jammu district.

A very clear picture of trend of agricultural development can be seen from Fig. 4.20. It is evident from the figure that maize, rice, wheat and total foodgrains show positive trend of development in area production as well as in productivity but in case of barley and pulses the conditions are little bit different as in barely area and production shows negative trend while productivity shows positive trend and in case of pulses production and productivity shows negative trend besides the area showing positive trend of development.

Trend of Agricultural Development in Doda

The cold climate as well as undulating landscape set limits for agricultural development in Doda. The rate of change in irrigation shows decline while intensity of landuse improves very little. Besides these unfavourable parameters the productivity of foodgrain crops increased. A brief account of agricultural development with reference to these crops is given below.

Dominance of maize in the cropping pattern is prominent as well as the improvement in agricultural productivity of this crop. It is evident from Tables 4.2, 4.3 and 4.4 that the rate of change in area, production and productivity of maize during 1954-55 to 1995-96 is found to be 45.21 per cent, 90.11 per cent and 30.92 per cent respectively. Such a high rate of change in production is a clear reflection and green revolution inputs applications in agricultural practices causes increase in productivity of maize.

Condition of rice is also seen improving besides the negative trend in irrigation. Area under rice has shown positive rate of change equal to 4.23 per cent while its production increased with the rate of change equal to 20.25 per cent. As a result the rate of change in productivity shows positive rate of change equal to 15.61 per cent from 1954-55 to 1995-96. In short the condition of rice showing satisfactory with positive trend of improvement.

The condition of wheat is such that its area is shrinking with the rate of change equal to –2.95 per cent but production and productivity showing positive trend with the rate of change equal to 13.24 per cent and 16.68 per cent respectively. Overall trend is remunerative but productivity should be increased along with the increase of area under the crop.

The agro-ecology of Doda is not suitable for barley crop as it is showing declining trend in area, production as well as in productivity. It is evident from Tables 4.2, 4.3 and 4.4

Fig. 4.21 : Trend line of Area, Production and Productivity of Cereals and Pulses (Doda-1954-55 to 1995-96)

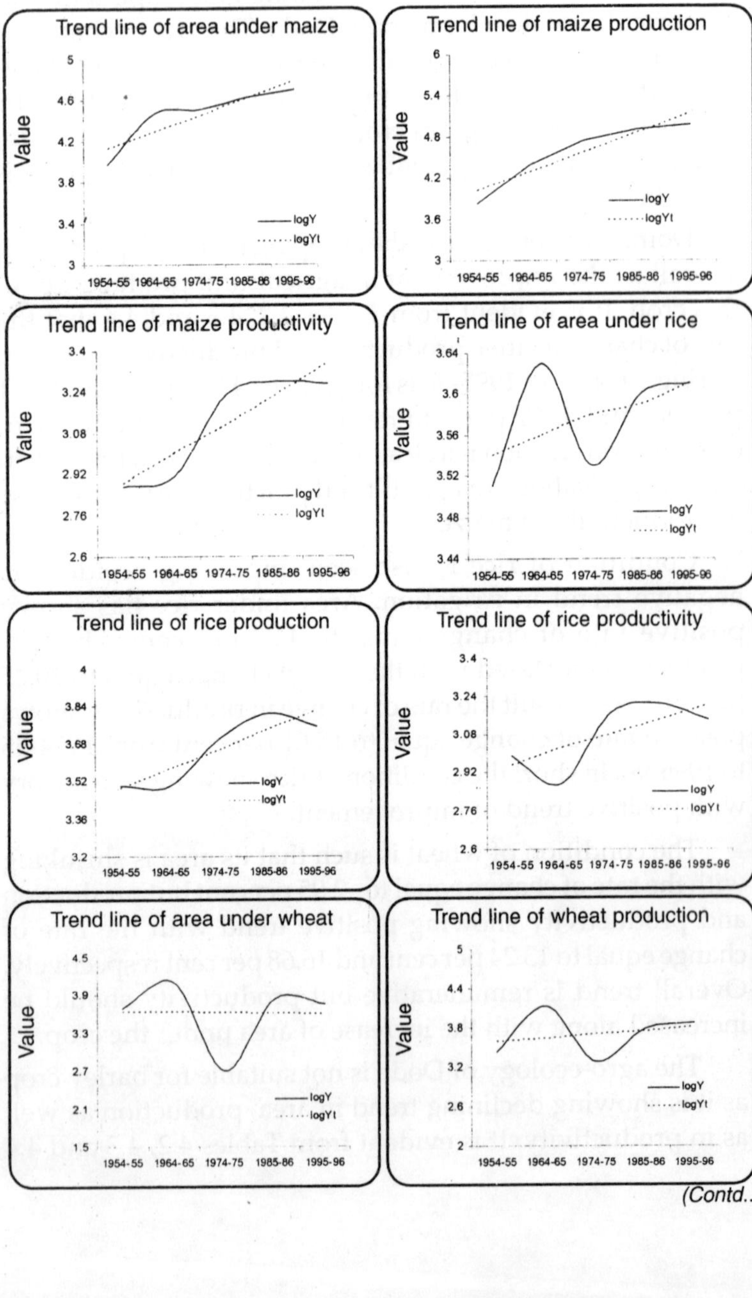

(Contd..)

Fig. 4.21 : *(Contd..)*

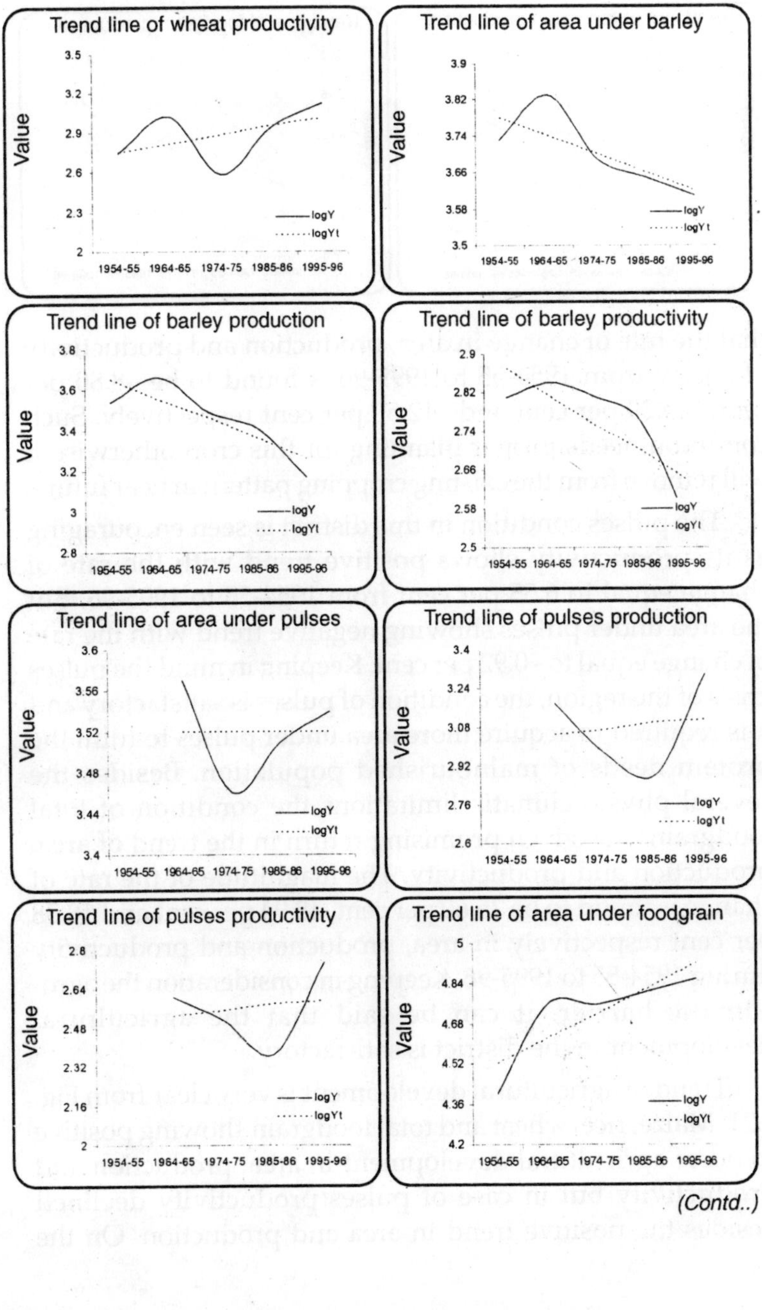

(Contd..)

Fig. 4.21 : *(Contd..)*

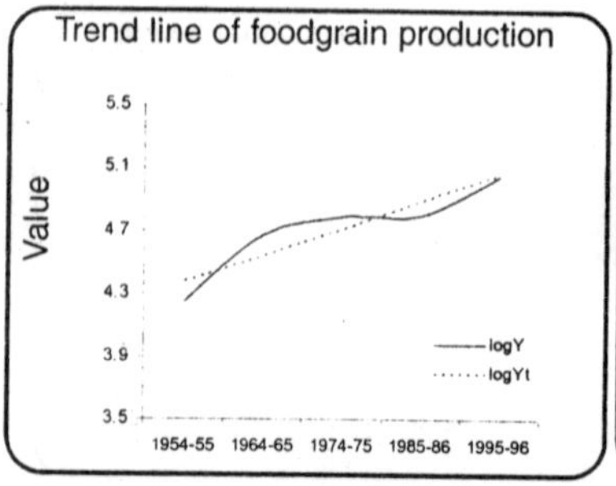

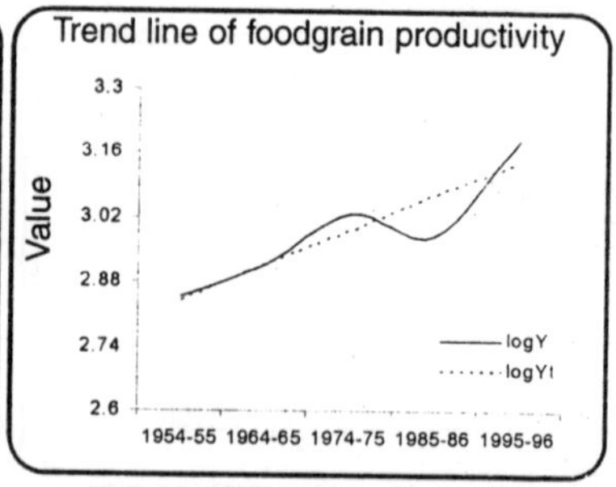

that the rate of change in area, production and productivity of barley from 1954-55 to 1995-96 is found to be –8.80 per cent, -20.20 per cent and –12.50 per cent respectively. Such condition needs proper planning for this crop otherwise it will remove from the existing cropping pattern in near future.

The pulses condition in this district is seen encouraging as its productivity shows positive trend with the rate of change equal to 6.55 per cent from 1954-55 to 1995-96. But the area under pulses showing negative trend with the rate of change equal to –0.92 per cent. Keeping in mind the pulses crisis of the region, the condition of pulses is satisfactory and it is required to acquire more area under pulses to fulfil the protein needs of malnourished population. Besides the several physio-climatic limitations the condition of total foodgrains has given promising return in the trend of area, production and productivity. The magnitude of the rate of change is found to be 25.60 per cent, 48.94 per cent and 18.58 per cent respectively in area, production and productivity during 1954-55 to 1995-96. Keeping in consideration the agro-climatic barrier, it can be said that the agricultural development in the district is satisfactory.

Trend of agricultural development is very clear from Fig. 4.21. Maize, rice, wheat and total foodgrain showing positive trend of agricultural development in area. production and productivity but in case of pulses productivity declined besides the positive trend in area and production. On the other hand in case of barley area, production and productivity shows negative trend in agricultural development.

Trend of Agricultural Development in Poonch

Poonch is an important district of Jammu region, where the agriculture practices are on every possible bit of land. Besides the slope factor and thin soil horizon the condition of total foodgrain production and productivity seems to be encouraging. Maize and wheat are the major crops of the cropping pattern, while rice, barley and pulses have insignificant share. The development of few crops taking into account are given below.

Maize is the second most dominant crop in terms of areal spreads besides its area has shown negative trend. The production and productivity of maize is giving promising return in the trend of agricultural development. The rate of change in area, production and productivity of maize during 1954-55 to 1995-96 is noted as –11.08 per cent, 30.02 per cent and 46.55 per cent respectively. Negative trend of area is seen due to the occupation of Indian land. Besides its production and productivity is improving by the help of modern agricultural efforts.

Rice is an insignificant crop in the cropping pattern of Poonch. Decline in the area will cause serious consequences in future. But the production and productivity of rice is improving due to the adoption of new agricultural measures of agricultural development. It is evident from Tables 4.2, 4.3 and 4.4 that the rate of change in area, production and productivity of rice from 1954-55 to 1995-96 is found to be –15.28 per cent, 25.03 per cent and 47.57 per cent respectively. It can be said that the development of productivity is satisfactory but it is required to increase the area of this crop for a balanced cropping pattern.

Fig. 4.22 : Trend line of Area, Production and Productivity of Cereals and Pulses (Poonch—1954-55 to 1995-96)

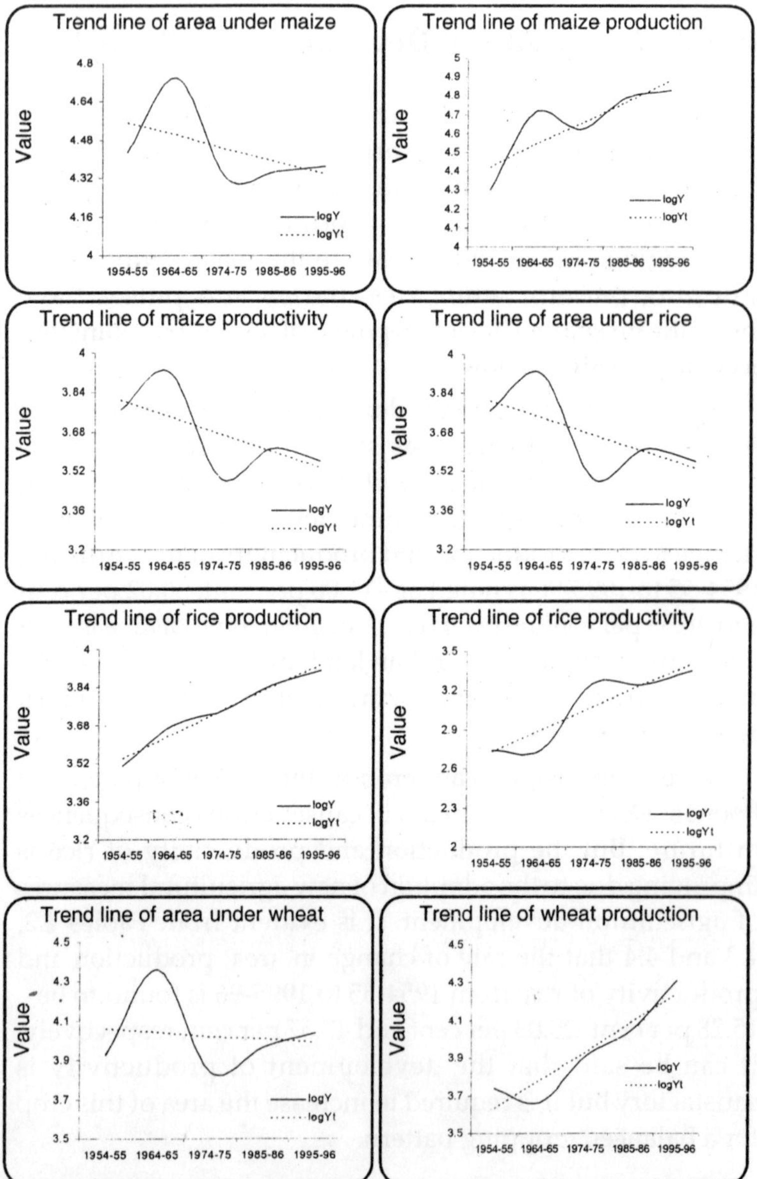

(Contd..)

Fig. 4.22 : *(Contd..)*

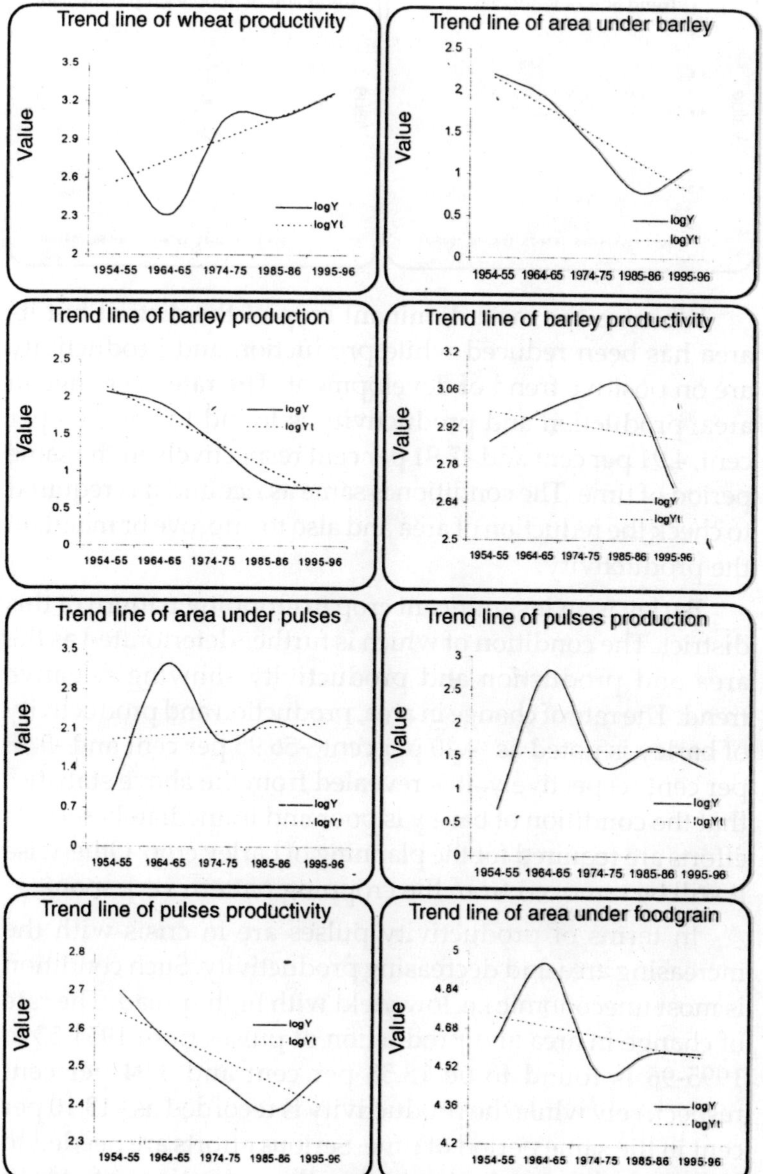

(Contd..)

Fig. 4.22 : *(Contd..)*

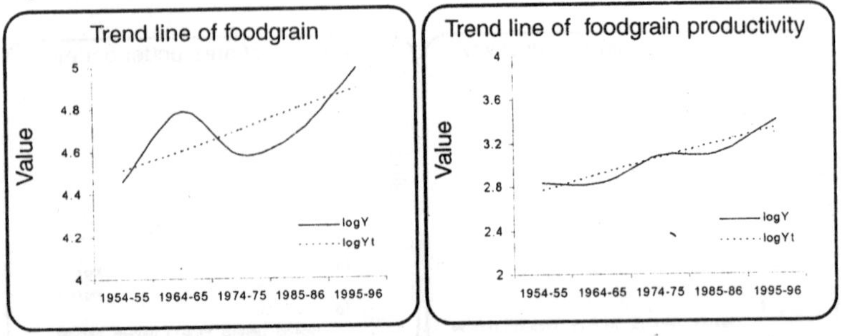

Wheat is the most dominant crop of the district but its area has been reduced while production and productivity are on positive trend of development. The rate of change in area, production and productivity is found to be –3.84 per cent, 4.91 per cent and 47.91 per cent respectively in the same period of time. The condition is same as rice and it is required to check the reduction of area and also to improve or maintain the productivity.

Barley is an insignificant crop in cropping pattern of this district. The condition of which is further deteriorated as the area and production and productivity showing negative trend. The rate of change in area, production and productivity of barley is noted as -4.40 per cent, -56.95 per cent and -5.38 per cent respectively. It is revealed from the above statistics that the condition of barley is poor and immediately serious efforts are required for the planning of barley crop. Otherwise it will be removed from the cropping pattern very soon.

In terms of productivity pulses are in crisis with the increasing area and decreasing productivity. Such condition is most uneconomic i.e. low yield with high spread. The rate of change in area and production of pulses from 1954-55 to 1995-96 is found to be 18.58 per cent and 3.04 per cent respectively while the productivity is recorded as –13.10 per cent in the same period of time. Serious efforts are needed to improve the productivity by the adoption of green revolutionary measures for a balanced cropping pattern.

The condition of overall foodgrain is encouraging besides its area has declined. But a significantly high positive change occurred in production and productivity of total foodgrains. The rate of change in area production and productivity from 1954-55 to 1995-96 is recorded as –9.64 per cent, 24.74 per cent and 38.36 per cent respectively. The decline in area is due to the encroachment of agricultural land by high rate of urbanization besides a large portion of Poonch district come under the illegal occupation.

Trend of agricultural development can be seen from Fig. 4.22 which shows that the area of maize, rice, wheat, barley and foodgrain giving negative trend while pulses give positive trend. Similar condition existing in case of production. But in case of productivity maize, rice, wheat and total foodgrains showing positive trend while barley and pulses are giving negative trend of agricultural development. It is revealed that proper, urgent and serious efforts are required for the development of barley and pulses otherwise serious consequences of cropping pattern will emerge in very near future.

Trend of Agricultural Development in Srinagar

Srinagar is situated in the heart of Kashmir valley, it is the capital of Jammu and Kashmir state and there is a huge urban agglomeration in these mountainous ranges. It is the most developed urban-pocket not only in the state but also in the region as a whole. The landscape of it is plain, so agricultural practices as well as other developmental work are easy to perform. As a result the level of agricultural development besides the overall development is very high. Availability and adoption of new agricultural technology is also high. Above all the agro-climatic conditions are favourable for a variety of food and non-food crops. A brief account of some food crops is given below.

Maize is the dominant crop of the region in terms of productivity while wheat is dominant in terms of area. But

Fig. 4.23 : Trend line of Area, Procuction and Productivity of Cerels and Pulses (Srinagar-1954-55 to 1995-96)

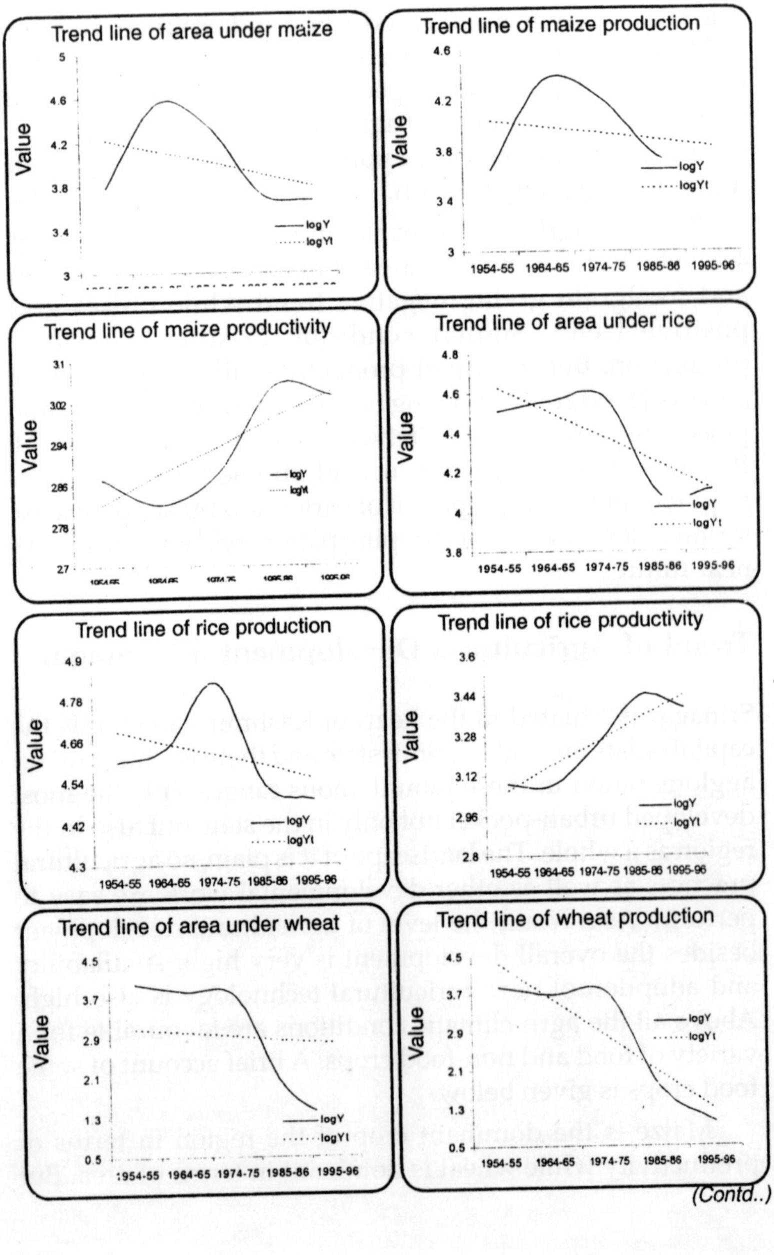

(Contd..)

Fig. 4.23 : *(Contd..)*

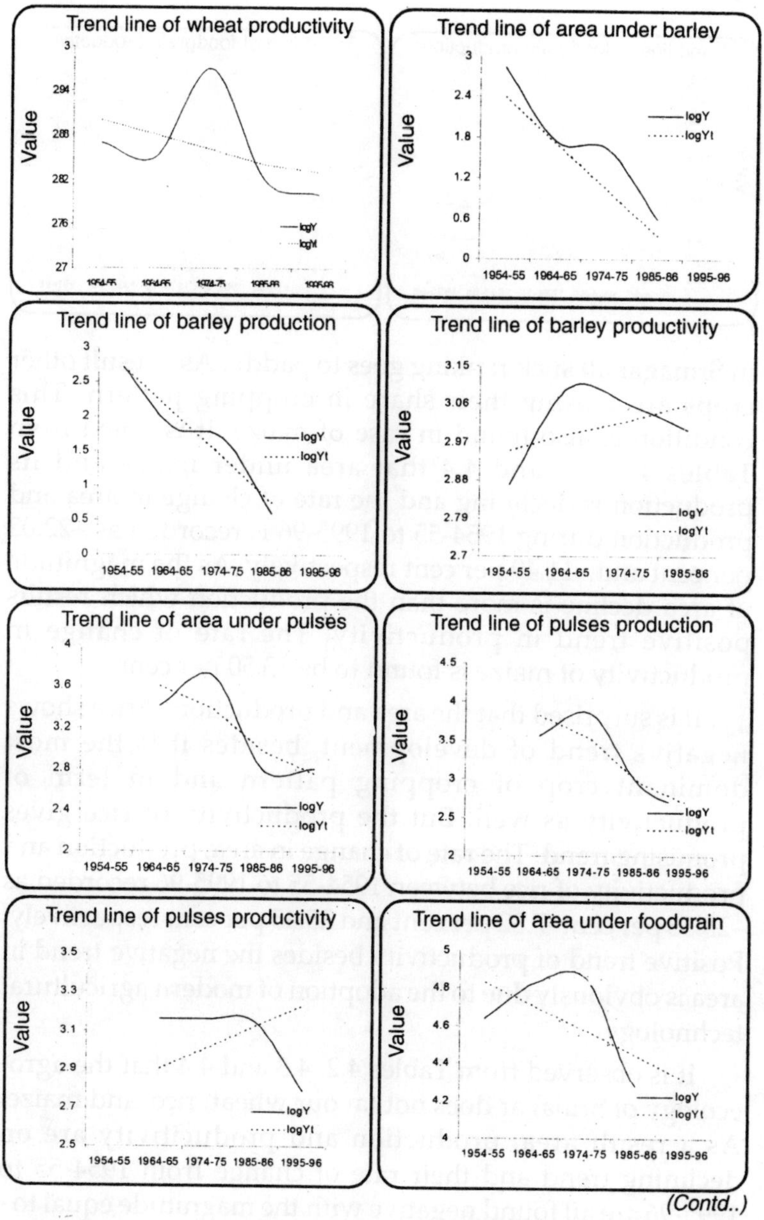

(Contd..)

Fig. 4.23 : *(Contd..)*

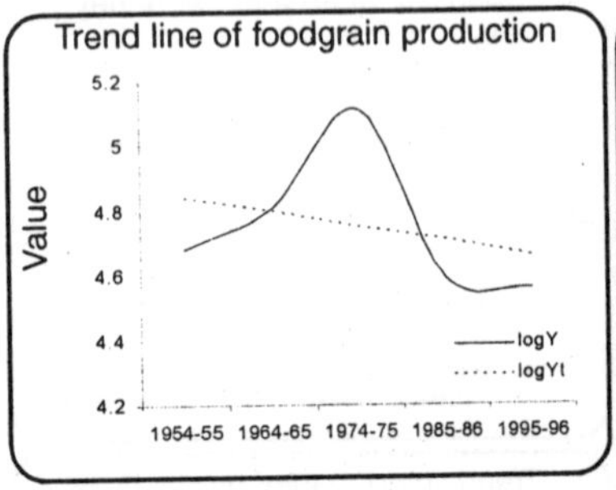

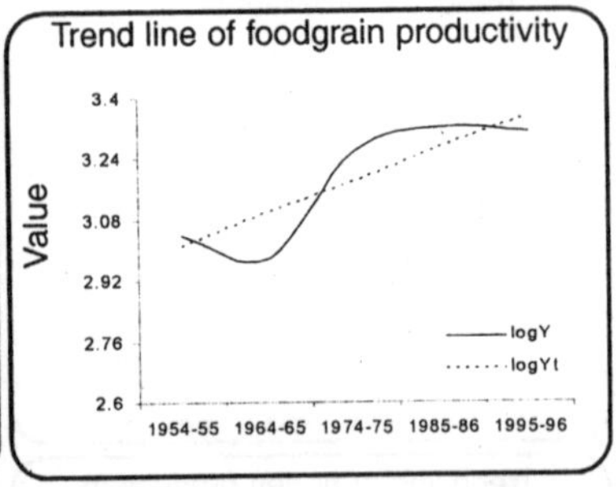

in Srinagar all such ranking goes to paddy. As a result other crops are loosing their share in cropping pattern. This condition is also found in case of maize. It is noted from Tables 4.2, 4.3 and 4.4 that area under maize and its production is declining and the rate of change in area and production during 1954-55 to 1995-96 is recorded as –22.02 per cent and –11.49 per cent respectively. As the magnitude of area decline is more than the production which results positive trend in productivity. The rate of change in productivity of maize is found to be 13.50 per cent.

It is surprised that the area and production of rice shows negative trend of development, besides it is the most dominant crop of cropping pattern and in term of productivity as well. But the productivity of rice gives promising trend. The rate of change in area, production and productivity of rice between 1954-55 to 1995-96 recorded as –25.53 per cent -7.32 per cent and 25.45 per cent respectively. Positive trend of productivity besides the negative trend in area is obviously due to the adoption of modern agricultural technology.

It is observed from Tables 4.2, 4.3 and 4.4 that the agro-ecology of Srinagar does not favour wheat, rice and maize. As a result area, production and producitivity are on declining trend and their rate of change from 1954-55 to 1995-96 are all found negative with the magnitude equal to -82.34 per cent, -82.02 per cent and -4.06 per cent respectively.

Condition of barley is more or less similar to wheat in terms of trend and rate of change which are negative. Barley is not an important foodgrain crop of the region and particularly in Kashmir Valley. Whatever, its share in the agriculture of Srinagar its condition is further deteriorating with the high rate of decline. The rate of change in area, production and productivity of barley during the same period of time is found to be –79.20 per cent, –78.42 per cent and –73.94 per cent respectively.

The condition of pulses is much poorer and its area is shrinking, production decline and productivity falling. It is evident from the given table that the rate of change in area, production and productivity of pulses is noted as –53.43 per cent, –71.02 per cent and –23.99 per cent respectively. It is the general condition of pulses crisis in the region with very few exceptions.

It is seen surprising that the area under foodgrain as well as its production showing a decline in it, due to the increasing population pressure on land resources, large areas of fertile agricultural land is urbanized for the growing population. As a result the agriculturally productive as well as potential lands are turned into urban colonies, causing shrinking net sown area and reduced production. The rate of change in area, production and productivity is found to be –21.26 per cent, -9.84 per cent and 21.90 per cent respectively.

It is revealed from Fig. 4.23 that the area and production of maize, rice, wheat, barley, pulses and total foodgrains are showing negative trend. In case of productivity it is found that maize, rice and foodgrain showing positive trend while wheat, barley and pulses showing negative trend. The negative trend of all crops taking into consideration is due to the encroachment of agricultural lands by sprawl of urbanization while increase in productivity is due to the adoption of modern agricultural technology like high yielding variety of seed, assured irrigation and chemical fertilizer, etc.

Trend of Agricultural Development in Anantnag

Anantnag is another important district of Kashmir valley where the agricultural practices are done with the adoption of modern agricultural inputs. Technological side of agriculture is showing positive trends of development. Producitivity development of some crops is discussed below.

Maize is the second most important crop next to rice in the cropping pattern. Out of the crops taking into account, it is the only one which is showing positive trend in area, production and in productivity as well. The positive rate of change has been noted in area, production and productivity of maize between 1954-55 to 1995-96, the magnitude of which is equal to 16.14 per cent, 28.53 per cent and 10.41 per cent respectively. The physiography is suitable for agricultural practices besides the adoption of modern inputs results in such an encouraging trend of development.

Rice is the most important crop not only in acreage but also in productivity as well. Though its area showing decline but the production is not. As a result the production showing promising return. The rate of change in area, production and productivity is found to be –13.70 per cent, 2.09 per cent and 18.30 per cent respectively. High rate of increase in production is due to the adoption of green revolution technology while the area reduces due to the division of the district to form a new district of Pulwama.

Though the share of wheat is very meagre both in area and in production as well. The rate of change further shows severe decline. The magnitude of which is such that –87.09 per cent and –86.26 per cent respectively. It is evident from Tables 4.2, 4.3 and 4.4 that besides the negative trend in area and production, the productivity of wheat showing positive trend with the magnitude of 6.41 per cent in the same period of time. It is required to increase the area under wheat so that a balance of cropping pattern could be maintained.

Fig. 4.24 : Trend line of Area, Production and Productivity of Cereals and Pulses (Anantnag-1954-55 to 1995-96)

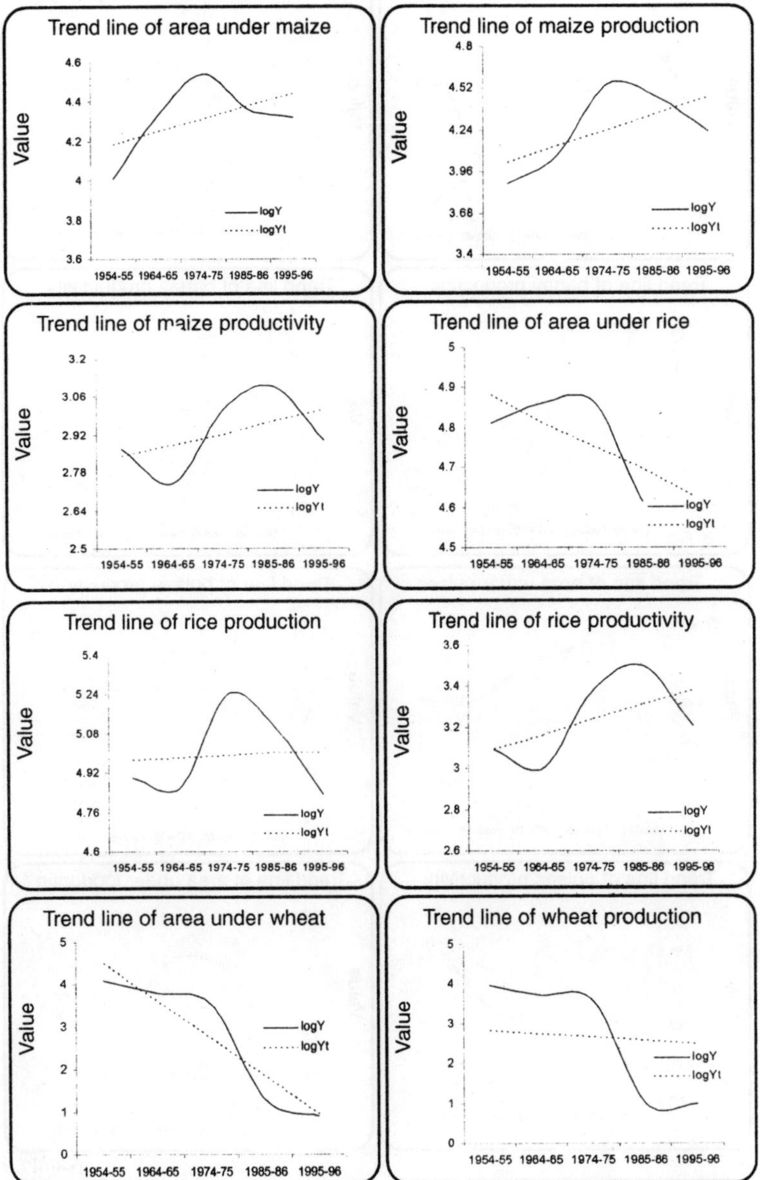

(Contd..)

Fig. 4.24 : *(Contd..)*

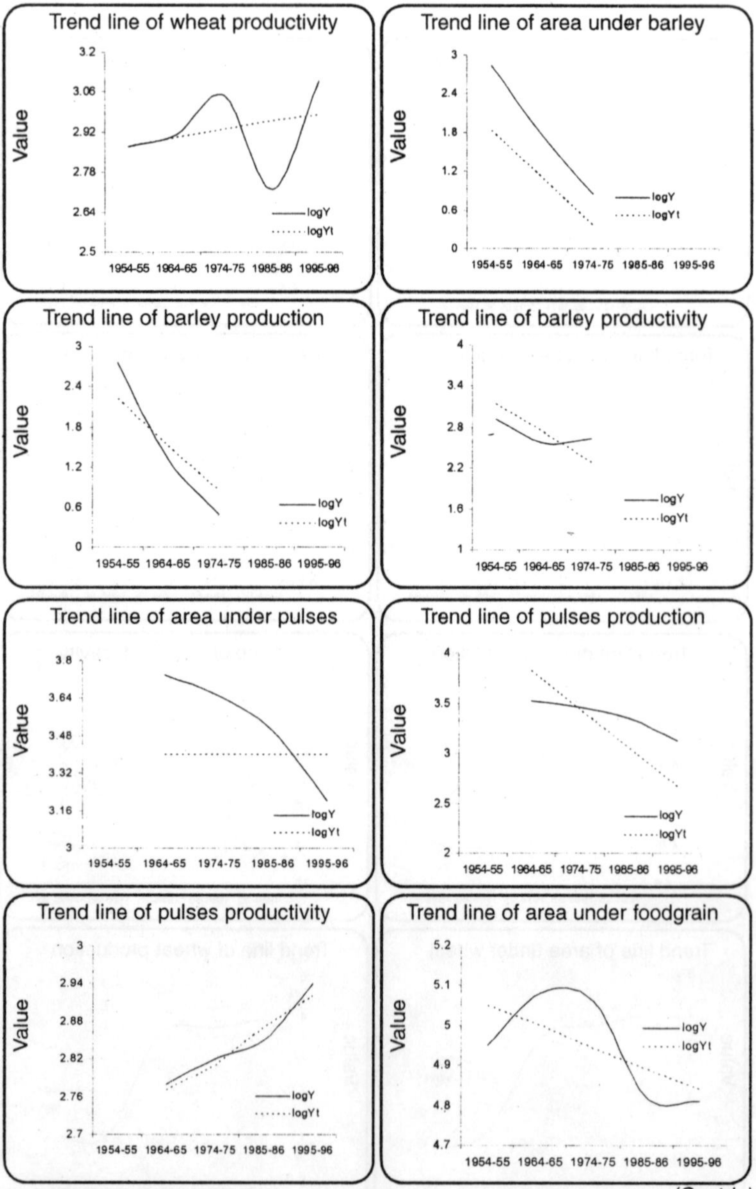

(Contd..)

Fig. 4.24 : *(Contd..)*

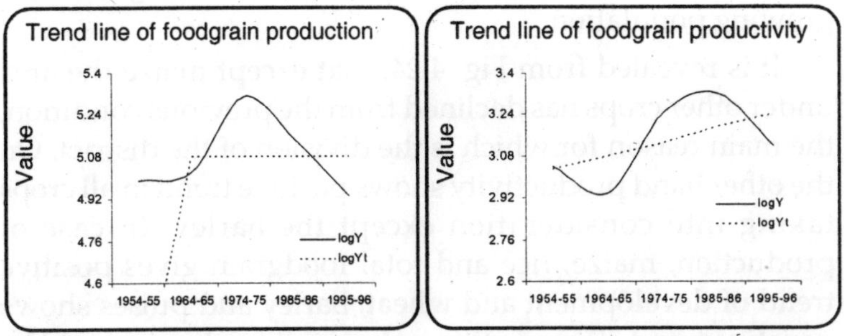

The condition of barley is seen in crisis as its area, production and productivity is showing negative trend of development, the magnitude of rate of change from 1954-55 to 1995-96 is recorded as –81.97 per cent, –73.27 per cent and –61.98 per cent respectively. Share of barley in the cropping pattern is only in fraction, and above this such a high rate of decline is matter of serious concern and if immediately serious efforts are not taken it will be removed from the cropping pattern very soon. It is now required not only to increase its area but its productivity too. At present its condition is not satisfactory.

Pulses which are sharing critical condition in most of the districts, showing the same condition here too. The area and production gives the negative trend but productivity increases. The rate of change from 1954-55 to 1995-96 in area, production and productivity of pulses is found to be –36.87 per cent, –27.01 per cent and 12.09 per cent respectively. Though decreasing area with increasing productivity is good but its share in the cropping pattern is very low which should be improved for balanced and sustainable agriculture.

The condition of total foodgrains is encouraging but the area under foodgrain is on decreasing trend due to the division of the district. The rate of change in area, production and productivity is recorded from Tables, 4.2, 4.3 and 4.4 as –11.49 per cent, 6.66 per cent and 12.98 per cent respectively. Decreasing trend of area with increasing production ultimately improve the productivity. The existing condition is not bad but the area should also be increased to overcome the feeding needs of growing population.

It is revealed from Fig. 4.24, that except maize the area under other crops has declined from the previous condition, the main reason for which is the division of the district. On the other hand productivity shows positive trend in all crops taking into consideration except the barley. In case of production, maize, rice and total foodgrain gives positive trend of development and wheat, barley and pulses shows negative trend. The condition of overall foodgrain could be improved further so that the balance of cropping should be maintained otherwise serious ecological consequences may be caused due to disbalanced cropping pattern.

Trend of Agricultural Development in Baramula

Baramula is the largest district of Kashmir valley, where rice, maize and pulses have significant share in cropping while wheat and barley occupying very meagre. On the basis of the productivity of foodgrain, the district is showing positive trend of agricultural development. A brief account of agricultural development with reference to a few crops is given below.

Maize, which is the second most dominant crop in the distict showing negative trend in area while production and productivity shows promising returns. It is evident from Tables 4.2, 4.3 and 4.4, that the rate of change in area, production and productivity of maize from 1954-55 to 1995-96 is noted as –0.23 per cent, 5.20 per cent and 4.95 per cent respectively. Though the area under maize decreased to a considerable extent but due to the adoption of modern agricultural technology, the production and productivity is on the rise.

Rice is the most dominant crop not only in areal spread but also in productivity and production as well. It is revealed from Tables 4.2, 4.3 and 4.4 that the area under rice crop

Fig. 4.25 : Trend line of Area, Production and Productivity of Cereals and Pulses (Baramula—1954-55 to 1995-96)

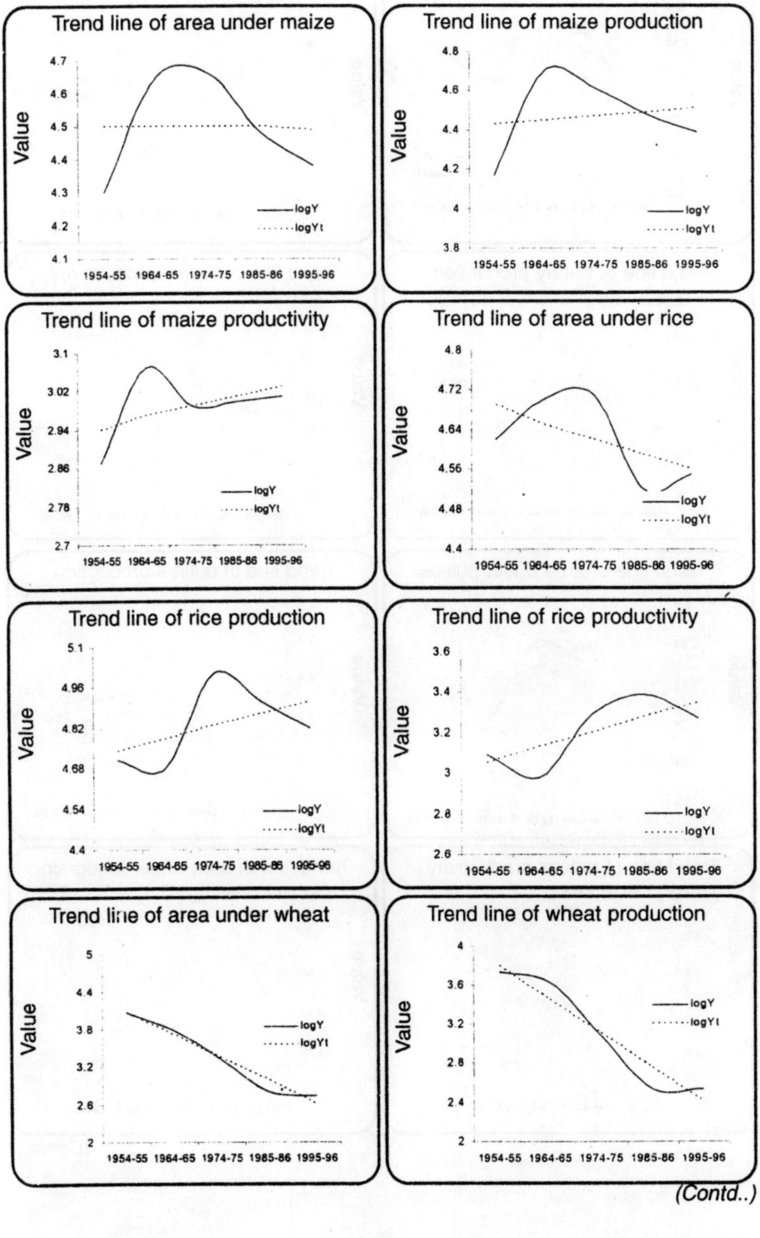

(Contd..)

Fig. 4.25 : *(Contd..)*

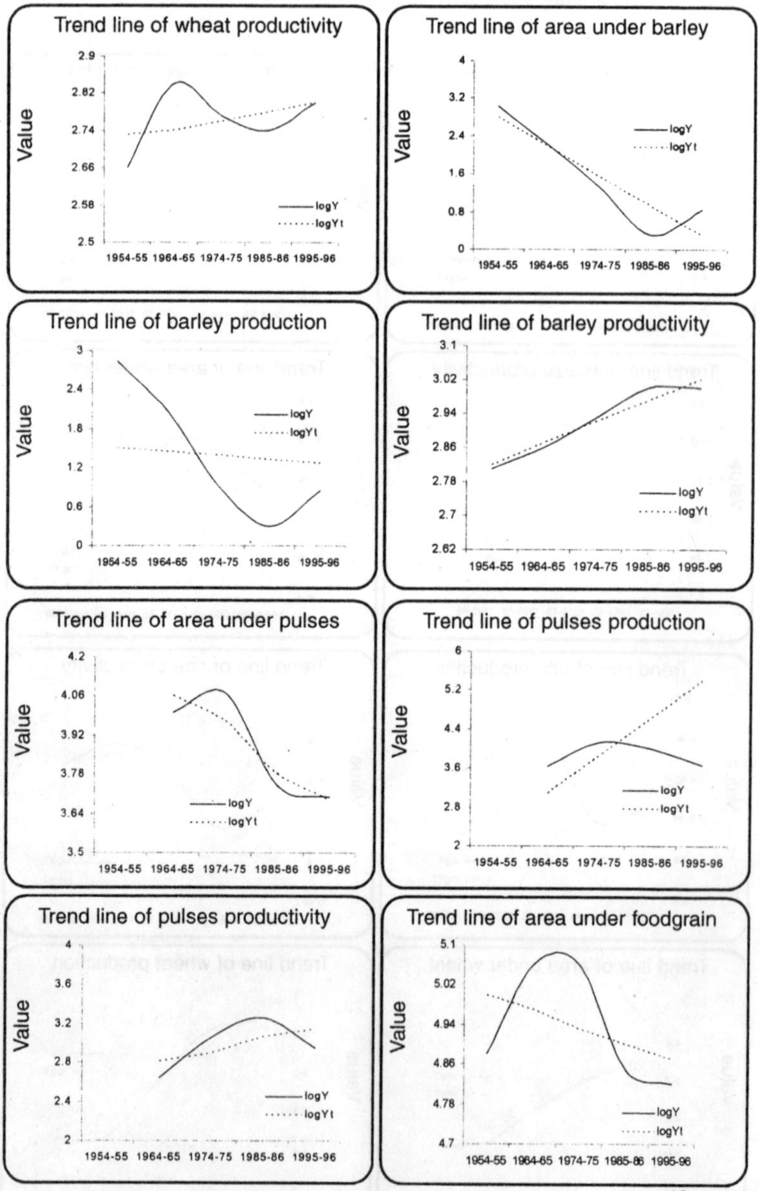

(Contd..)

Fig. 4.25 : *(Contd..)*

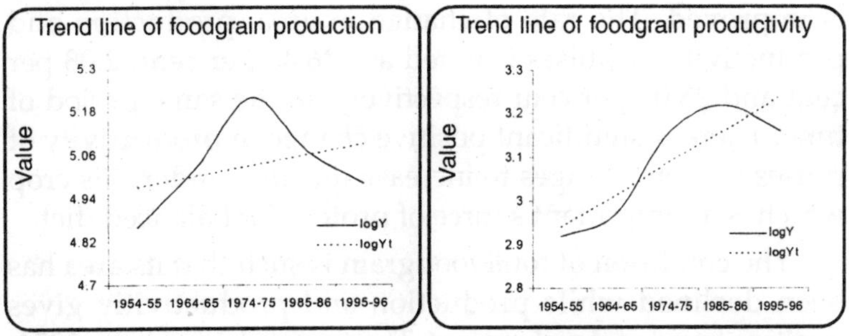

shows negative trend with the rate of change equal to –7.10 per cent. On the other hand, production and productivity gives positive trend with the rate of change equal to 10.66 per cent and 19.12 per cent respectively. Suitable landscape, plenty of water and modern agricultural measures play important role in improving the agriculture besides the reduction in area under this crop.

Wheat is the most dominant crop of the north-western Himalaya but recorded very little share in the cropping pattern of Kashmir valley. The condition become worst when this little area shows negative trend. The rate of change in area is noted as –56.85 per cent while in production it is found to be –55.02 per cent besides it, the productivity gives positive trend with the rate of change equal to 4.23 per cent. Such a decline in area is a matter of serious concern and it should be checked soon otherwise wheat will be removed from the cropping scenario of this district in near future.

Barley, which is an insignificant crop of the region shows negative trend in area and in production as well, but the productivity is improved. It is noted from Tables 4.2, 4.3 and 4.4 that the rate of change in area, production and productivity of barley is found to be –76.29 per cent, –73.27 per cent and 12.72 per cent. On the basis of the trend of productivity, it is analyzed that the area under barley will increase which will give promising return in production too.

Pulses, which are in crisis, in the region as a whole again

gives negative trend in area but production and productivity is improved. The rate of change in area, production and productivity of pulses is noted as –26.40 per cent, 2.08 per cent and 28.05 per cent respectively, in the same period of time. There is significant positive change in productivity of pulses and encourages to increase the area under this crop which is an important source of protein for balanced diet.

The condition of total foodgrain is such that its area has been declined while production and productivitiy gives positive trend of development. The rate of change from 1954-55 to 1995-96 is found to be –7.32 per cent, 6.90 per cent and 18.85 per cent for area, production and productivity respectively. The overall area under foodgrain reduced due to the population pressure and heavy urbanization encroached the fertile agricultural land for non-agricultural purposes. But the improvement in the productivity is due to the adoption of new agricultural technology. The trend of agricultural development is very clear from Fig. 4.25, in which the trend of area is seen negative for rice, wheat, barley, pulses and total foodgrain and only maize gives positive trend in it. In case of production of maize, rice, pulses and total foodgrain, there is positive trend of development while wheat and barley show negative trend. But the actual development of agriculture is shown from the productivity which gives positive trend in all the crops taking into account.

Between Page No. 234-235

1. Farm yard manure is the main source of soil fertility in this mountainous region. Which is the demand of eco-friendly approach of agricultural development.

2. Livestock are important sources for household income and fertility of soil as well in the region.

3. Urbanization enveloped the hill ranges of Shimla. In ancient time no doubt the forest covered these ranges in the same manner. Can we call them urban hills or urban havoc?

4. Unwanted Hydrophytes is a big threat for Shikara sailing in Dal lake besides they consume water and causes shrinkage of it.

5. Because of human ignorance towards ecological balance uncontrolled forest cutting in the region results in degraded forest (a case of Rajouri district)

6. A view of scattered population settlement with their agricultural fields to look upon. (Salwa and Banola village of Mendhar, Poonch)

7. Households on agriculture fields is a common practice in the region. (Shahadra Sharief, Rajouri)

8.Terraced farming is the ultimate way of agriculture practices in mountainous region. Sowing of paddy is in progress.

9. Destruction of forest to fulfill the fuel needs of the household reached to the young trees
and young ones are engaged in this un-ecological act of deforestation

10. Conventional method of fodder reserve. Fodder for off-seasen is generally kept on trees. In this way it is not damaged by soil moisture and canopy of the tree also protect it from snow.

11. Edible crops gaining importance in the region because of nutrition value in balanced diet. Mustard field at Barkot (Uttarkashi)

12. Population pressure forced encroachment of fertile and limited agricultural land for urbanization (a case of Rajouri district)

13. Urban encroachment of agricultural land results in barren lands and degraded forest.

14. Urban development is a major threat for the ecology of most of the tourist centers in hilly regions. Nainital (lake) is shrinking and landslides are common phenomena in the region.

15. Wide scope for horticultural development with a variety of fruits and vegetables in the region. A variety of large size lemon in Karn Prayag (Chamoli)

16. New Tehri is developed as a compensation for the resident of Old Tehri which will submerge when the Tehri dam start its functioning.

5

Inequalities in Agricultural Development and Agricultural Potential of the region

Concept of Agricultural Development

In order to attain rapid progress in agriculture and rural development, several strategies have been evolved and executed time to time. These developmental strategies for agriculture are designed not only to achieve a certain level of output but also to maintain nutritionally desirable composition of output. If we go through the Five Year Plan for the pattern of agricultural development during the post-independentce, it is found that the emphasis of agricultural development has been made.

In the First Five Year Plan the emphasis was on to fill the vacuum of foodgrain production created by the partition of country. Considerable stress is laid on the conservation of existing resources, production of foodgrain as well as commercial crops.

In the Second-Five Year Plan, main emphasis has not been on agriculture in India as a whole but on heavy industries, diversification of agriculture, development of dairying and horticulture was on the top priority to boost the agricultural economy.

The basic objective of the Third Five Year Plan was to provide sound foundations for sustained economic growth. The first priority necessarily belongs to agriculture and agricultural production has to be increased to the highest levels of feasibility. Development of fertilizers, improved seeds and credit along with the provision of extension services reaching down to the village level are measures undertaken directly to increase production.

Fourth Five Year Plan focussed on the availability of water in adequate quantities and at the right time is one of the basic determinant of agricultural productivity. Climatic conditions are favourable to vegetative growth throughout the year. If assured water supply and good seeds are available high yields can be achieved by the application of fertilizers, manures and improved methods of farming. Foodgrain Policy Committee (1966) postulated three objectives of food policy (a) to achieve self-reliance in production, (b) to ensure equitable distribution, and (c) to bring about price stability in the context of both production and distribution. The emphasis has to be attached with the role of agricultural technology as a major input of agriculture production. Entirely new crop rotations have been made possible by the development of short duration varieties of paddy, maize, jowar and bajra suited to different agro-climatic conditions. Among other crops included in the rotation are barley, ragi, oilseeds, potato and vegetables.

In the Fifth Five Year Plan, besides other measures of agricultural development, special livestock development programme through small or marginal farmers and agricultural labourers started. Production oriented projects such as intensive cattle development projects, intensive poultry production-cum-marketing centres, sheep and wool extension centres and fluid milk plants and milk product are the target of achievement in this plan.

In eighties during the Sixth Five Year Plan, the concept of agricultural development has changed a little with the consideration of ecological phenomenon. Here, the focus is not only on the increase of productivity but to avoid the problems of waterlogging, salinization, erosion ec. in irrigated areas and to evaluate and stabilize production in un-irrigated areas through water harvesting and conservation. Also eco-friendly crops are encouraged to suit different weather conditions i.e., eco-development.

During Seventh Five Year Plan, the focus of regional development was emphasized on sustainable agriculture and self-sufficiency in food. For food self-sufficiency, adequate and balanced attention must be paid to cereals, oilseeds, pulses, fruits, vegetables and protective food like milk, egg, meat and fish. The development of ecology and environment from three viewpoints (a) eco-restoration, (b) eco-preservation, and (c) eco-development.

In the Eighth Five Year Plan agro-climatic regional planning approach was initiated to formulate the integrated development plan for agricultural and allied sectors differentiated by homogenous agro-climatic regions. The emphasis is on development of resources and their optimum utilization in an integrated and sustainable manner. Special focus was laid on productive sectors of hill economies especially in modernizing agricultural practices and small scale industries at household, cottage and village levels.

The developmental aims of Ninth Five Year Plan (1997 to 2002), was to ensure the food and nutrition for all. Food security implied a situation where everyone has access at all times to the food needed for an active and healthy life. Thus the essential elements of food security are : (a) adequate availability of food, (b) efficient distribution through public distribution system, and (c) availability of adequate purchasing power in the hands of the people.

It is found by the main points, that through nine five year

plans since independence till now the mode of agricul-tural development has changed time to time according to prevailing agro-climatic conditions and demand of the time.

Regional pattern of agricultural productivity analyses their socio-economic condition and spatial variations of level of regional development. Agricultural productivity is viewed as measure of efficiency with which regional development can be measured. As variations in the agricultural productivity reflect disparities in the use of agricultural resources in a region and the level of development.

Regional development is interpreted as intra-regional development in the overwhelmingly agriculture situation which can only be achieved by developing agriculture and reducing gaps between regions with regard to the efficiency of agricultural system. Progress in agriculture also leads to the modernization and social development and better level of living through education and generating inclination to consume other than agricultural goods.

The measurement of agricultural productivity and levels of development involve multiple variables and fall in the domain of multivariate analysis. By its very definition there are many socio-economic and cultural variables which together define the level of development.

The term agricultural productivity denotes the existing level of performance of a piece or unit of land which may differ inevitably from one area to another causing regional variation in development. The aggregate performance of crop yield per unit of area is the primary determinant to demarcate the agriculturally weaker or efficient region. Sometimes productivity is incorrectly and interchangeably been used with production. Production refers to the volume of output while productivity signifies the output in relation to resources expanded. Production can be increased by employing more resources without increasing productivity. On the other hand productivity can be increased without increasing production by reducing the resource. Therefore, agricultural productivity can be defined as a measure of efficiency with which an agricultural production system employs land, labour, capital and other resources.

As an indicator of regional development, focussed attention of scholars from different disciplines attempt to measure and quantify agricultural productivity in India and abroad as well with different angles.

Since agriculture is the only means of livelihood and mainstay of the economy in this region, imbalances in agricultural productivity may be taken as the major indicator of the regional development.

Measures of Agricultural Development

The meaning of development is different to different thoughts. However, it is accepted that the objective of development is to meet one's basic needs like food, shelter, clothing, education, health, sanitation, clean drinking water, transport and so on. For regional development of an area, the people should have basic necessities of life, a good environment which can lead to satisfactory life and is also contributing to the onward march. Thus it is a multifarious and multidimensional development of the economy by exploiting the optimum level of resources in man, material, land and water on the basis of head initiative and derive. It is the interaction of all economic and social activities (agriculture, industry, infrastructure, transport, labour and education) with physical environment as it implies planning in all its aspects needed for modernization and betterment of living.

In this section main emphasis will be given on how far the physical limitations have overcome by the introduction of new agricultural technology in the region. Because it is observed that some areas of the region which are poorer than other in physical terms but their yield level is much higher due to new agricultural technology consisting of high yielding varieties irrigation, fertilizer consumption and farm machinery which augmenting the regional inequalities in agricultural development. The existing level of inequalities may be due to several factors as regional differences in the size of land holdings, tenurial systems, technological factors, management factors, sociological factors, availability of required infrastructures like electricity and transport, and differences in natural resources like water. But in this study inequalities in agricultural development discussed on the basis of farm machinery (wooden plough, iron plough, pumping sets, sugarcane crushers, tractors, wheat threshers, rice threshers, and shellers), fertilizer consumption, irrigation and intensity of landuse. As the landscape for agricultural technology is very tough and not permit a number of modern machinery on its terraced farm and sloppy nature. Because of these barrier the value of agricultural machinery is very low i.e. per thousand or per ten thousand hectare of agricultural inputs comes in fractions with a few exceptions. So it is decided that to find out the trend in the level of development, a number of such parameters have been taken into account, which in general taken as per 1000 or per 10,000.

Spatial Variation of Irrigation Status

Irrigation is one of the most important factors for the agricultural development of any region. Assured irrigation is a key factor for the development of agricultural productivity because it is a base for other inputs. Water is such a crucial factor which alone is responsible to enhance the agricultural productivity as well as to create inequalities in agricultural development. During 1961-63, the total irrigated land in the region was just 16 per cent which increased to 18 per cent in 1971-74. A considerable increased found during 1982-85 when it became 29 per cent and then

30 per cent in 1995-96. So it is revealed that overall situation of irrigation is improved with a growth of 87.5 per cent from 1961-63 to 1992-95.

A comparative analysis of irrigation development during 1982-85 and 1992-95 can be done from Table 5.1. It is revealed from the given table that overall irrigation in the region became 29.84 per cent from 29.05 per cent or in terms of growth it increased by 2.72 per cent. The status of irrigation

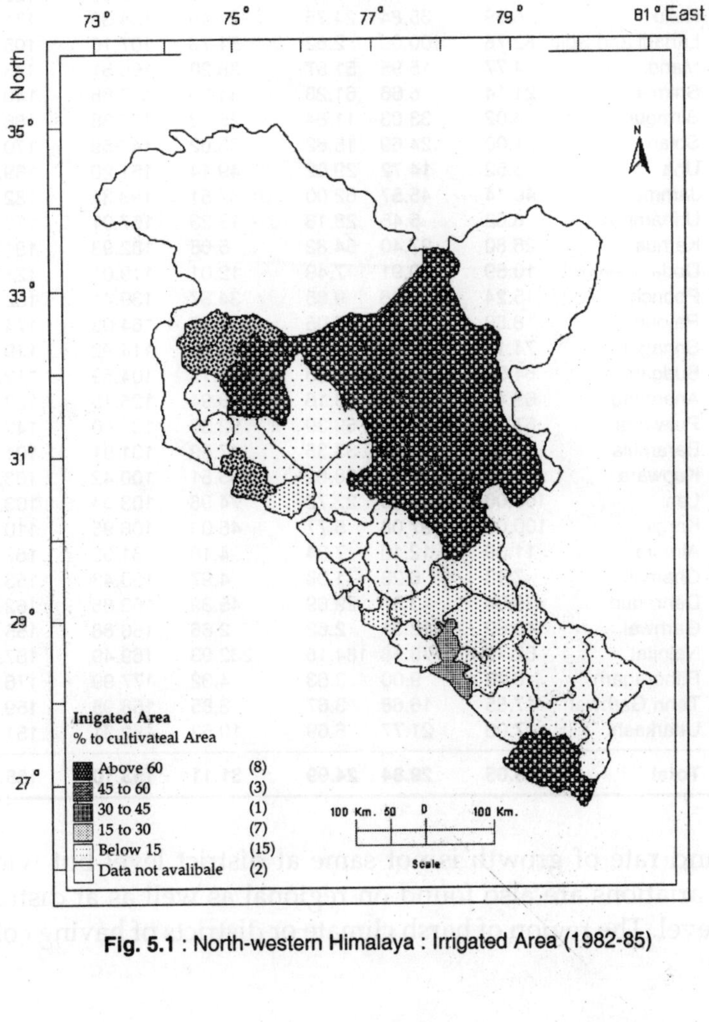

Fig. 5.1 : North-western Himalaya : Irrigated Area (1982-85)

Table 5.1 : Spatial variations in Levels of irrigation (%), Fertilizer consumption (kg./hectare) and Intensity of Agricultural Landuse (%)

Districts	*Irrigation*		*Fertilizer Consumption*		*Intensity of Landuse*	
	1982-85	*1992-95*	*1985*	*1995*	*1982-85*	*1992-95*
Bilaspur	9.23	9.44	7.56	33.09	189.08	190.01
Chamba	10.38	10.45	4.74	13.72	153.25	142.04
Hamirpur	4.18	4.44	19.21	33.77	191.92	196.33
Kangra	2.53	29.68	26.81	26.81	178.24	181.65
Kinnaur	21.54	55.69	0.27	7.46	132.35	122.41
Kullu	5.69	35.84	24.75	27.46	154.38	171.48
Lahaul and Spiti	82.78	100.00	2.62	51.78	107.16	105.53
Mandi	4.77	15.95	51.57	28.20	169.51	173.30
Shimla	21.14	6.66	61.28	34.73	147.65	140.76
Sirmour	4.02	33.03	11.84	25.03	177.38	186.01
Solan	21.00	24.69	15.62	35.62	147.59	170.42
Una	5.52	14.72	29.02	49.14	151.20	169.94
Jammu	46.14	45.57	62.00	37.51	193.31	182.69
Udhampur	6.62	5.45	28.13	13.23	157.01	168.01
Kathua	26.80	27.40	64.82	5.66	182.93	191.41
Doda	10.59	10.91	7.49	12.01	119.01	121.37
Poonch	15.24	14.06	9.65	34.37	139.41	132.49
Rajouri	8.58	8.93	17.05	19.47	164.03	174.42
Srinagar	74.14	76.12	116.54	137.65	114.92	119.36
Budgam	64.30	65.20	56.35	71.17	104.69	112.41
Anantnag	69.48	72.79	63.16	83.94	125.19	128.13
Pulwama	67.77	67.79	69.39	82.36	132.00	149.32
Baramula	49.82	50.54	33.35	47.59	101.01	101.15
Kupwara	50.56	48.50	28.62	65.51	100.42	103.48
Leh	100.00	99.80	22.72	74.98	103.04	103.97
Kargil	100.00	97.08	5.77	45.00	108.95	110.35
Almora	11.39	12.10	7.34	4.18	161.95	167.68
Chamoli	7.13	6.28	1.08	4.97	153.43	153.72
Dehradun	38.04	43.07	29.69	45.39	153.85	162.47
Garhwal	10.69	12.31	2.62	2.66	150.88	155.18
Nainital	62.70	79.18	184.16	232.93	169.49	167.93
Pithoragarh	8.49	9.00	3.63	4.32	177.99	176.60
Tehri Garhwal	17.85	16.68	3.67	3.85	158.95	159.40
Uttarkashi	17.06	21.77	5.69	10.53	152.31	151.15
Total	**29.05**	**29.84**	**24.99**	**31.11**	**153.10**	**155.45**

and rate of growth is not same at district level but wide variations are also found on regional as well as at district level. The region of harsh climate or districts of having cold desert like climate noted with highest level of irrigation like Leh and Kargil have almost 100 per cent area under irrigation whereas in Lahaul and Spiti 82.78 per cent area was under assured irrigation during 1982-85 which become 100 per cent during 1992-95. Next to it, all the districts of Kashmir valley have good status of irrigation i.e., more than 50 per cent with little improvement in all the districts except in Kupwara where irrigation declined by –4.07 per cent. Highest growth rate during this period recorded in Himachal region where a number of districts show more

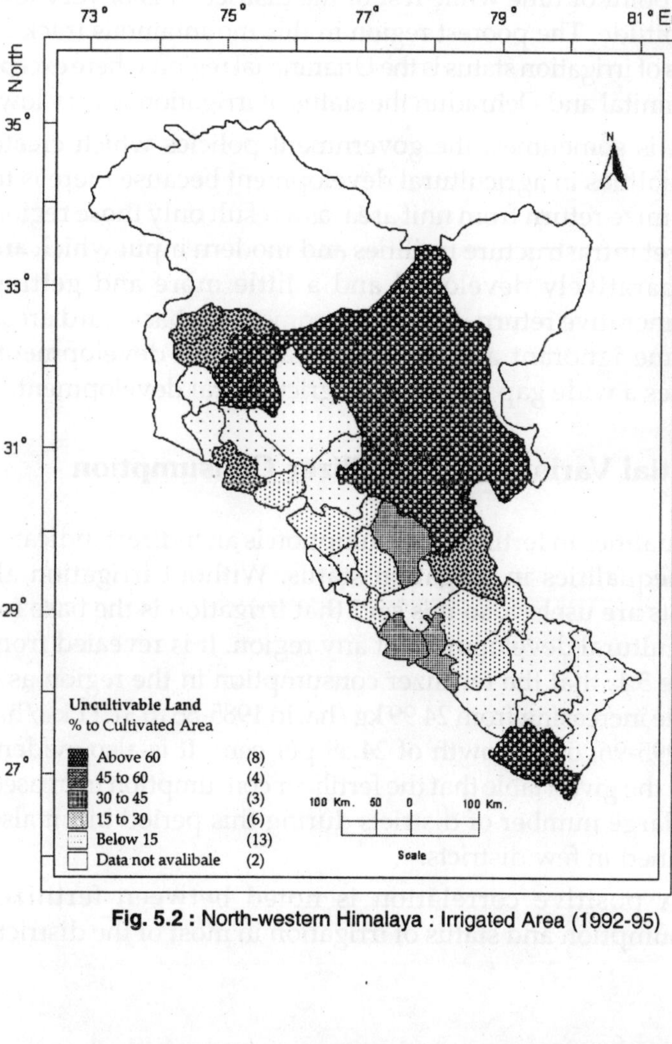

Fig. 5.2 : North-western Himalaya : Irrigated Area (1992-95)

than 100 per cent growth in irrigation. Though the status of irrigation development in this region is not very good but a considerable improvement from very low status can easily be noticed while in case of Kashmir valley improvement in irrigation is very low but its status is already very high. In Himachal region only Shimla shows a decline of –68.49 per cent in irrigation i.e. from 21.14 per cent to 6.66 per cent. Irrigation status of Jammu region comes next, the highest level of irrigation is noted in Jammu where it is about 45 per cent at both point of time while rest of the districts it is of very low magnitude. The poorest region in this mountainous track in terms of irrigation status is the Uttaranchal region where except in Nainital and Dehradun the status of irrigation is very low.

It is sometimes, the government policies which create inequalities in agricultural development because there is to maximize return from unit area, as a result only those region can get infrastructure facilities and modern input which are comparatively developed and a little more and getting remunerative return. As a result sometimes backward areas become ignorant and developed areas get development, creates a wide gap in levels of agricultrural development.

Spatial Variation of Fertilizer Consumption

Inequalities in fertilizer consumption is an indirect indicator of inequalities in irrigation status. Without irrigation all inputs are useless, so it is said that irrigation is the base for agricultural development of any region. It is revealed from Table 5.1, that the fertilizer consumption in the region as a whole increasing from 24.99 kg/ha. in 1985-86 to 31.11 kg/ha. in 1995-96, or a growth of 24.49 per cent. It is also evident from the given table that the fertilizer consumption increased in a large number of districts during this period but it also declined in few districts.

A positive correlation is noted between fertilizer consumption and status of irrigation in most of the districts

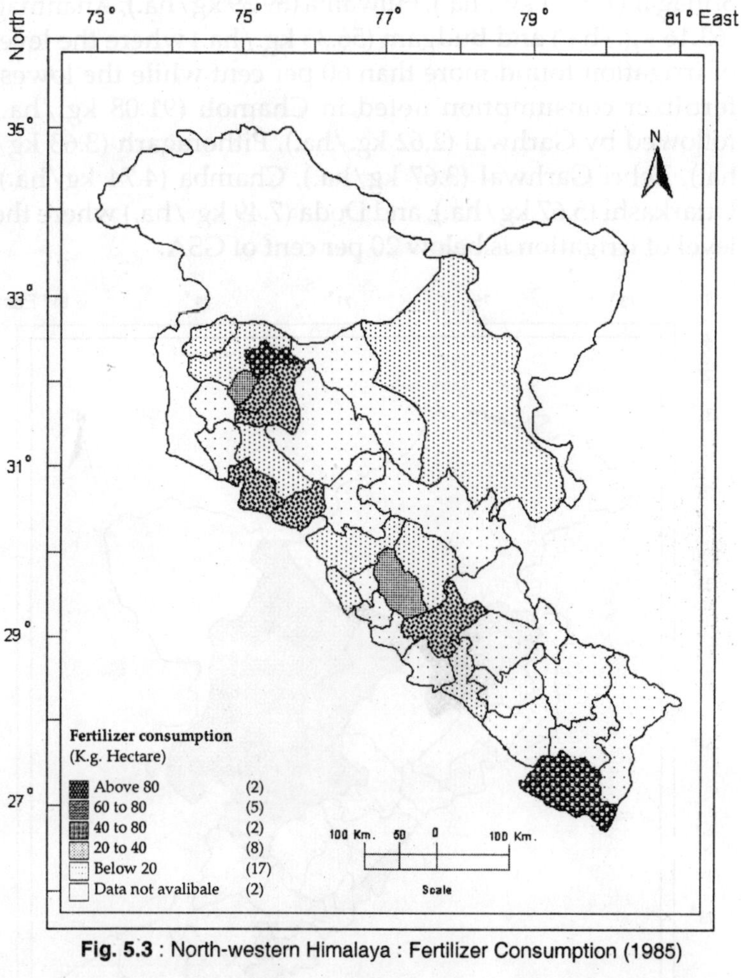

Fig. 5.3 : North-western Himalaya : Fertilizer Consumption (1985)

with few exceptions like in Mandi, Kathua, Poonch, Kupwara, Almora and Chamoli, the per hectare fertilizer consumption decreased while percentage of irrigated area increased, leaving those districts of cold desert where irrigation reached to its saturation level.

During 1985-86, highest fertilizer consumption per hectare is recorded in Nainital (184.16 kg/ha.) followed by

Srinagar (116.54 kg./ha.), Pulwama (69.39 kg/ha.), Anantnag (63.16 kg./ha.) and Budgam (56.35 kg./ha.) where the level of irrigation found more than 60 per cent while the lowest fertilizer consumption noted in Chamoli (91.08 kg./ha.) followed by Garhwal (2.62 kg./ha.), Pithoragarh (3.63 kg/ha.), Tehri Garhwal (3.67 kg/ha.), Chamba (4.74 kg/ha.), Uttarkashi (5.67 kg/ha.), and Doda (7.49 kg./ha.) where the level of irrigation is below 20 per cent of GSA.

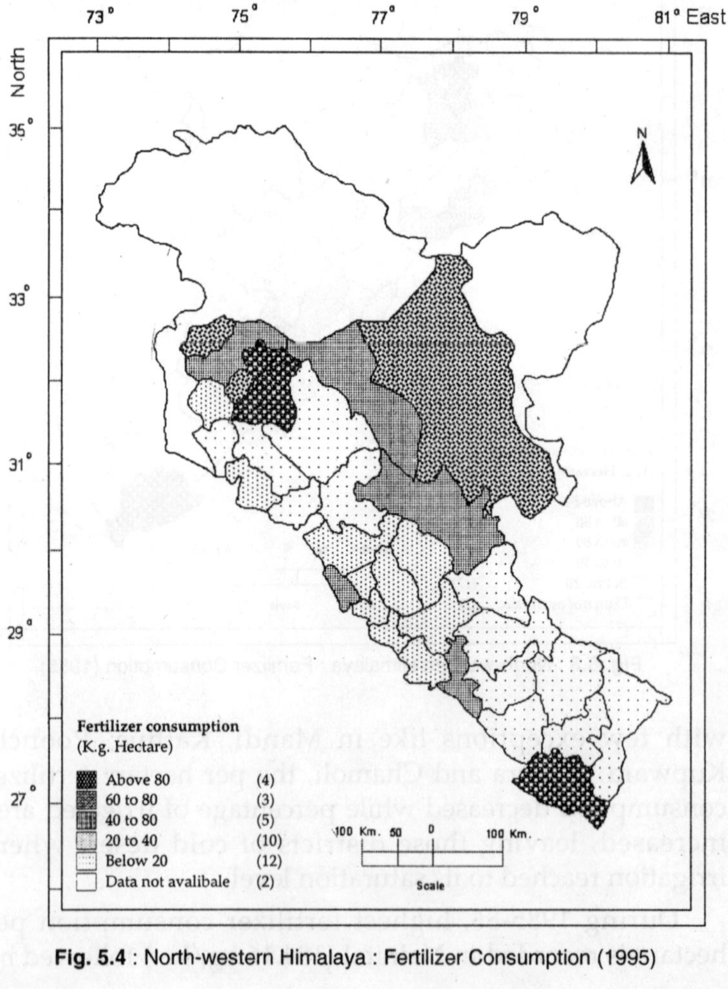

Fig. 5.4 : North-western Himalaya : Fertilizer Consumption (1995)

In 1995-96, the condition of fertilizer consumption is more or less same with slight improvement, the highest fertilizer consumption again found in Nainital (232.93 kg/ha.) followed by Srinagar (137.65 kg./ha.), Anantnag (83.94 kg/ha.) come ahead of Pulwama (82.36 kg/ha.) and Budgam (71.17 kg./ha.) while the lowest fertilizer consumption is found in Garhwal (2.66% kg./ha.), Tehri Garhwal (3.85 kg./ha.), Almora (4.18 kg./ha.), Chamoli (4.97 kg./ha.), Pithoragarh (4.32 kg./ha.). The pattern of irrigation level also found with positive correlation.

It is also revealed that fertilizer consumption per hectare is highest in Kashmir valley region while lowest in Uttaranchal. Himachal region is found more developed than Jammu region in fertilizer intake which reflect inequalities in level of agricultural development in the region.

Spatial Variation of Intensity of Agricultural Landuse

It indicates the efficiency of agricultural land in such a way that "a unit of land used for agriculture how much more than unity?" It is an indicator of agricultural development which is the outcome of all other parameters used or adopted in combination for the agricultural development.

It is evident from Table 5.1, that the landuse intensity of north-western Himalaya has been increased from 153.10 per cent in 1982-95 to 155.45 per cent in 1992-95. Inequalities in agricultural development can also be assessed by the analysis of spatial variations of landuse intensity in the region. It also reflects the intensification of the adoption of agricultural technology or input in an areal unit. It is found that the intensity of landuse in the region ranges up to 196.33 per cent. It is surprising to note that the district of Kashmir valley where landscape and climate is suitable besides the adoption of agricultural technology is also high, the intensity of landuse found lower than the rest of the region. If we categorize the intensity of land into five categories i.e. less than 120 per cent, 120 per cent - 140 per cent, 140 per cent - 160 per cent, 160 per cent - 180 per cent and more than 180 per cent it is found that during 1982-85, eight districts found below 120 per cent of landuse intensity namely Leh, Kargil, Lahaul and Spiti, Kupwara, Baramula, Badgam, Srinagar and Doda. While in 1992-95, except Doda all the above districts remain in this category. It is interesting to note that in term of landuse intensity districts of Kashmir valley

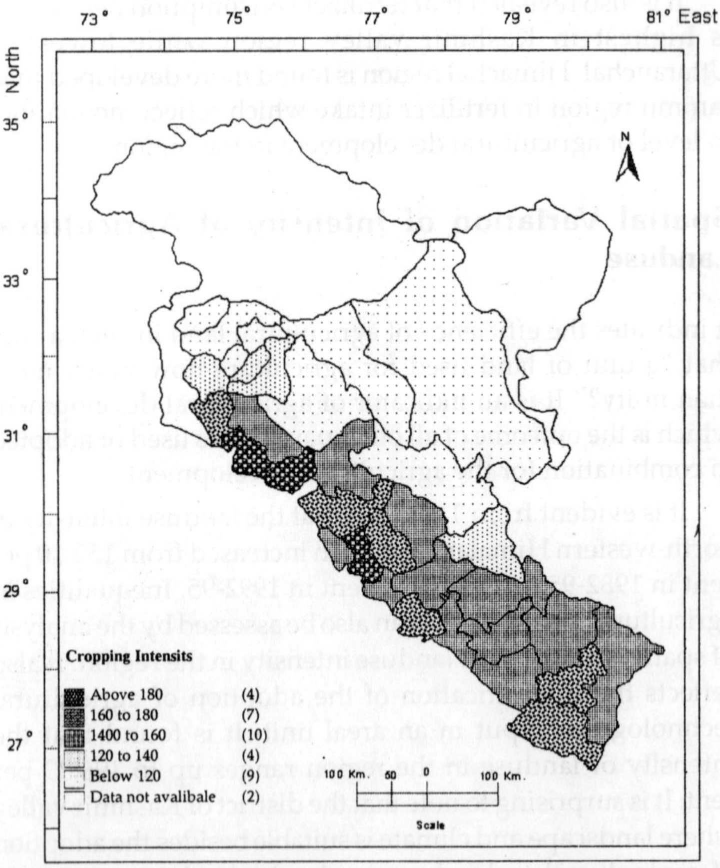

Fig. 5.5 : North-western Himalaya : Intensity of Agricultural Landuse (1982-85)

come along the districts of cold deserts. In the next category (i.e. 120-140%), only four districts were found in 1982-85 namely Pulwama, Anantnag, Poonch, and Kinnaur and in 1992-95, the number of districts were same but Doda come in this category in place of Pulwama which jump into the next higher category. In the medium category of landuse intensity i.e. 140 per cent to 160 per cent eleven districts were found in 1982-85 namely Uttarkashi, Tehri Garhwal, Garhwal, Dehradun, Chamoli, Udhampur, Solan, Una, Shimla and Chamba which mostly belong to Uttaranchal and Himachal region. But in 1992-95 only

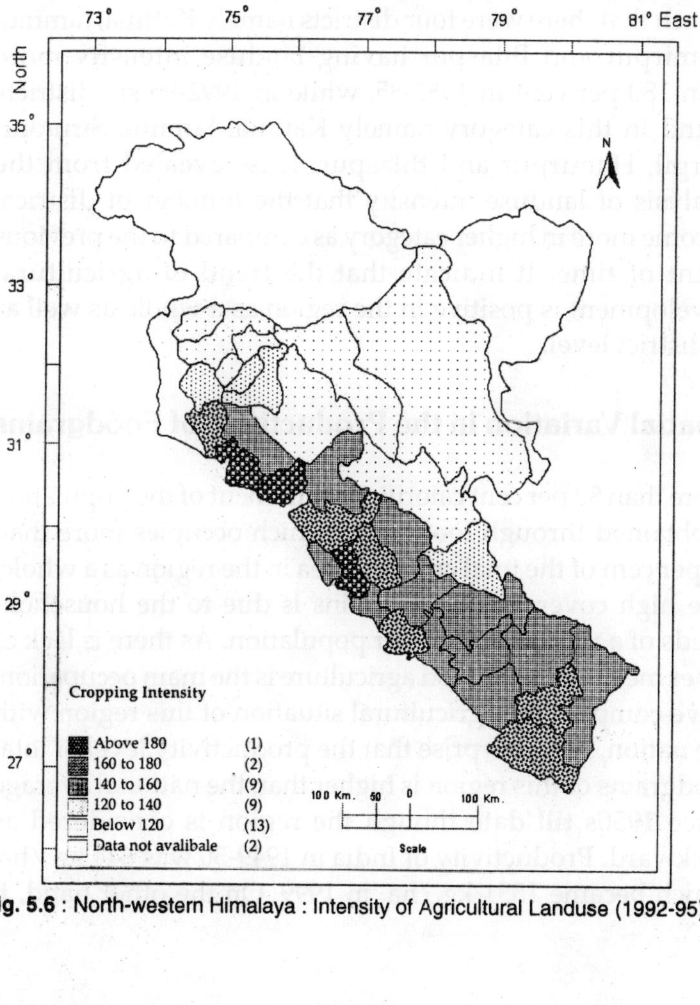

Fig. 5.6 : North-western Himalaya : Intensity of Agricultural Landuse (1992-95)

seven districts were remained in it namely Uttarkashi, Tehri Garhwal, Garhwal, Chamoli, Pulwama, Shimla and Chamba. In the next category seven districts were found during 1982-85, namely, Pithoragarh, Nainital, Almora, Rajouri, Sirmour, Mandi and Kangra while the number of district increased in this category to ten namely Pithoragarh, Nainital, Dehradun, Almora, Rajouri, Udhampur, Una, Solan, Mandi and Kullu. The condition is seen improving as Dehradun, Udhampur, Una, Solan and Kullu districts jump to higher level of landuse intensity. The condition of highest level of landuse intensity is such that there were four districts namely Kathua, Jammu, Hamirpur and Bilaspur having landuse intensity more than 180 per cent in 1982-85, while in 1992-95 six districts found in this category namely Kathua, Jammu, Sirmour, Kargil, Hamirpur and Bilaspur. It is revealed from the analysis of landuse intensity that the number of districts become more in higher category as compared to the previous point of time. It indicate that the trend of agricultural development is positive in the region as a whole as well as at district level.

Spatial Variation in the Productivity of Foodgrains

More than 80 per cent calorific requirement of the population is obtained through foodgrains which occupies more than 90 per cent of the total cropped area in the region as a whole. The high coverage of foodgrains is due to the household needs of a vast section of our population. As there is lack of other means of livelihood agriculture is the main occupation. If we compare the agricultural situation of this region with the nation, it is a surprise that the productivity level of total foodgrains of this region is higher than the national average since 1950s till date though the region is considered as backward. Productivity of India in 1949-50 was 644 kg./ha. which became 1551 kg./ha. in 1999. On the other hand, it was 747 kg./ha. in 1954-55 and increased to 808, 1262, 1422 and 1614 kg./ha. in 1964-65, 1974-75, 1985-86 and 1995-96 respectively. The high productivity level of the region shows the agricultural potentiality of the region. And if special weightage is given, the changes for prosperity is bright enough to feed others too. It is evident from Table 5.2, that during 1954-1955 except Srinagar (1107 kg./hect.) and Anantnag (1100 kg./hect.) no district reached the productivity level even above 900 kg./ha. The situation in

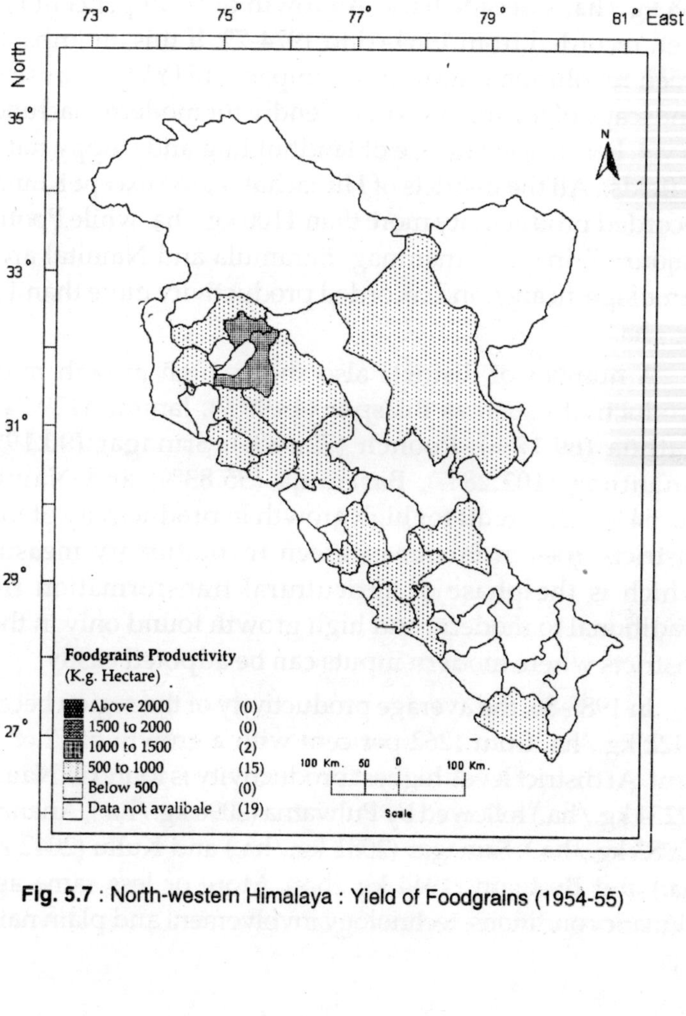

Fig. 5.7 : North-western Himalaya : Yield of Foodgrains (1954-55)

1964-65 improved and eight districts recorded more than 900 kg./ha. productivity namely, Bilaspur, Chamba, Kinnaur, Mandi, Shimla, Sirmour, Srinagar and Baramula. A large number of districts are also found where data was not available at said point of time, because, some of the districts formed later and in this study administrative setup is taken according to the population census of 1991.

During 1974-75, the improvement in productivity is very clear, the prouctivity of the region became 1262 kg./ha. from 808 kg./ha. a decade back. A growth of 56.19 per cent have been recorded from 1964-65 to 1974-75. If it is an impact of green revolution it must be an impact of HYVS because the landscape of the region is not friendly for modern machinery as yet, because small size of landholding and sloppy nature of fields.. All the districts of Himachal region except Kinnaur recorded productivity more than 1100 kg./ha. while Poonch, Rajouri, Srinagar, Anantnag, Baramula and Nainital are of same significance and recorded productivity more than 1100 kg./ha.

A number of districts also shows high growth rate in productivity such as Bilaspur (43.99%), Jammu (129.74%), Kathua (69.17%), Poonch (72.98%), Srinagar (90.19%), Anantnag (102.28%), Baramula (55.83%) and Nainital (57.54%). The credit for high growth in productivity of these districts goes in favour of green revolutionary measures which is the phase of agricultural transformation from traditional to modern, and high growth found only in those districts where modern inputs can be adpoted easily.

In 1985-86, the average productivity of the region became 1422 kg./ha. from 1262 per cent with a growth of 12.68 per cent. At district level, highest productivity is found in Nainital (2234 kg./ha.) followed by Pulwama (2096 kg./ha.), Anantnag (2083 kg./ha.), Srinagar (2081 kg./ha.) and Kullu (2042 kg./ha.) and Budgam (1944 kg./ha.). More or less same agro-climatic conditions, technology involvement and plain nature

Table 5.2 : Changing pattern of productivity of pulses and foodgrains

Districts	*Changing Pattern of Total Pulses Productivity*					*Changing Pattern of Total Foodgrain Productivity*				
	1954-55	*1964-65*	*1974-75*	*1985-86*	*1995-96*	*1954-55*	*1964-65*	*1974-75*	*1985-86*	*1995-96*
Bilaspur	807	932	447	290	333	591	1032	1486	1267	1320
Chamba	N.A.	N.A.	272	500	371	581	971	1263	1538	1909
Hamirpur	N.A.	N.A.	327	254	333	N.A.	N.A.	1365	1243	1583
Kangra	N.A.	N.A.	326	643	325	N.A.	N.A.	1254	1493	1515
Kinnaur	N.A.	N.A.	160	N.A.	1111	911	1022	708	751	844
Kullu	N.A.	N.A.	315	458	583	N.A.	N.A.	1144	2042	1894
Lahaul and Spiti	N.A.	N.A.	2450	N.A.	1454	N.A.	N.A.	2147	N.A.	1454
Mandi	265	1013	925	535	355	618	965	1244	1518	1893
Shimla	297	1086	223	130	387	685	958	1258	1409	1371
Sirmour	359	1202	425	742	318	634	987	1329	1409	1867
Solan	N.A.	N.A.	456	390	292	N.A.	N.A.	1128	1542	1664
Una	N.A.	N.A.	1006	975	266	N.A.	N.A.	1368	1172	1732
Jammu	646	364	364	363	99	691	427	981	1239	1600
Udhampur	587	340	359	367	396	643	705	918	865	2043
Kathua	645	350	283	216	226	758	613	1037	1130	1638
Doda	N.A.	411	324	236	566	706	823	1073	960	1534
Poonch	500	366	291	239	305	692	692	1197	1286	2567
Rajouri	N.A.	N.A.	432	386	384	N.A.	N.A.	1178	985	2105
Srinagar	N.A.	1442	1444	1403	620	1107	958	1822	2081	2025
Budgam	N.A.	N.A.	N.A.	1257	1055	N.A.	N.A.	N.A.	1944	2172
Anantnag	N.A.	601	654	704	867	1100	877	1774	2083	1314
Pulwama	N.A.	N.A.	N.A.	884	439	N.A.	N.A.	N.A.	2098	1794
Baramula	N.A.	434	1129	1819	900	825	926	1443	1673	1451
Kupwara	N.A.	N.A.	N.A.	854	500	N.A.	N.A.	N.A.	1549	1038

Table 5.2 : *(Contd..)*

Leh	N.A.	174	674	493	445	722	725	881	721	656
Kargil	N.A.	N.A.	N.A.	500	768	N.A.	N.A.	N.A.	656	608
Almora	N.A.	N.A.	N.A.	550	610	N.A.	N.A.	N.A.	1035	1111
Chamoli	N.A.	N.A.	N.A.	569	521	N.A.	N.A.	N.A.	1103	1396
Dehradun	600	330	443	726	618	645	606	678	1505	1537
Garhwal	N.A.	N.A.	N.A.	552	513	N.A.	N.A.	N.A.	1039	1192
Nainital	691	383	462	772	969	718	869	1369	2234	2641
Pithoragarh	N.A.	N.A.	N.A.	601	631	N.A.	N.A.	N.A.	1203	1438
Tehri Garhwal	N.A.	N.A.	N.A.	637	685	N.A.	N.A.	N.A.	1229	1337
Uttarkashi	N.A.	N.A.	N.A.	592	579	N.A.	N.A.	N.A.	1318	1649
Total	666	465	561	635	511	747	808	1262	1422	1674

N.A. : Not Available

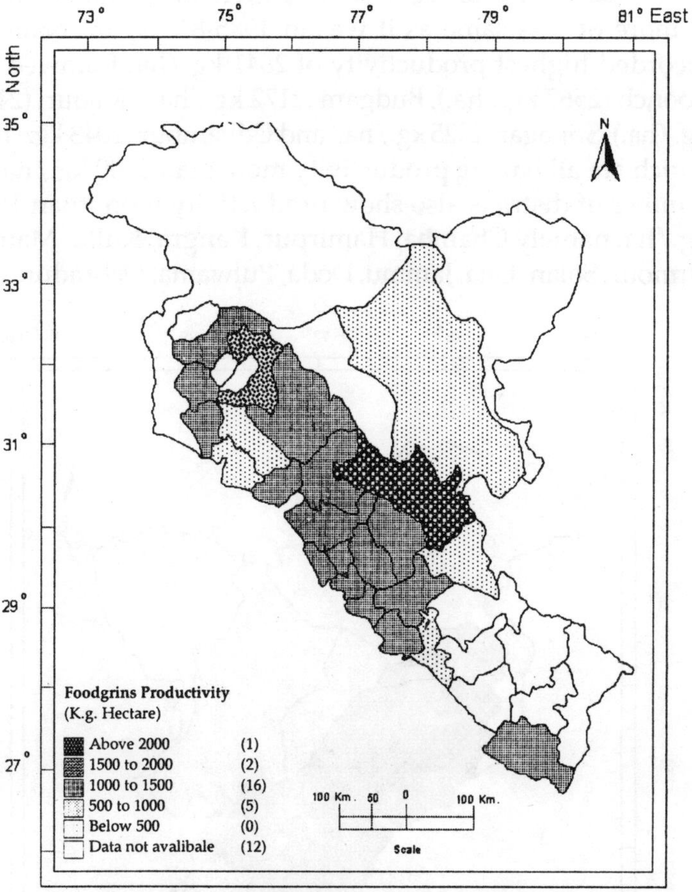

Fig. 5.8 : North-western Himalaya : Yield of Foodgrains (1974-75)

of landscape are responsible for high productivity in these districts. On the other hand, low productivity of Kinnaur (751 kg./ha.), Udhampur (865 kg./ha.), Doda (960 kg./ha.), Leh (721 kg./ha.) and Kargil (656 kg./ha.) is the response of harsh climate of the region on agricultural development.

Finally, the agricultural situation of foodgrain in 1995-96 recorded a growth of 17.72 per cent in productivity. The productivity of foodgrain in the region became 1674 kg./ha.

from 1422 kg./ha. of 1985-86. The pattern of spatial variation is more or less same as it was in 1985-86. Again Nainital recorded highest productivity of 2641 kg./ha. followed by Poonch (2567 kg./ha.), Budgam (2172 kg./ha.), Rajouri (2105 kg./ha.), Srinagar (2025 kg./ha.) and Udhampur (2043 kg./ha.) which are all having productivity more than 2000 kg./ha. A number of districts also show productivity more than 1500 kg./ha. namely Chamba, Hamirpur, Kangra, Kullu, Mandi, Sirmour, Solan, Una, Jammu, Doda, Pulwama, Dehradun and

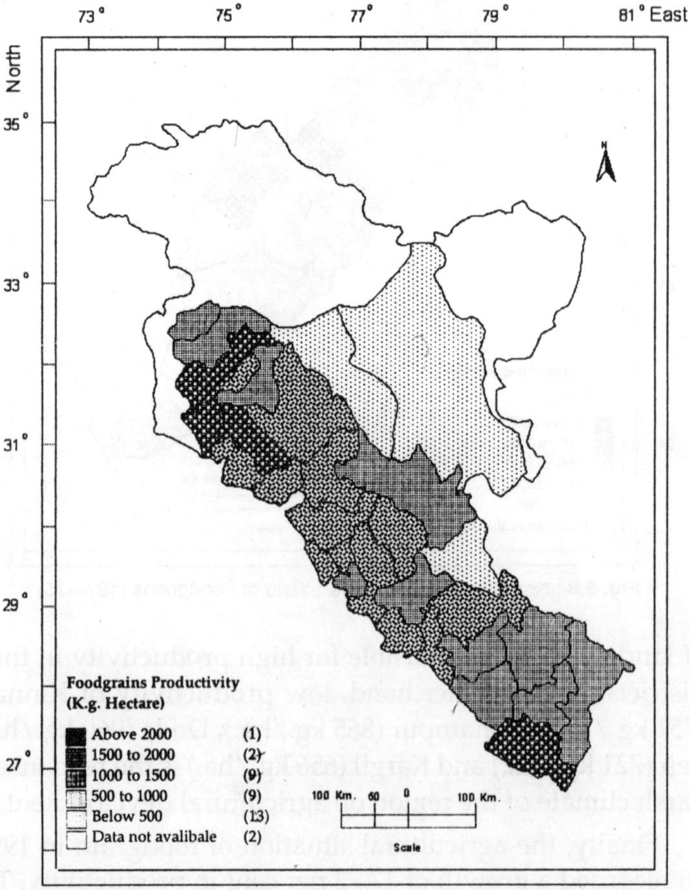

Fig. 5.9 : North-western Himalaya : Yield of Foodgrains (1995-96)

Uttarkashi. Low productivity of Kinnaur (844 kg./ha.), Leh (656 kg./ha.) and Kargil (608 kg./ha.) is an adverse effect of harsh climate on the agriculture.

At district level, there are wide variations in the level of productivity. Highest growth during 1985-86 to 1995-96 is recorded in Udhampur, (136.18%) where productivity became 2043 kg./ha. from a low of 865 kg/ha., Rajouri (113.71 kg/ha.), Poonch (99.61%) and Doda (59.79%) are other districts of significant growth. Agro-climatic similarities are found in these high growth districts of Rajouri-Poonch and Udhampur-Doda.

On the other hand, Kupwara (-32.99%), Anantnag (-36.92%), Pulwama (-14.49%) and Baramula (-13.27%) districts experienced negative growth. The productivity decreased in these districts while most of the conditions are favourable for agricultural practices except one which must be the cause of poor agricultural development in the valley districts i.e. militancy. It engulfed most of the areas in Kashmir valley in last decade resulted in negative or poor agro-economic development.

Spatial Variation of Productivity of Pulses

Pulses are not only the cheap source of protein for the majority of population but also maintains the nitrogen level in soil fertility. Trend of pulses production against the other cereal crops is noted discouraging. It is evident from the scenario of India that the area and production of pulses remained almost stagnant but the population increases with unprecedented growth rate which causes severe decline in availability from 70.95 gram/day in 1950 to 36.98 gm/day in 1995-96. The share of pulses in total foodgrain production which was 16.55 per cent in 1950-51 came down to 7.84 per cent during 1994-95.

The situation of pulses production, spread and

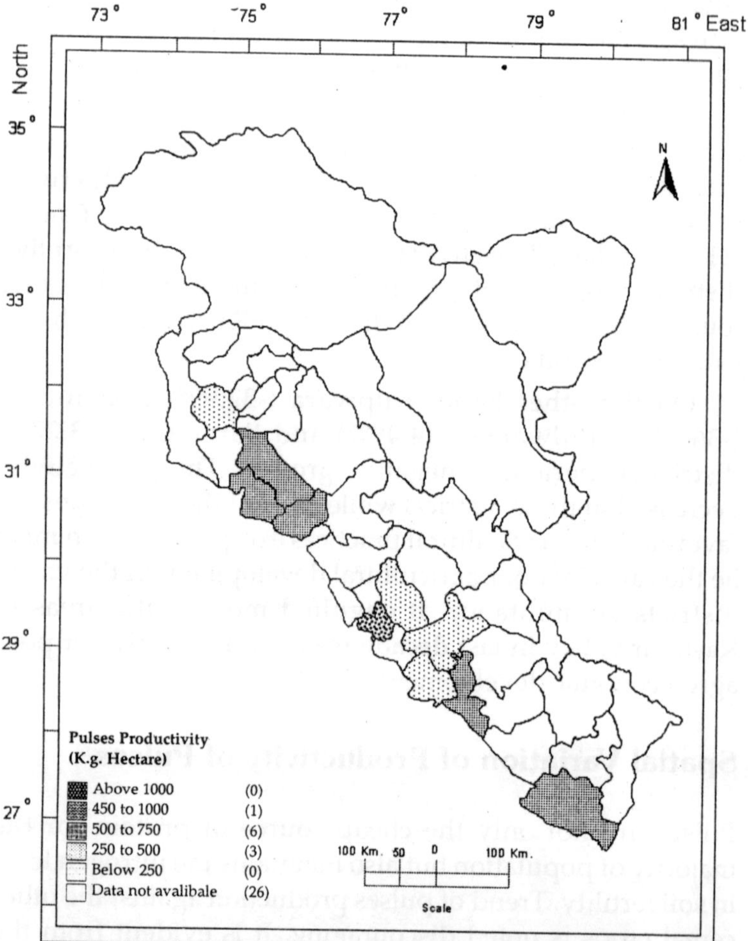

Fig. 5.10 : North-western Himalaya : Yield of Pulses (1954-55)

productivity is also on declining trend in the study region. It is evident from the analysis of wheat, rice, maize, barley and pulses that, out of these crops acreage of pulses was 4.25 per cent in 1954-55 which increased to 7.10 per cent and 7.80 per cent in 1964-65 and 1974-75 but afterward severe decline and areal percentage became only 3.30 per cent and 4.07 per cent in 1985-86 and 1995-96 respectively.

Table 5.3 : Changing pattern of percentage share of area and Production of Pulses

Districts	*1954-55*		*1964-65*		*1974-75*		*1985-86*		*1995-96*	
	area	*production*	*area*	*production*	*area*	*production*	*area*	*production*	*area*	*production*
Bilaspur	11.34	15.48	7.50	7.22	14.04	4.23	1.55	0.36	3.06	0.77
Chamba	N.A.	N.A.	N.A.	N.A.	6.07	1.23	N.A.	N.A.	6.12	1.15
Hamirpur	N.A.	N.A.	N.A.	N.A.	7.52	1.67	0.25	0.05	0.84	0.18
Kangra	N.A.	N.A.	N.A.	N.A.	5.35	1.39	0.46	0.20	2.25	0.48
Kinnaur	N.A.	N.A.	N.A.	N.A.	5.11	0.64	N.A.	N.A.	25.00	23.81
Kullu	N.A.	N.A.	N.A.	N.A.	7.14	1.80	0.06	0.01	7.83	2.31
Lahaul and Spiti	N.A.	N.A.	N.A.	N.A.	10.47	10.89	N.A.	N.A.	55.00	55.17
Mandi	0.55	0.23	0.50	0.53	4.49	3.25	0.12	0.04	3.04	0.56
Shimla	0.47	0.19	0.65	0.74	6.29	1.02	0.04	0.00	10.36	2.82
Sirmour	4.23	2.37	3.05	3.73	8.17	2.56	0.88	0.46	6.53	1.10
Solan	N.A.	N.A.	N.A.	N.A.	17.48	7.00	5.81	2.02	7.00	1.23
Una	N.A.	N.A.	N.A.	N.A.	15.57	11.45	1.85	1.54	2.29	0.35
Jammu	3.63	3.35	10.21	9.51	9.34	3.01	7.44	2.85	4.49	0.23
Udhampur	0.38	0.35	7.66	3.67	6.49	1.89	4.82	1.35	2.62	0.50
Kathua	2.86	2.38	8.64	5.02	8.48	2.35	6.65	1.32	5.53	0.69
Doda	0.00	0.00	6.46	3.22	6.60	1.42	5.26	0.78	5.16	1.80
Poonch	0.02	0.02	1.97	1.04	0.26	0.04	0.51	0.05	0.19	0.02
Rajouri	N.A.	N.A.	N.A.	N.A.	2.42	0.85	1.49	0.34	1.19	0.21
Srinagar	N.A.	N.A.	3.17	4.87	7.21	7.62	3.45	2.16	3.62	1.11
Budgam	N.A.	N.A.	N.A.	N.A.	N.A.	N.A.	4.86	3.38	4.03	3.02
Anantnag	N.A.	N.A.	5.18	3.53	3.85	1.35	4.77	1.41	2.44	1.60
Pulwama	N.A.	N.A.	N.A.	N.A.	N.A.	N.A.	5.28	1.97	2.85	0.70
Baramula	N.A.	N.A.	8.95	4.04	10.55	8.47	7.90	8.24	7.62	4.71
Kupwara	N.A.	N.A.	N.A.	N.A.	N.A.	N.A.	2.78	1.22	1.83	0.88

(Contd.)

Table 5.3 : *(Contd.)*

Leh	N.A.	N.A.	16.78	4.51	53.85	51.61	44.13	34.41	45.62	29.39
Kargil	N.A.	N.A.	N.A.	N.A.	N.A.	N.A.	99.84	99.37	100.00	100.00
Almora	N.A.	N.A.	N.A.	N.A.	N.A.	N.A.	3.28	1.77	3.12	1.64
Chamoli	N.A.	N.A.	N.A.	N.A.	N.A.	N.A.	1.05	0.54	1.50	0.57
Dehradun	8.98	8.28	8.52	4.73	7.80	5.01	4.28	2.15	3.98	1.58
Garhwal	N.A.	N.A.	N.A.	N.A.	N.A.	N.A.	4.84	2.62	4.98	2.25
Nainital	18.05	17.15	19.18	8.92	9.05	3.05	3.63	1.24	2.69	0.98
Pithoragarh	N.A.	N.A.	N.A.	N.A.	N.A.	N.A.	8.27	4.23	7.14	3.25
Tehri Garhwal	N.A.	N.A.	N.A.	N.A.	N.A.	N.A.	5.16	2.68	4.73	2.39
Uttarkashi	N.A.	N.A.	N.A.	N.A.	N.A.	N.A.	3.69	1.62	5.72	2.00
Total	4.25	3.70	7.10	4.08	7.80	3.30	3.51	1.48	4.07	1.19

N.A. : Not Available

The condition of production became worst than ever. The percentage share of pulses in major foodgrains noted 3.70 per cent, 4.08 per cent, 3.30 per cent, 1.48 per cent and 1.19 per cent at the same point of time respectively (Table 5.3).

The productivity of pulses in the region is also not static and left behind a considerable gap when compared to other cereals. There is continuous increase found in the productivity of total foodgrain i.e. from 747 kg./ha. to 1674 kg./ha. But in case of pulses it recroded a decrease from 666 kg./ha. in 1954-55 to 511 kg./ha. in 1995-96 (Table 5.2). A clear picture of trend of area, production and productivity of pulses can be seen in Fig. 4.1. Where trend line curves of area and production shows increasing trends while productivity shows decreasing trend. It is noted from Tables 4.2, 4.3 and 4.4 that the magnitude of rate of change in area, production and productivity from 1954-55 to 1995-96 is 15.08 per cent, 12.72 per cent and -2.05 per cent respectively.

The spatial variations in area, production and productivity of pulses found different from the regional average. It is evident from Table 5.3 that Bilaspur and Nainital were the most prominent districts of pulse's cultivation in 1954-55 where pulses occupies 11.34 per cent and 18.05 per cent areal spread and contribute 15.48 per cent and 17.15 per cent in total foodgrain production. The situation in 1964-65 was such that Jammu (10.21%), Leh (16%) and Nainital (19.18%) were the districts where pulses occupied more than 10 per cent areal spread but except Jammu (9.51%) production share was very low in Leh (i.e. 4.51%) and Nainital (8.92%). Spatial analysis of productivity shows that Mandi (1013 kg./ha.), Shimla (1086 kg./ha.), Sirmour (1202 kg./ha.) and Lahaul & Spiti (1442 kg./ha.) are the prominent districts besides the area under pulses in these districts is very low.

During 1974-75, Bilaspur (14.04%), Lahaul and Spiti (10.47%), Solan (17.48%), Una (15.57%), Baramula (10.55%) and Leh (53.85%) are the districts where pulses occupied more than 10 per cent area under this crop but in case of

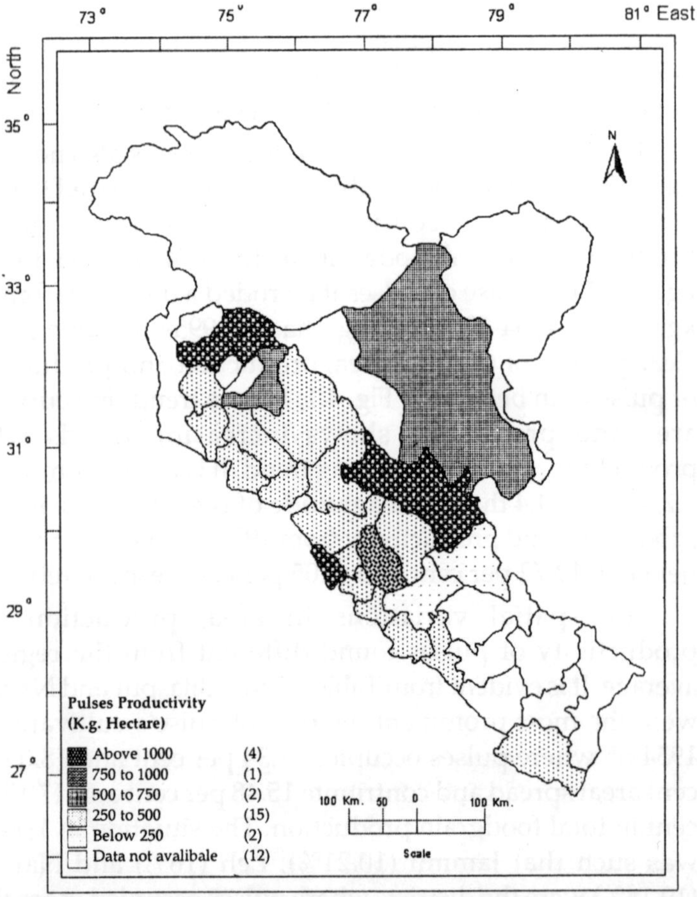

Fig. 5.11 : North-western Himalaya : Yield of Pulses (1974-75)

production only Lahaul and Spiti (10.89%), Una (11.45%) and Leh (51.61%) have more than 10 per cent share. Lahaul and Spiti and Leh are the cold desert regions where cultivation is done with most advance methods to overcome the climatic vagaries and no crop is more remunerative than pulses in such circumstance as a result it occupies a commendable position in these districts of cold deserts. The productivity shows a different situation, Lahaul and Spiti (2450 kg./ha.)

which shows a great chance of productivity increase in other districts if the technological skills are provided. Una (1006 kg./ha.), Srinagar (1444 kg./ha.) and Baramula (1129 kg./ ha.) are other districts of significance.

The condition of pulses in this region during 1985-86 and 1995-96 became more critical as area under pulses increased a little bit while production slightly decline to make the condition worst. Except Leh (44.13%) no district could reach 10 per cent average while in production except these arid districts no one had share even more than 5 per cent. In 1995-96 the condition was more or less same only Kinnaur (25%), Lahaul and Spiti (55%), Leh (45.62%) were the prominent districts in acreage as well as production (Table 5.3). It is noted that all these districts come under cold deserts and reason for high acreage in these districts is advanced methods of agriculture and economic value of pulses as discussed above. Productivity of pulses in Kinnaur, Lahual and Spiti was 1111 kg./ha. and 1454 kg./ha. while other arid districts show average condition but Budgam a plain district of Kashmir valley recorded 1257 kg./ha. and 1055 kg./ha. during 1985-86 and 1995-96 respectively. It indicates that possibilities for high productivity in other districts of valley are also bright if required weightage is given to it.

Spatial Variation of Productivity of Wheat

Wheat is one of the most significant foodgrain crops of Indian dietary system. It occupies about 33 per cent of the total foodgrain production of the country. Its production has increased from 6 million tonnes in 1950 to 68 million tonnes in 1996-97.

In the study region, wheat is also the most dominant crop of the region since independence. It occupied 39.89 per cent area and contributed 36.77 per cent in production during 1954-55. Its share declined to 31.15 per cent and production shared to 26.57 per cent in 1964-65. Since the green revolution its condition improve year after year and average share became

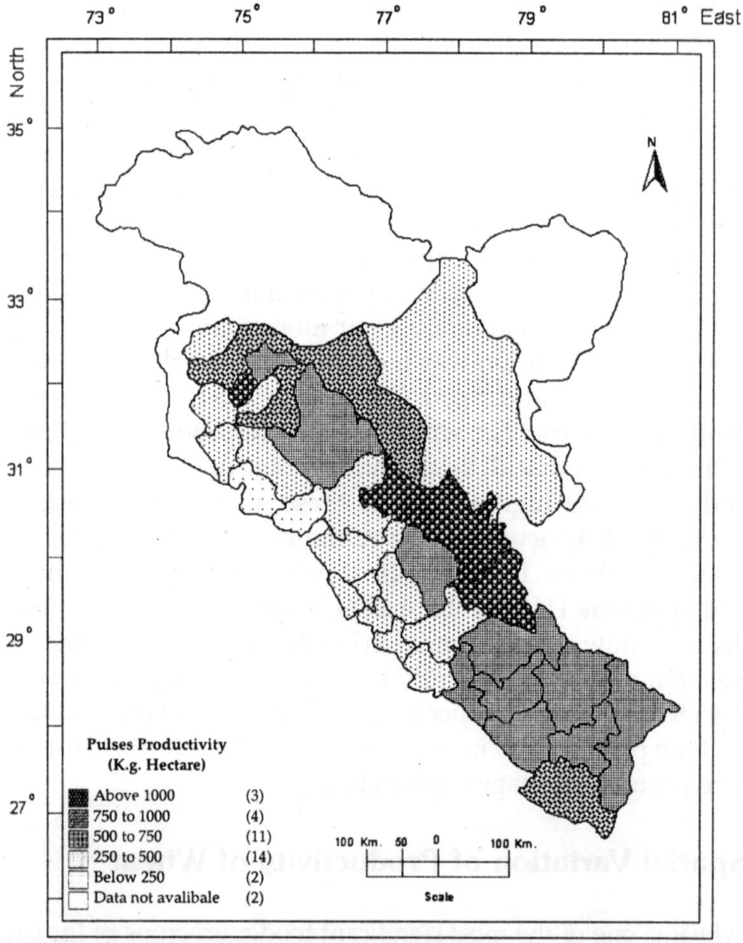

Fig. 5.12 : North-western Himalaya : Yield of Pulses (1995-96)

33.70 per cent, 40.83 per cent and 40.84 per cent while production share became 27.97 per cent, 34.33 per cent and 41.12 per cent in 1974-75, 1985-86 and 1995-96 respectively (Table 5.4).

It is evident from the spatial variation that all the districts of the region are wheat dominated except the districts of Kashmir valley where paddy is the dominant crop. In 1995-96, it is noted that in 20 districts out of 34 wheat occupied more than 40 per cent area while production share found more than 40 per cent in 16 districts. Besides Kinnaur (16.67%), Lahaul and Spiti (10%), Doda (8.52%), all the six districts of Kashmir valley have negligible area under wheat cultivation while in Leh and Kargil no area devoted for wheat cultivation. On the other hand in pre-green revolution period of 1954-55 and 1964-65 wheat occupied 75.47 per cent and 64.82 per cent area respectively in Leh.

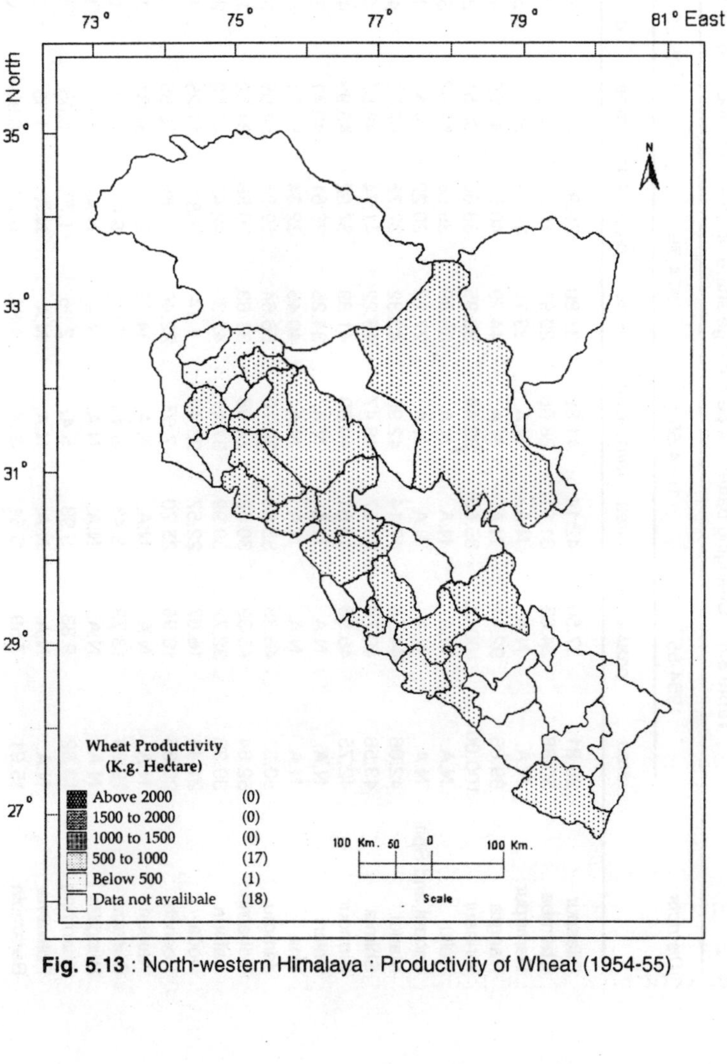

Fig. 5.13 : North-western Himalaya : Productivity of Wheat (1954-55)

Table 5.4 : Changing pattern of percentage share of Area and Production of Wheat

Districts	1954-55		1964-65		1974-75		1985-86		1995-96	
	area	*production*	*area*	*production*	*area*	*production*	*area*	*production*	*area*	*production*
Bilaspur	36.81	37.50	42.14	41.88	37.80	30.91	48.41	37.53	46.43	40.59
Chamba	30.88	34.63	31.91	28.86	29.91	16.27	35.12	28.94	34.09	23.45
Hamirpur	N.A.	N.A.	N.A.	N.A.	39.10	31.52	50.14	51.38	49.37	49.60
Kangra	89.86	92.41	N.A.	N.A.	44.79	40.51	47.62	47.71	47.67	53.02
Kinnaur	100.00	100.00	35.63	32.44	33.95	33.60	35.51	21.77	16.67	14.29
Kullu	N.A.	N.A.	N.A.	N.A.	39.89	35.56	44.68	37.14	46.09	50.44
Lahaul and Spiti	N.A.	N.A.	N.A.	N.A.	27.23	20.22	N.A.	N.A.	10.00	13.79
Mandi	42.06	44.58	47.14	42.95	43.32	37.78	47.76	42.11	46.69	37.50
Shimla	43.56	40.70	49.57	46.47	45.22	41.74	49.57	45.24	43.44	39.91
Sirmour	42.75	45.56	43.61	40.90	41.89	37.92	45.98	40.31	43.47	39.87
Solan	N.A.	N.A.	N.A.	N.A.	34.25	26.61	40.41	35.88	40.10	45.03
Una	N.A.	N.A.	N.A.	N.A.	40.45	32.34	50.71	46.81	48.93	52.95
Jammu	50.33	46.38	55.58	36.10	56.64	59.83	54.65	37.18	52.87	57.38
Udhampur	52.64	45.39	30.68	24.96	31.63	20.88	34.50	17.61	39.39	31.65
Kathua	39.33	32.72	39.99	37.17	42.38	42.45	45.10	28.41	50.89	53.29
Doda	21.44	16.67	22.57	28.49	1.14	2.88	11.25	6.12	8.52	6.84
Poonch	20.35	18.96	25.70	7.53	22.46	14.85	25.65	13.95	28.49	21.07
Rajouri	N.A.	N.A.	N.A.	N.A.	34.39	22.44	36.31	24.90	41.53	36.87
Srinagar	19.89	13.79	8.43	6.43	7.92	5.41	0.41	0.12	0.09	0.03
Budgam	N.A.	N.A.	N.A.	N.A.	N.A.	N.A.	5.11	1.49	2.00	1.25
Anantnag	13.92	9.33	5.93	5.47	2.96	1.77	0.03	0.01	0.01	0.01
Pulwama	N.A.	N.A.	N.A.	N.A.	N.A.	N.A.	1.35	0.59	0.90	0.43
Baramula	15.91	7.59	5.51	3.95	1.84	0.77	0.90	0.29	0.85	0.37

(Contd.)

Table 5.4 : (Contd.)

Kupwara	N.A.	N.A.	N.A.	N.A.	N.A.	N.A.	0.06	0.01	0.19	0.14
Leh	75.47	75.48	64.82	74.64	N.A.	N.A.	N.A.	N.A.	N.A.	N.A.
Kargil	N.A.	N.A.	N.A.	N.A.	N.A.	N.A.	N.A.	N.A.	N.A.	N.A.
Almora	N.A.	N.A.	N.A.	N.A.	N.A.	N.A.	55.40	54.17	55.29	55.22
Chamoli	N.A.	N.A.	N.A.	N.A.	N.A.	N.A.	52.22	49.07	51.98	53.04
Dehradun	42.21	38.16	40.94	32.48	44.57	35.84	42.24	45.87	42.05	47.92
Garhwal	N.A.	N.A.	N.A.	N.A.	N.A.	N.A.	57.87	59.13	52.43	58.68
Nainital	30.23	32.97	26.97	31.52	37.47	35.42	46.95	48.96	49.31	48.74
Pithoragarh	N.A.	N.A.	N.A.	N.A.	N.A.	N.A.	44.66	48.57	45.41	50.63
Tehri Garhwal	N.A.	N.A.	N.A.	N.A.	N.A.	N.A.	59.83	61.79	57.67	60.95
Uttarkashi	N.A.	N.A.	N.A.	N.A.	N.A.	N.A.	55.40	49.23	48.99	53.73
Total	39.89	36.77	31.15	26.57	33.70	27.97	40.83	34.23	40.84	41.12

N.A.= Not Available.

It is evident from Table 5.5, that productivity of wheat in 1954-55 was 706 kg./ha. which slightly declined to 689 kg./ha. in 1964-65, afterward gradual increase is noted as a result productivity became 1100 kg./ha., 1776 kg./ha. and 1758 kg./ha. in 1974-75, 1985-86 and 1995-96 respectively.

District level analysis proves that during 1954-55 no district of the region reached the productivity level of 1000 kg./ha. Most of districts ranges between 600-700 kg./ha. while Kangra with 801 kg./ha. and Kinnaur with 911 kg./ha. are the highest productivity districts of the study region at this point of time. The overall condition of productivity is declining but 9 districts reached the productivity level of more than 800 kg./ha. Chamba with 1140 kg./ha and Doda with 1040 kg./ha. are the top productivity districts.

Impact of green revolution is clearly shown in 1974-75 when the productivity of wheat became 1100 kg./ha. from 689 kg./ha. The growth of 59.65 per cent in productivity is no doubt the result of green revolutionary measures where because of high yielding variety of seeds, 49.20 per cent area increased during the decade while the growth in production during the same period was recorded as 138.10 per cent.

Highest productivity of the region is found in Lahaul and Spiti (1750 kg./ha) which is the result of scientific agriculture on very small area followed by Nainital (1297 kg./ha.), Shimla (1265 kg./ha.), Kinnaur (1265 kg./ha.) Sirmour (1228 kg./ha.) and Bilaspur (1215 kg./ha.), where productivity level is noted more than 1200 kg./ha. under reasonable acreage. A number of districts also show more than 1000 kg./ha. productivity namely Hamirpur, Kangra, Kullu, Mandi, Una, Jammu, Kathua, Poonch and Anantnag. The impact of green revolution is clear in a number of ways like increase in areal spread, production and productivity in the study region as a whole and at district level too.

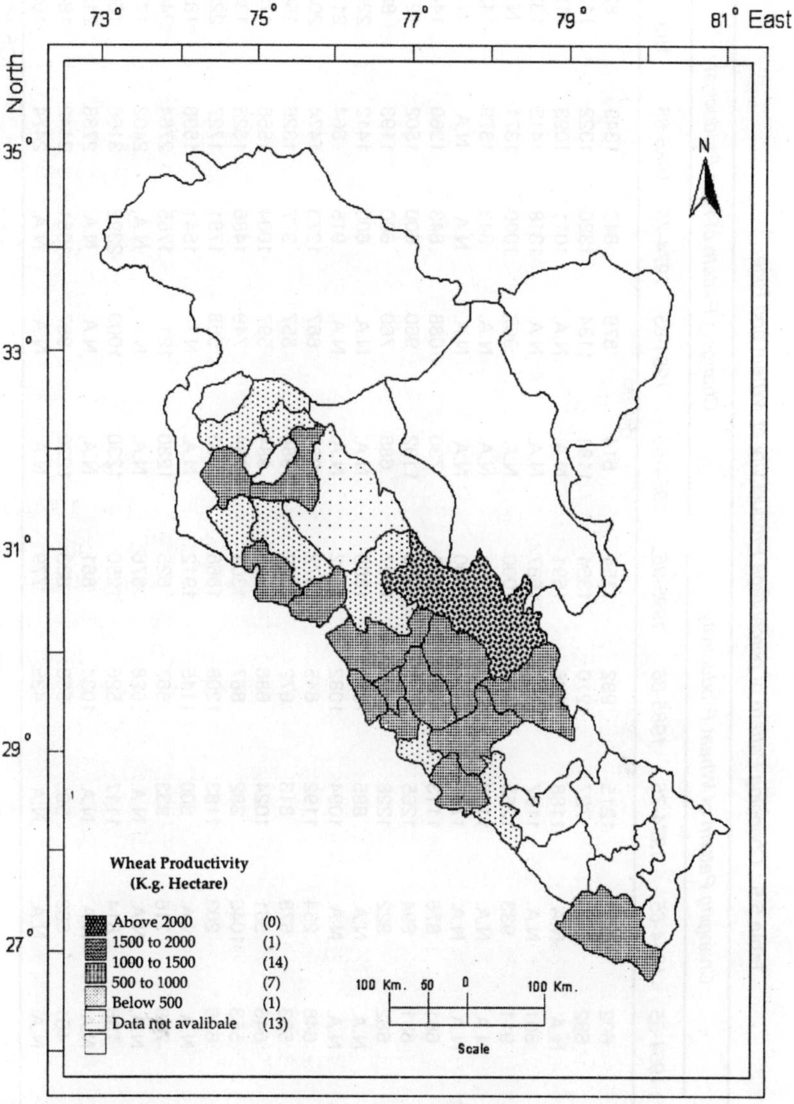

Fig. 5.14 : North-western Himalaya : Productivity of Wheat (1974-75)

Table 5.5 : *Changing Pattern of percentage Productivity of Wheat and Rice*

Districts	Changing Pattern of Wheat Productivity					Changing Pattern of Rice Productivity				
	1954-55	1964-65	1974-75	1985-86	1995-96	1954-55	1964-65	1974-75	1985-86	1995-96
Bilaspur	602	962	1215	982	1154	514	876	840	1349	826
Chamba	682	1140	727	1270	1354	1181	1134	1320	1322	1129
Hamirpur	N.A.	N.A.	1188	1026	1591	N.A.	N.A.	1011	1098	1187
Kangra	801	N.A.	1137	1496	1687	N.A.	N.A.	1318	1415	1392
Kinnaur	911	933	1265	456	1000	N.A.	967	1000	1371	N.A.
Kullu	N.A.	N.A.	1113	1659	2165	N.A.	N.A.	843	1379	1428
Lahaul and Spiti	N.A.	N.A.	1750	N.A.	2000	N.A.	N.A.	N.A.	N.A.	N.A.
Mandi	681	876	1113	1352	1541	730	1088	843	1360	1445
Shimla	681	894	1265	1325	1308	1192	960	600	1502	1273
Sirmour	682	922	1228	1245	1727	685	760	600	1193	889
Solan	N.A.	N.A.	885	1378	1868	N.A.	N.A.	600	1412	2256
Una	N.A.	N.A.	1094	1082	1872	N.A.	N.A.	915	864	2176
Jammu	646	254	1192	645	2072	984	667	1273	1474	2036
Udhampur	553	578	813	672	1672	984	557	977	1326	1588
Kathua	646	561	1024	686	1893	984	597	1004	1558	1735
Doda	553	1040	382	867	1304	984	749	1486	1623	1380
Poonch	646	203	1183	1208	1866	553	568	1781	1727	2221
Rajouri	N.A.	N.A.	800	1145	1912	N.A.	N.A.	1541	1598	1852
Srinagar	738	716	933	667	625	1230	1214	1755	2764	2443
Budgam	N.A.	N.A.	N.A.	528	876	N.A.	N.A.	N.A.	2402	1772
Anantnag	738	814	1117	526	1250	1230	1009	2392	3166	1605
Pulwama	N.A.	N.A.	N.A.	1034	861	N.A.	N.A.	N.A.	2788	2128
Baramula	461	689	585	553	633	1230	957	1951	2442	1844
Kupwara	N.A.	N.A.	N.A.	435	779	N.A.	N.A.	N.A.	2474	1625

(Contd...)

Table 5.5 : (*Contd.*)

Leh	738	747	N.A.	N.A.	N.A.	N.A.	N.A.	N.A.	N.A.	N.A.
Kargil	N.A.	N.A.	N.A.	N.A.	N.A.	N.A.	N.A.	N.A.	N.A.	N.A.
Almora	N.A.	N.A.	N.A.	997	1161	N.A.	N.A.	N.A.	1072	1231
Chamoli	N.A.	N.A.	N.A.	1042	1396	N.A.	N.A.	N.A.	1184	1345
Dehradun	588	471	556	1571	1773	653	1015	769	1378	1736
Garhwal	N.A.	N.A.	N.A.	1043	1273	N.A.	N.A.	N.A.	1040	1032
Nainital	793	963	1297	2356	2639	675	942	1468	2399	2917
Pithoragarh	N.A.	N.A.	N.A.	1280	1547	N.A.	N.A.	N.A.	1115	1362
Tehri Garhwal	N.A.	N.A.	N.A.	1264	1433	N.A.	N.A.	N.A.	1250	1403
Uttarkashi	N.A.	N.A.	N.A.	1197	1819	N.A.	N.A.	N.A.	1643	1649
Total	706	689	1100	1276	1758	997	936	1607	1882	2532

N.A. : Not Available.

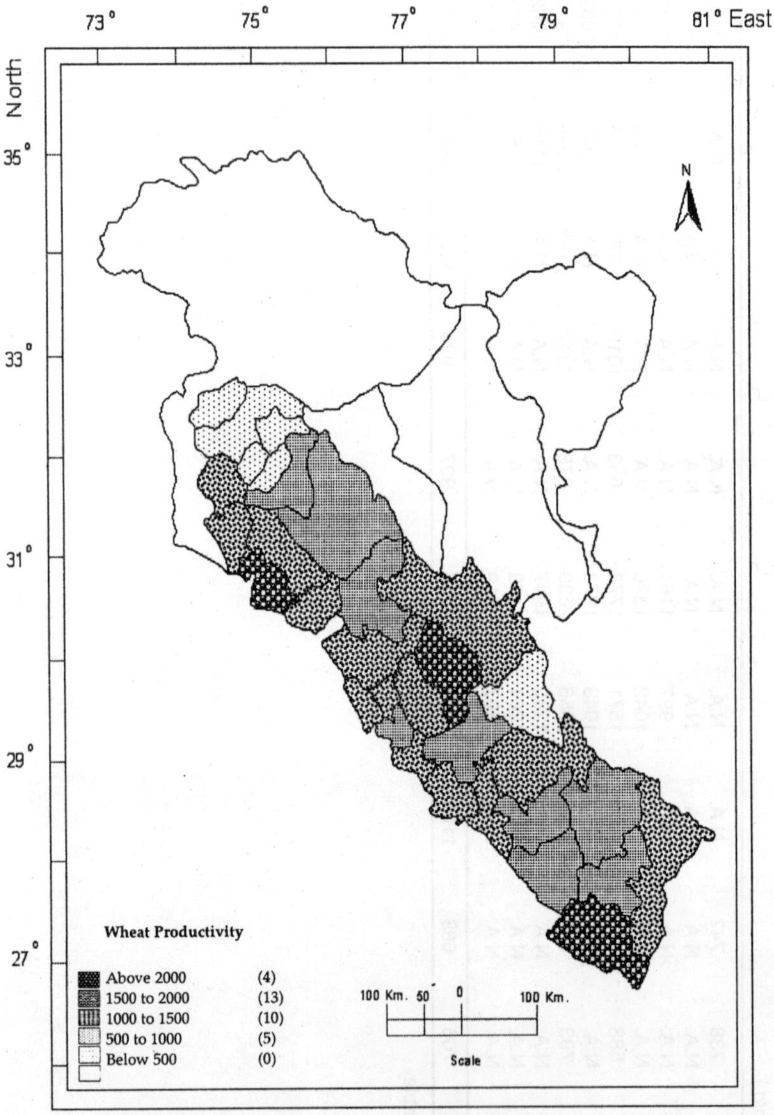

Fig. 5.15 : North-western Himalaya : Productivity of Wheat (1995-96)

The condition of wheat in the study region further improved during 1985-86, as overall productivity of the region became 1216 kg./ha. with a growth of 16 per cent which is not at par with the growth of the previous decade. Dehradun recorded highest growth with 182.35 per cent, Doda came next with 126.96 per cent growth. Nainital (81.65%), Chamba (74.69%) and Solan (55.71%) are other districts of significant growth. On the other hand, productivity decreased in some of the districts like Kinnaur (-63.95%), Jammu (-45.89%) and Anantnag (-52.91%). Anantnag is a paddy dominant district while Kinnaur is a cold deserts need controlled environment all the time but the decrease in productivity of Jammu is a matter of great concern. In terms of productivity. Nainital recorded highest i.e. 2356 kg./ha. Kullu and Dehradun are also considered significant with productivity more than 1500 kg./ha. A large number of districts fall between 1000 to 1500 kg./ha. productivity level namely Chamba, Hamirpur, Kangra, Mandi, Shimla, Sirmour, Solan, Una, Poonch, Rajouri, Pulwama, Chamoli, Garhwal, Pithoragarh, Tehri Garhwal and Uttarkashi. It shows that majority of the districts come close to regional average.

In 1995-96, the overall productivity of the region became 1758 kg./ha. with an increase of 37.77 per cent from the previous point of time. At district level, Nainital recorded highest productivity with 2639 kg./ha. while Kullu, Lahaul and Spiti of Jammu experienced more than 2000 kg./ha. A large number of districts namely Hamirpur, Kangra, Mandi, Sirmour, Solan, Una, Udhampur, Kathua, Poonch, Rajouri, Dehradun, Pithoragarh and Uttarkashi come under the productivity level between 1500 to 2000 kg./ha. Wheat is totally absent in Leh and Kargil while Kupwara, Baramula, Pulwama, Budgam and Srinagar fall below 1000 kg./ha. productivity, all belongs to valley of Kashmir where Paddy and Maize are the dominant crops while wheat is an insignificant crop.

Spatial Variation of Productivity of Rice

Rice is the second most dominant crop of the region in terms of productivity but occupied next to wheat in areal expansion. Its productivity increased to 1839 kg./ha. in 1995-96 from 997 kg./ha. during 1954-55 showing a growth of 84.45 per cent in this period of time. Its percentage share in area is found more or less stagnant i.e. 26.74 per cent in 1954-55 and 26.02 per cent in 1995-96 while a minor decline was found in production i.e. 34.81 per cent to 27.42 per cent in the same period of time.

Before discussing the variation in productivity, we briefly discuss the situation of area and production of rice in the region. It is evident from Table 5.6 that, in 1954-55 rice is the most dominant crop in all the districts of Kashmir valley. In Srinagar, Anantnag and Baramula rice occupied 66.38 per cent, 73.61 per cent and 55.68 per cent area out of total foodgrains.

Kathua with 31.35 per cent and Nainital with 32.55 per cent are also significant districts in terms of areal expansion and production as well out of total foodgrain. In 1964-65, the pattern of spatial variation of area and production is more or less same with very minor fluctuations. A slight decline is observed in area and production of valley districts while Nainital and Dehradun show a slight increase.

In 1974-75, the impact of green revolution measures is prominently seen in production of rice with a growth of 113 per cent from 1964-65, while in area it is only 24.23 per cent. Besides the growth in magnitude of area and production, the percentage share in total foodgrain shows a decline which indicates that other crops of the region benefitted more with green revolution technology than rice and occupied more and contributed more in production. In case of productivity a growth of 71.68 per cent is recorded in the same time period.

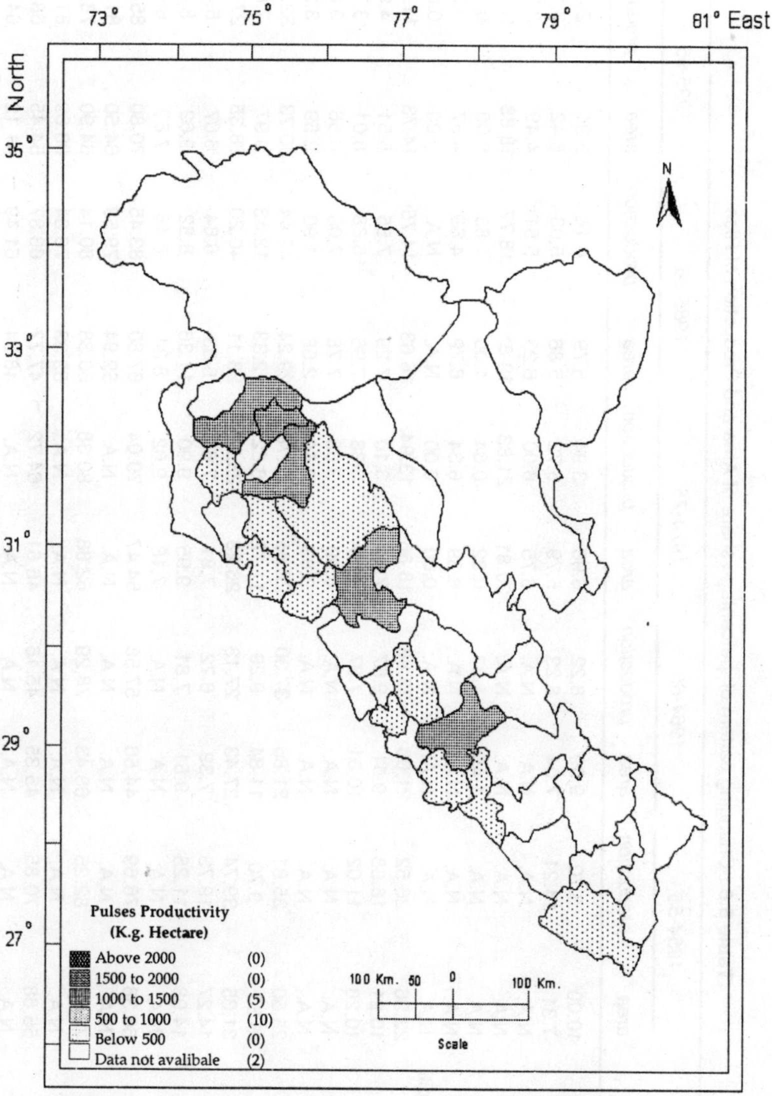

Fig. 5.16 : North-western Himalaya : Productivity of Rice (1954-55)

Table 5.6 : Changing pattern of percentage share of Area and Production of Rice

Districts	*1954-55*		*1964-65*		*1974-75*		*1985-86*		*1995-96*	
	area	*production*	*area*	*production*	*area*	*production*	*area*	*production*	*area*	*production*
Bilaspur	10.00	8.70	9.09	8.22	6.96	3.94	5.79	6.16	3.91	2.45
Chamba	7.31	14.21	7.04	8.23	5.79	5.72	5.86	5.03	5.42	3.11
Hamirpur	N.A.	N.A.	N.A.	N.A.	8.75	6.00	6.23	5.51	4.49	3.37
Kangra	N.A.	N.A.	N.A.	N.A.	20.81	21.83	19.81	18.77	18.88	17.32
Kinnaur	N.A.	N.A.	1.45	1.37	0.82	0.64	1.00	1.85	0.00	0.00
Kullu	N.A.	N.A.	N.A.	N.A.	8.39	6.34	6.79	4.69	4.57	3.30
Lahaul and Spiti	N.A.	N.A.	N.A.	N.A.	0.00	0.00	N.A.	N.A.	0.00	0.00
Mandi	23.36	26.52	21.08	23.86	18.86	13.94	16.63	14.75	14.73	11.09
Shimla	10.21	16.68	9.51	9.57	7.08	3.10	7.29	7.55	5.51	4.93
Sirmour	10.28	11.02	10.31	7.97	7.64	3.38	7.96	6.68	8.01	3.78
Solan	N.A.	N.A.	N.A.	N.A.	6.99	3.68	7.76	7.05	6.66	9.03
Una	N.A.	N.A.	N.A.	N.A.	3.83	2.56	2.58	1.90	2.59	3.26
Jammu	25.50	35.81	21.86	37.30	22.85	25.78	29.24	45.44	30.73	32.76
Udhampur	6.33	9.70	11.84	9.29	11.12	8.82	12.33	12.43	9.97	7.62
Kathua	31.35	39.74	27.43	27.13	26.28	25.79	28.11	40.20	28.35	27.22
Doda	14.27	19.73	7.39	6.72	7.81	7.68	6.53	6.64	6.07	5.15
Poonch	14.08	11.25	9.51	7.81	9.95	9.90	11.36	8.82	9.62	8.46
Rajouri	N.A.	N.A.	N.A.	N.A.	7.18	9.02	8.84	8.46	7.69	6.61
Srinagar	66.38	76.69	44.56	57.58	54.47	70.04	67.80	83.45	70.80	85.42
Budgam	N.A.	N.A.	N.A.	N.A.	N.A.	N.A.	59.94	79.59	64.50	81.16
Anantnag	73.61	82.25	68.43	78.29	62.88	80.38	60.38	80.14	64.90	78.62
Pulwama	N.A.	N.A.	N.A.	N.A.	N.A.	N.A.	65.33	76.84	68.89	81.78
Baramula	55.68	70.85	45.35	45.16	46.61	64.72	47.77	66.87	54.45	68.92
Kupwara	N.A.	N.A.	N.A.	N.A.	N.A.	N.A.	48.34	61.49	41.12	64.40

(Contd.)

Table 5.6 : *(Contd.)*

Leh	N.A.	N.A.	N.A.	N.A.	N.A.	N.A.	N.A.	N.A.	N.A.	N.A.
Kargil	N.A.	N.A.	N.A.	N.A.	N.A.	N.A.	N.A.	N.A.	N.A.	N.A.
Almora	N.A.	N.A.	N.A.	N.A.	N.A.	N.A.	32.96	34.66	34.15	36.18
Chamoli	N.A.	N.A.	N.A.	N.A.	N.A.	N.A.	41.76	44.60	41.99	41.29
Dehradun	24.29	24.38	24.44	41.72	26.52	31.53	22.47	21.41	27.93	31.16
Garhwal	N.A.	N.A.	N.A.	N.A.	N.A.	N.A.	32.13	32.74	31.12	28.22
Nainital	32.55	30.19	38.78	44.32	44.20	55.90	44.78	47.57	44.51	48.63
Pithoragarh	N.A.	N.A.	N.A.	N.A.	N.A.	N.A.	35.87	34.01	34.49	33.85
Tehri Garhwal	N.A.	N.A.	N.A.	N.A.	N.A.	N.A.	25.56	26.11	28.54	29.54
Uttarkashi	N.A.	N.A.	N.A.	N.A.	N.A.	N.A.	37.60	45.87	41.93	41.67
Total	26.74	34.81	27.06	31.33	24.38	29.54	26.32	32.54	26.02	27.42

***N.A* : Not Available.**

The average productivity of the region became 1607 kg./ha. from 936 kg./ha. (Table 5.4). But variations at district level are prominent with highest productivity noted in Anantnag (2392 kg./ha.), followed by Baramula (1951 kg./ha.), Poonch (1781 kg./ha.) and Srinagar (1755 kg./ha.). Some districts also experienced negative growth in productivity namely Bilaspur (-4.11%), Mandi (-22.52%), Shimla (-21.05%) and Dehradun (-24.24%). It is investigated that agro-ecology of the above districts is not as favourable as for other foodgrain crops like maize and wheat due to which rice did not get equal weightage and declined its productivity, which can be overcome by the use of modern technology.

The condition of rice productivity in 1985-86, improved further with a growth of 17.11 per cent and became 1882 kg./ha. while it was only 1607 kg./ha. a decade back. A number of districts noted high positive growth in productivity namely Nainital (63.42%), Dehradun (79.19%), Srinagar (57.49%) Kathua (55.10%), Solan (135.33%), Sirmour (98.58%) and Bilaspur (60.60%), all the above districts show a growth of more than 50 per cent in productivity which shows slightly late impact of green revolution in these districts.

As a result, the highest productivity found in Anantnag i.e., 3166 kg./ha. followed by Pulwama (2788 kg./ha.) Srinagar (2764 kg./ha.), Kupwara (2474 kg./ha.), Baramula (2442 kg./ha.) and Budgam (2402 kg./ha.). All these districts belong to paddy dominated Kashmir valley, besides only Nainital with 2399 kg./ha. found above 2000 kg./ha. productivity level. In Leh, Kargil and Lahaul and Spiti rice is totally absent where agro-ecology does not permit natural agriculture because of cold desertic climate.

The condition of rice productivity in the region shows a decline from 1882 kg./ha. to 1839 kg./ha. in 1995, similarly share in area and production also recorded a slight decline out of total foodgrains.

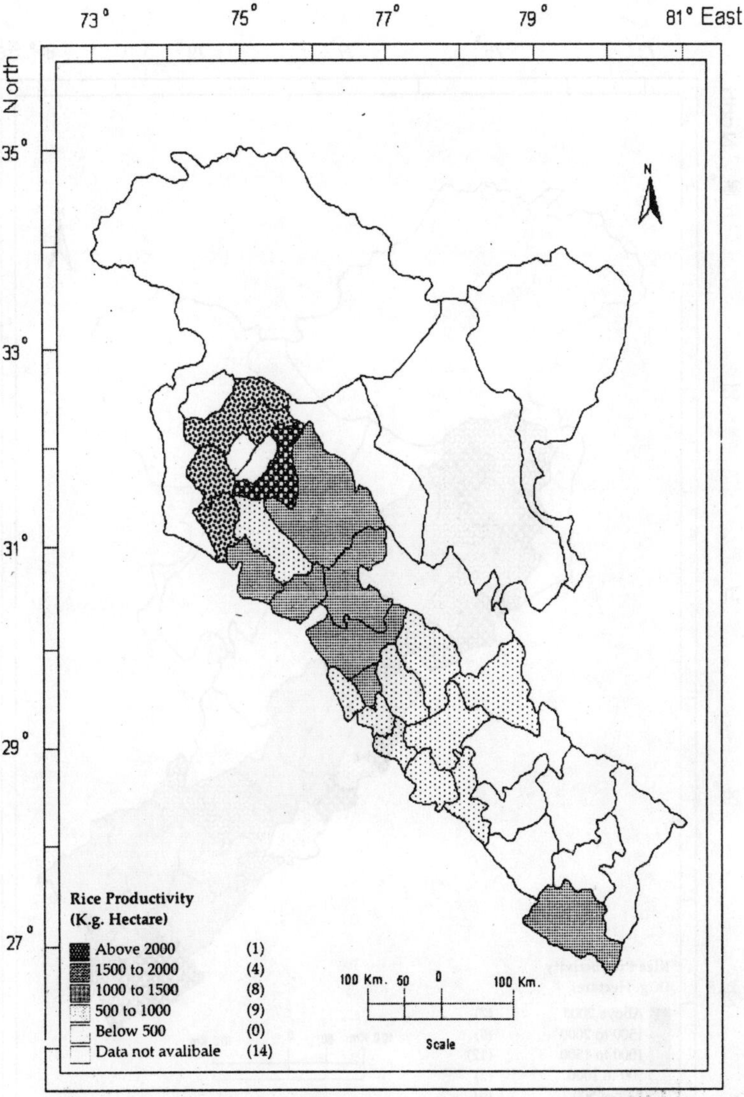

Fig. 5.17 : North-western Himalaya : Productivity of Rice (1974-75).

The reason for declining share of area, production and productivity can be caused by militancy, because paddy dominates in the valley and occupy a significant share in

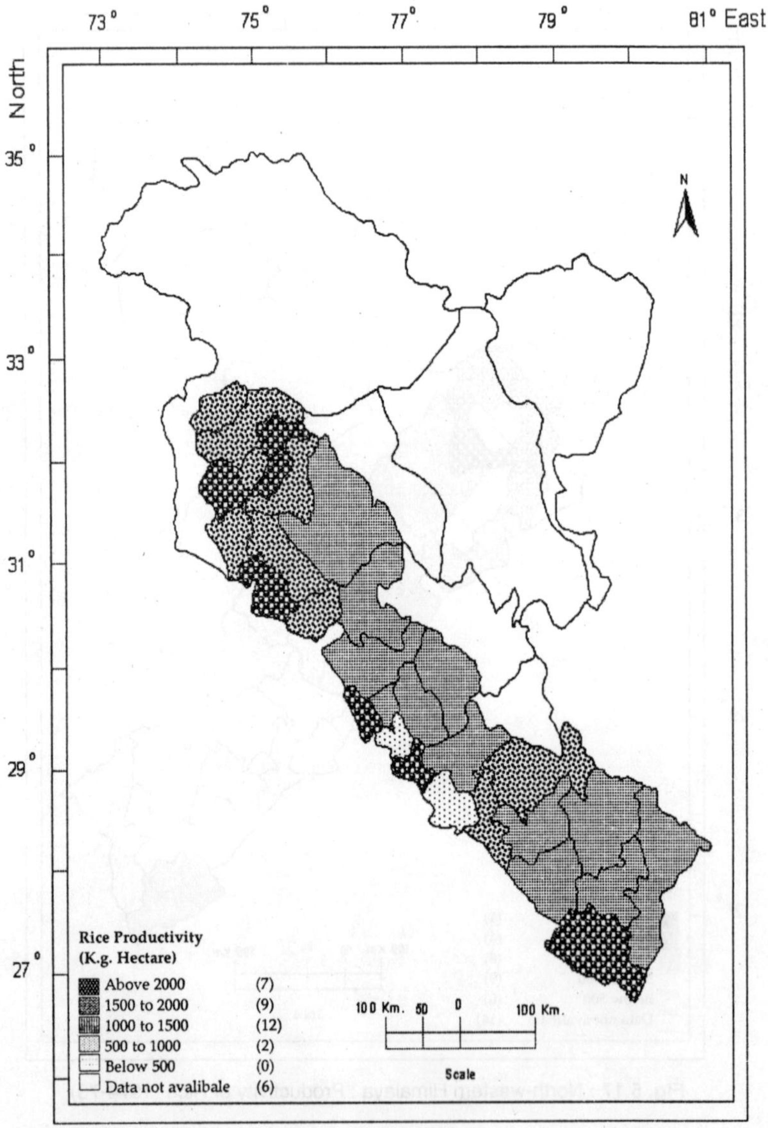

Fig. 5.18 : North-western Himalaya : Productivity of Rice (1995-96)

area as well as production. Besides it, rice is a dominant crop both in area and production in number of districts. It is evident from Table 5.6 that rice occupied 70.80 per cent of total foodgrain area in Srinagar and contributes 85.42 per cent in total production. Other districts of significance are Budgam which occupied 64.50 per cent area and contribute 81.16 per cent in production, in Anantnag it occupied 64.90 per cent area and contribute 78.62 per cent production, in Pulwama it occupied 68.89 per cent area and 71.78 per cent production, in Baramula it occupied 54.45 per cent area and contribute 68.89 production. All these dominant districts belong to Kashmir valley where agro-ecology is most favourable for rice than other region of north-western Himalaya, but to some extent militancy caused decline in productivity in the last decade.

The pattern of productivity decline can be shown in 13 districts, all the districts of Kashmir valley in particular show significant decline as compared to other districts. On the other hand, a number of districts from Himachal and Uttaranchal region show positive growth in rice productivity. Kinnaur, Lahaul and Spiti, Leh and Kargil are districts of cold desert like climate and rice is totally absent in these districts.

Spatial Variation of Productivity of Maize

The credit for high productivity increase goes to the growth in production i.e. 637.43 per cent. Agro-ecology of the region is most favourable for maize crops than other part of the country. As a result it responded best and is the most dominant crop of the region in terms of productivity. The productivity of maize in the region as a whole increased by 178.6 per cent from 1954-55 to 1995-96 while area increased by 164.66 per cent in the same period of time. But highest growth is found during 1964-65 to 1974-75 i.e., 77.77 per cent. Obviously it is an impact of green revolutionary measures. The growth in the production of maize was also found tremendous during this period i.e. 144.95 per cent while in area it increased by only 37.90 per cent.

In 1954-55 when the average productivity of the region was just 674 kg./ha. Shimla ranked highest with 762 kg./ha. while Dehradun ranked second with 743 kg./ha. It was the period of First Five Year Plan and data of a number of districts was not available. On the basis of available data for all the districts of Jammu Kashmir shows same productivity of 738 kg./ha. which shows some doubts in secondary data. Percentage share of area and production is also low i.e. 22.32 per cent and 19.63 per cent out of total foodgrain.

In 1964-65, the average productivity of Maize increased to 886 kg./ha. with growth of 31.45 per cent. At district level it is found that Kinnaur district of Himachal region ranked first with 1347 kg./ha. productivity. Also Bilaspur, Mandi, Shimla and Sirmour, districts of H.P. fall above 1000 kg./ha. of productivity. On the other hand in Himachal region only Baramula shows high productivity i.e. 1166 kg./ha.

In 1974-75, passed through the period of green revolution and its impact is clearly reflected in the growth of productivity which became 1575 kg./ha. against 886 kg./ha. of 1964-65. This increase of productivity is the result of growth in production i.e. 144.97 per cent in the same period of time. Highest productivity of maize is recorded in Bilaspur i.e. 2210 kg./ha. while Hamirpur and Poonch also come above 2000 kg./ha. of productivity level. It is also revealed from Table 5.7 that productivity of maize in almost all the districts of Jammu and Himachal region is above 1500 kg./ha. except Udhampur (1240 kg./ha.) and Rajouri (1488 kg./ha.) where it is close to the existing productivity level. Kashmir and Uttaranchal region showed low productivity of this crop because agro-ecology of these regions is more friendly and remunerative for paddy and rice cultivation.

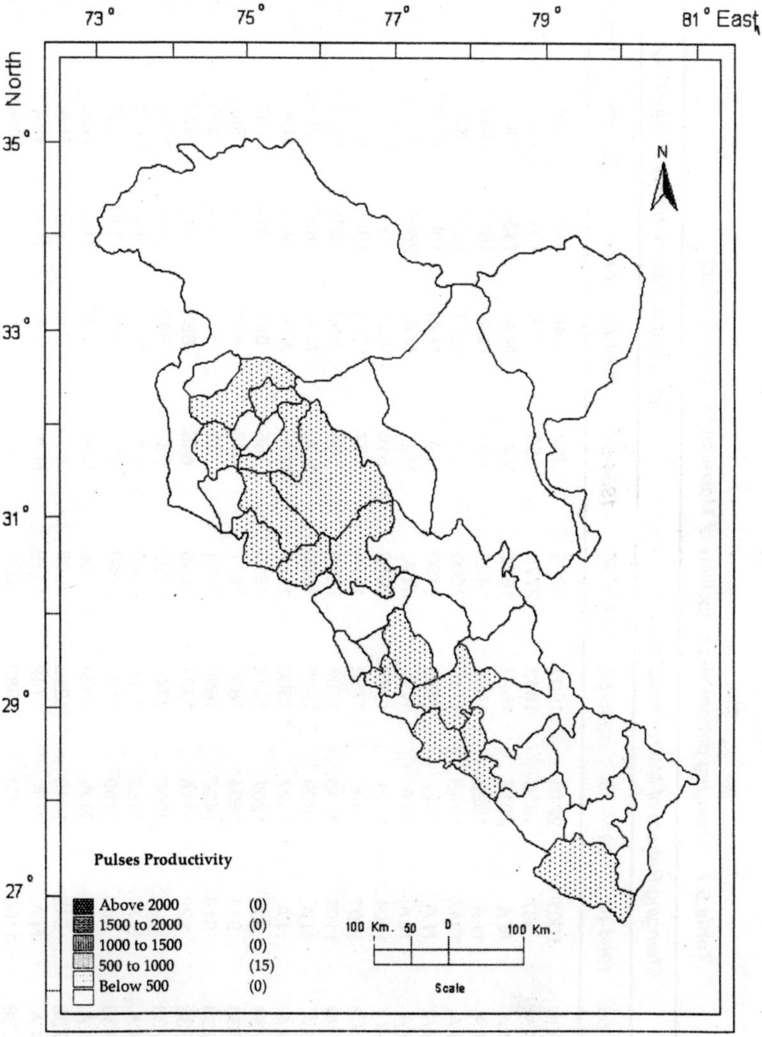

Fig. 5.19 : North-western Himalaya : Productivity of Maize (1954-55)

Table 5.7 : Changing pattern of productivity of Maize and Barley (Kg./hac)

Districts	*Changing Pattern of Maize Productivity*					*Changing pattern of Barley Productivity*				
	1954-55	*1964-65*	*1974-75*	*1985-86*	*1995-96*	*1954-55*	*1964-65*	*1974-75*	*1985-86*	*1995-96*
Bilaspur	538	1001	2210	1609	1595	802	1046	1107	1120	1400
Chamba	538	987	1936	1843	2778	359	1045	1013	1203	1219
Hamirpur	N.A.	N.A.	2016	1523	1638	N.A.	N.A.	1136	1121	1500
Kangra	N.A.	N.A.	1628	1581	1432	583	N.A.	1014	1071	1167
Kinnaur	N.A.	1347	1848	1740	2000	N.A.	1063	1326	647	1000
Kullu	N.A.	N.A.	1822	2655	2130	N.A.	N.A.	1067	1398	1900
Lahaul and Spiti	N.A.	N.A.	N.A.	N.A.	N.A.	N.A.	N.A.	2605	N.A.	1286
Mandi	547	1004	1821	1963	2902	516	1007	1069	1150	1067
Shimla	762	1001	1995	1752	1971	537	1079	1326	1040	1505
Sirmour	628	1106	1943	1708	2650	536	952	826	1170	965
Solan	N.A.	N.A.	1809	1926	1626	N.A.	N.A.	827	778	1353
Una	N.A.	N.A.	1827	1300	1627	N.A.	N.A.	N.A.	N.A.	N.A.
Jammu	738	529	1207	1616	1555	65	597	656	682	500
Udhampur	738	911	1694	1896	2637	646	470	667	867	1299
Kathua	738	818	1420	1690	2404	738	557	769	939	810
Doda	738	822	1716	1921	1859	646	688	640	573	363
Poonch	738	922	2000	2745	2882	737	942	1087	1166	500
Rajouri	N.A.	N.A.	1488	2076	2445	N.A.	N.A.	N.A.	1000	500
Srinagar	738	666	759	1131	1067	738	1273	1139	1000	N.A.
Budgam	N.A.	N.A.	N.A.	934	696	N.A.	N.A.	N.A.	N.A.	N.A.
Anantnag	738	548	1019	1263	802	830	368	428	N.A.	N.A.
Pulwama	N.A.	N.A.	N.A.	1742	1119	N.A.	N.A.	N.A.	N.A.	N.A.
Baramula	738	1166	982	989	1021	646	727	857	1000	1000
Kupwara	N.A.	N.A.	N.A.	1486	630	N.A.	N.A.	N.A.	972	1523

(Contd.)

Table 5.7 : *(Contd.)*

Leh	N.A.	N.A.	N.A.	N.A.	N.A.	738	735	737	742	897
Kargil	N.A.	N.A.	N.A.	N.A.	N.A.	N.A.	N.A.	N.A.	N.A.	N.A.
Almora	N.A.	N.A.	N.A.	1221	1187	N.A.	N.A.	N.A.	1120	1040
Chamoli	N.A.	N.A.	N.A.	916	1168	N.A.	N.A.	N.A.	1343	1612
Dehradun	743	344	848	1677	1043	846	905	1053	1011	1741
Garhwal	N.A.	N.A.	N.A.	879	1055	N.A.	N.A.	N.A.	1390	1086
Nainital	729	845	854	1013	1218	807	726	457	1443	1445
Pithoragarh	N.A.	N.A.	N.A.	1329	1219	N.A.	N.A.	N.A.	1426	1409
Tehri Garhwal	N.A.	N.A.	N.A.	1189	1044	N.A.	N.A.	N.A.	1231	1086
Uttarkashi	N.A.	N.A.	N.A.	1347	1093	N.A.	N.A.	N.A.	1326	1803
Total	674	886	1575	1710	1878	220	882	1027	1111	1199

N.A. : Not Available.

As a result of green revolutionary measures and new technology implications in agricultural development, the percentage share of production of maize in the total foodgrain of the region became 36.71 per cent and average became 30.91 per cent. It shows that Maize was the most dominant crop of the region in areal expansion as well as in production but left behind in productivity with rice.

During 1985-86, it is evident from Table 5.7 and Table 5.8, that the average productivity of maize in the region became 1710 kg./ha., but the percentage share of area as well as production declined to 26.29 per cent and 29.53 per cent respectively. Table 5.8, reveals that the highest area occupied by maize recorded in Doda (69.44%) followed by Poonch with 62.47 per cent and Rajouri 53.26 per cent. These districts also contribute considerable share in production i.e. 83.84 per cent, 77.17 per cent and 66.23 per cent respectively. Areal expansion of more than 40 per cent in area is noted in the districts of Bilaspur, Chamba, Hamirpur, Sirmour, Solan, Una, Udhampur, Rajouri, Baramula and Kupwara. All the said districts are also contributing more than 40 per cent share in production except Baramula and Kupwara where because of low productivity it is not contributing as significant as in area. Ladakh and Uttaranchal districts are insignificant for this crop, only Dehradun shows a continuous increase as a result percentage share of area and production became 19.34 per cent and 22.41 per cent respectively while Nainital shows a continuous decline in it.

Table 5.7, revealed that the highest productivity of the region found in Poonch district i.e. 2745 kg./ha. while Kullu ranked second i.e. 2655 kg./ha. with the growth of 37.25 per cent and 45.72 per cent respectively. The highest growth in productivity is found in Dehradun i.e. 97.76 per cent where productivity became 1677 kg./ha. from a low of 848 kg./ha.

It is said earlier that agro-climate of Himachal and Jammu region is more suitable for maize than other region of north-western Himalaya and it is proved by the prevailing

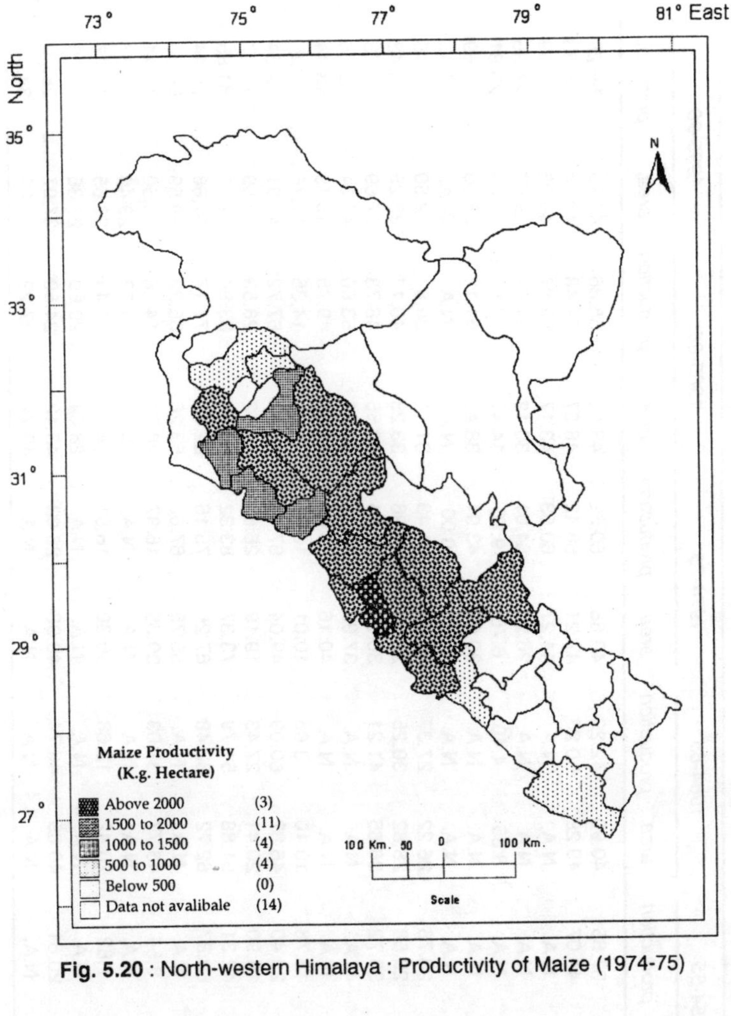

Fig. 5.20 : North-western Himalaya : Productivity of Maize (1974-75)

conditions of productivity level, it is noted that all the districts except Una of Himachal Pradesh show high productivity i.e. more than 1500 kg./ha. and come under the upper category of productivity.

Finally in 1995-96, productivity of maize increased to 1878 kg./ha. with a growth of 9.82 per cent. A number of districts noted more than 2500 kg./ha. of productivity. Highest productivity

Table 5.8 : Changing Pattern of percentage share of Area and Production of Maize

Districts	*1954-55*		*1964-65*		*1974-75*		*1985-86*		*1995-96*	
	area	*production*	*area*	*production*	*area*	*production*	*area*	*production*	*area*	*production*
Bilaspur	41.35	37.65	40.90	42.29	40.68	60.52	43.58	55.36	45.75	55.28
Chamba	49.78	44.07	49.29	50.23	47.21	68.42	48.07	57.48	48.78	68.83
Hamirpur	N.A.	N.A.	N.A.	N.A.	44.27	60.53	43.17	52.87	45.02	46.59
Kangra	N.A.	N.A.	N.A.	N.A.	26.25	34.01	30.12	31.89	29.63	27.96
Kinnaur	N.A.	N.A.	3.35	4.40	6.75	9.76	14.47	33.80	13.89	23.81
Kullu	N.A.	N.A.	N.A.	N.A.	30.09	43.91	38.43	51.12	35.00	37.69
Lahaul and Spiti	N.A.	N.A.	N.A.	N.A.	0.00	0.00	N.A.	N.A.	N.A.	N.A.
Mandi	27.91	23.76	26.22	27.37	29.03	41.43	31.11	39.81	32.50	49.15
Shimla	28.33	29.58	28.82	30.25	28.86	41.98	33.28	40.17	32.25	43.43
Sirmour	36.38	35.72	36.65	41.21	36.90	52.84	40.55	48.73	37.69	53.03
Solan	N.A.	N.A.	N.A.	N.A.	37.98	60.32	43.26	53.66	43.34	42.36
Una	N.A.	N.A.	N.A.	N.A.	40.16	53.65	44.86	49.75	46.19	43.44
Jammu	13.08	13.78	10.10	13.66	10.01	10.71	8.42	14.35	11.76	9.58
Udhampur	25.57	29.40	46.82	60.09	49.04	67.47	46.99	67.72	46.91	59.54
Kathua	19.88	18.90	20.41	27.43	19.19	26.65	18.42	28.59	13.59	18.07
Doda	40.81	42.31	51.88	51.79	73.37	83.32	69.64	83.84	74.17	84.85
Poonch	65.16	69.37	62.72	83.48	67.26	75.16	62.47	77.17	61.68	70.44
Rajouri	N.A.	N.A.	N.A.	N.A.	55.75	67.68	53.26	66.23	49.59	56.30
Srinagar	12.32	8.54	43.77	31.03	30.35	16.87	28.31	14.26	25.50	13.44
Budgam	N.A.	N.A.	N.A.	N.A.	N.A.	N.A.	30.09	15.53	29.46	14.57
Anantnag	11.67	7.82	20.42	12.68	30.30	16.51	34.83	18.44	32.65	19.77
Pulwama	N.A.	N.A.	N.A.	N.A.	N.A.	N.A.	28.04	20.60	27.36	17.09
Baramula	26.99	20.61	40.05	46.74	40.98	26.03	43.42	24.60	37.07	26.00
Kupwara	N.A.	N.A.	N.A.	N.A.	N.A.	N.A.	48.72	37.22	56.82	34.52

(Contd...)

Table 5.8 : *(Contd.)*

Leh	N.A.	N.A.	N.A.	N.A.	N.A.	N.A.	0.00	0.00	0.00	0.00
Kargil	N.A.	N.A.	N.A.	N.A.	N.A.	N.A.	0.00	0.00	0.00	0.00
Almora	N.A.	N.A.	N.A.	N.A.	N.A.	N.A.	2.01	2.40	2.33	2.39
Chamoli	N.A.	N.A.	N.A.	N.A.	N.A.	N.A.	0.58	0.46	0.73	0.63
Dehradun	16.96	19.36	19.84	11.50	15.34	18.82	19.34	22.41	21.84	14.64
Garhwal	N.A.	N.A.	N.A.	N.A.	N.A.	N.A.	3.02	2.60	3.54	3.28
Nainital	15.01	15.05	13.48	13.84	8.78	5.46	3.89	1.75	2.84	1.30
Pithoragarh	N.A.	N.A.	N.A.	N.A.	N.A.	N.A.	4.56	5.16	6.50	5.72
Tehri Garhwal	N.A.	N.A.	N.A.	N.A.	N.A.	N.A.	2.89	2.81	4.26	3.28
Uttarkashi	N.A.	N.A.	N.A.	N.A.	N.A.	N.A.	1.54	1.54	2.45	1.61
Total	22.32	19.63	30.91	33.89	30.91	36.71	26.29	29.53	26.48	28.49

(Contd.)

N.A : Not Available

is found in Mandi i.e. 2902 kg./ha. followed by Poonch (2882 kg./ha.), Chamba (2778 kg./ha.), Shimla (2650 kg./ha.) and Udhampur (2637 kg./ha.), besides these in a number of districts productivity level found between 2000-2500 kg./ha., namely Kinnaur, Kullu, Kathua and Rajouri. All these districts belong to Himachal and Jammu region, and

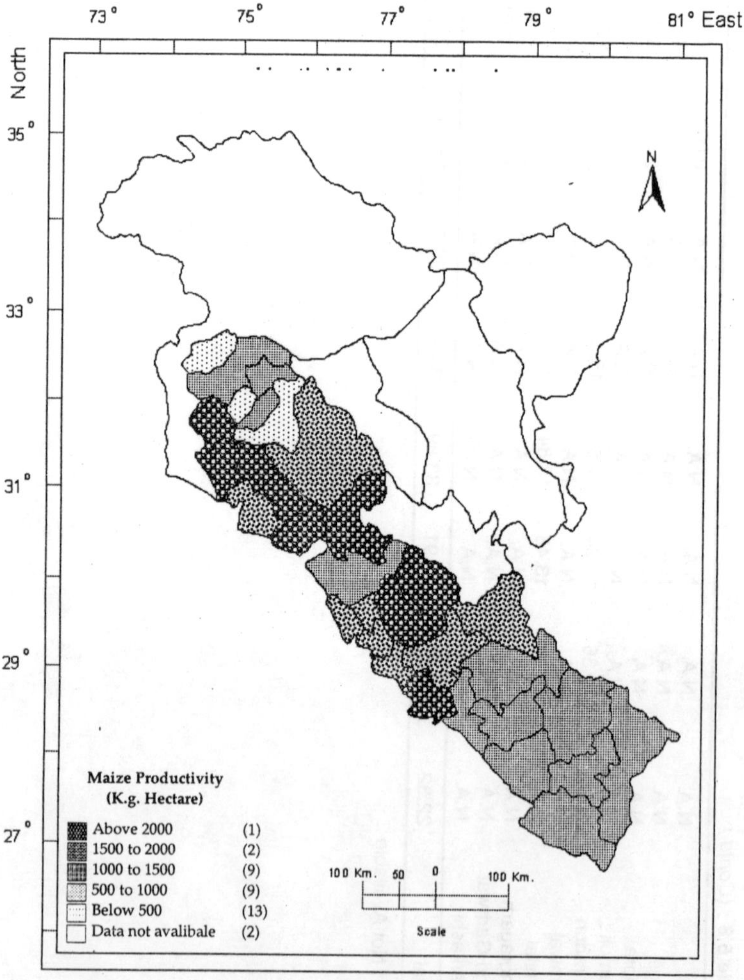

Fig. 5.21 : North-western Himalaya Productivity of Maize (1995-96)

remaining districts of maize productive region come between 1500-2000 kg./ha. of productivity. And all the districts of Kashmir valley and Uttaranchal records lower than 1500 kg./ha. productivity which are not eco-friendly for maize. The percentage of share area and production with minor fluctuations followed the same trend as it was in 1985-86.

Spatial Variation of Productivity of Barley

Barley is an insignificant crop but hardy in nature which can survive in harsh climatic condition also. Its share in area remained almost stagnant with a slight decline from 1954-55 to 1995-96 i.e., 4.25 per cent and 4.07 per cent respectively but percentage share production decreased significantly i.e., from 3.70 per cent to 1.19 per cent. Though its productivity is not at par with wheat, rice and maize but in terms of growth it is the most significant crop which showed a growth of 445 per cent in productivity i.e. 220 kg./ha. in 1954-55 to 1199 kg./ha. in 1995-96. The growth is continuous without any decline. Such a high growth in productivity is because of the production increased during this period by 77.37 per cent while area shrinked by –15.20 per cent.

As said earlier that it is a hardy crop which can survive in harsh and extremely cold climate where other crops can not, it is evident from Table 5.9, that during 1954-55 it occupied highest share in Leh (i.e. 24.53%) in area and 24.52 per cent in production. Doda and Udhampur are the next important districts during this period. The productivity on an average is very low and only 4 districts namely Dehradun, Anantnag, Nainital and Bilaspur could reach 800 kg./ha. productivity level.

In 1964-65, Kinnaur occupied 59.57 per cent of total area under foodgrain and contributed 61.78 per cent of production. The next prominent district of barley is Leh where it occupied 18.40 per cent area and 20.85 per cent production. In terms of productivity Srinagar ranked first with 1273 kg./ha. while Shimla (1079 kg./ha.), Kinnaur (1063 kg./ha.), Bilaspur (1046 kg./ha.), Chamba and Mandi (1007 kg./ha.), are the districts where productivity level is also noted more than 1000 kg./ha. (Table 5.7).

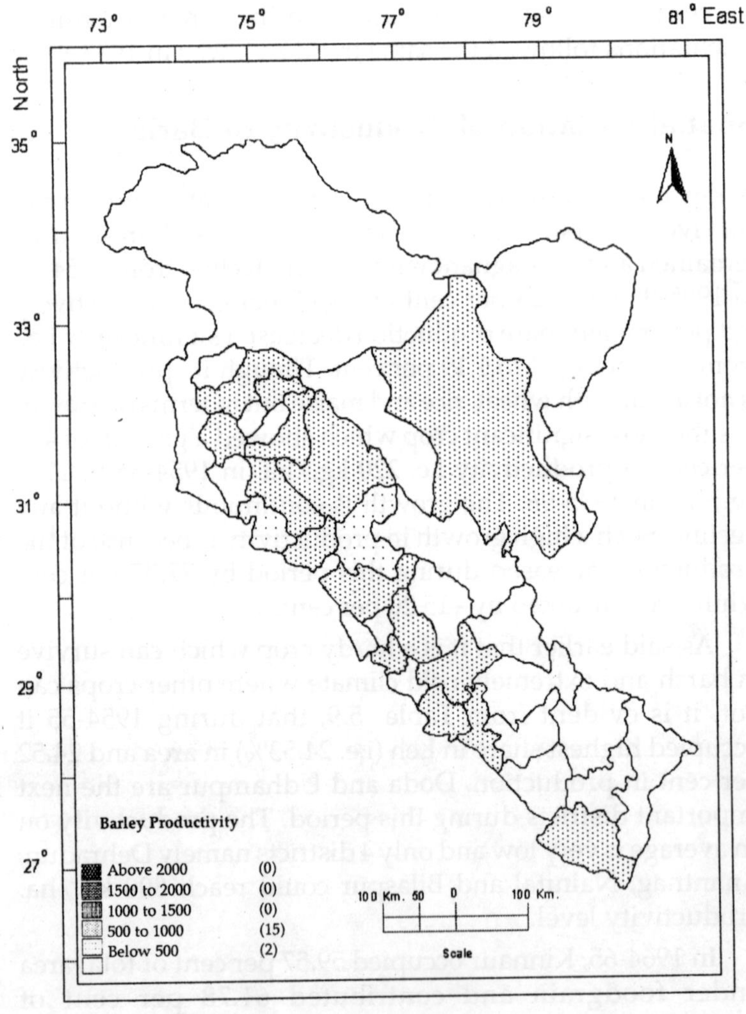

Fig. 5.22 : North-western Himalaya : Productivity of Barley (1954-55)

Table 5.9 : Changing pattern of percentage share of Area and Production of Barley

Districts	*1954-55*		*1964-65*		*1974-75*		*1985-86*		*1995-96*	
	area	*production*	*area*	*production*	*area*	*production*	*area*	*production*	*area*	*production*
Bilaspur	0.49	0.67	0.36	0.39	0.53	0.39	0.67	0.59	0.85	0.90
Chamba	12.02	7.10	11.76	12.68	11.02	8.35	10.94	8.55	5.59	3.46
Hamirpur	N.A.	N.A.	N.A.	N.A.	0.36	0.27	0.21	0.19	0.28	0.27
Kangra	10.14	7.59	N.A.	N.A.	2.79	2.25	1.99	1.43	1.57	1.21
Kinnaur	0.00	0.00	59.57	61.78	53.37	55.36	49.02	42.58	44.44	38.10
Kullu	N.A.	N.A.	N.A.	N.A.	14.49	12.38	10.04	7.03	6.52	6.26
Lahaul and Spiti	N.A.	N.A.	N.A.	N.A.	62.30	68.89	N.A.	N.A.	35.00	31.03
Mandi	6.12	4.91	5.05	5.29	4.30	3.60	4.39	3.29	3.04	1.69
Shimla	17.44	12.85	11.45	12.96	12.55	12.15	9.82	7.04	8.44	8.92
Sirmour	6.36	5.33	6.39	6.18	5.41	3.29	4.63	3.82	4.30	2.21
Solan	N.A.	N.A.	N.A.	N.A.	3.30	2.40	2.76	1.38	2.90	2.36
Una	N.A.	N.A.	N.A.	N.A.	N.A.	N.A.	N.A.	N.A.	N.A.	N.A.
Jammu	7.46	0.69	2.25	3.44	1.16	0.67	0.25	0.18	0.16	0.04
Udhampur	15.07	15.16	3.00	1.99	1.72	0.93	1.36	0.90	1.11	0.69
Kathua	6.59	6.26	3.53	3.26	3.67	2.76	1.73	1.49	1.64	0.73
Doda	23.47	21.29	11.71	9.78	11.09	4.71	7.32	2.63	6.08	1.36
Poonch	0.39	0.41	0.10	0.13	0.07	0.05	0.02	0.01	0.03	0.01
Rajouri	N.A.	N.A.	N.A.	N.A.	0.26	0.00	0.10	0.06	N.A.	N.A.
Srinagar	1.42	0.98	0.07	0.09	0.06	0.05	0.02	0.01	N.A.	N.A.
Budgam	N.A.	N.A.	N.A.	N.A.	N.A.	N.A.	N.A.	N.A.	N.A.	N.A.
Anantnag	0.79	0.60	0.05	0.02	0.01	0.00	N.A.	N.A.	N.A.	N.A.
Pulwama	N.A.	N.A.	N.A.	N.A.	N.A.	N.A.	N.A.	N.A.	N.A.	N.A.
Baramula	1.42	0.95	0.14	0.10	0.02	0.01	N.A.	N.A.	0.01	0.01
Kupwara	N.A.	N.A.	N.A.	N.A.	N.A.	N.A.	0.10	0.05	0.04	0.06

(Contd.)

Table 5.9 : *(Contd.)*

Leh	24.53	24.52	18.40	20.85	46.15	48.39	55.87	65.59	54.38	70.61
Kargil	N.A.	N.A.	N.A.	N.A.	N.A.	N.A.	0.16	0.63	0.00	0.00
Almora	N.A.	N.A.	N.A.	N.A.	N.A.	N.A.	6.37	7.00	5.11	4.57
Chamoli	N.A.	N.A.	N.A.	N.A.	N.A.	N.A.	4.38	5.30	3.80	4.47
Dehradun	7.56	9.83	6.26	9.57	5.77	8.80	11.66	8.15	4.20	4.70
Garhwal	N.A.	N.A.	N.A.	N.A.	N.A.	N.A.	2.14	2.91	7.93	7.57
Nainital	4.17	4.63	1.59	1.40	0.51	0.17	0.75	0.48	0.65	0.35
Pithoragarh	N.A.	N.A.	N.A.	N.A.	N.A.	N.A.	6.64	8.04	6.45	6.55
Tehri Garhwal	N.A.	N.A.	N.A.	N.A.	N.A.	N.A.	6.56	6.60	4.79	3.84
Uttarkashi	N.A.	N.A.	N.A.	N.A.	N.A.	N.A.	1.77	1.74	0.91	0.99
Total	6.80	5.09	3.78	4.13	3.21	2.48	3.04	2.22	2.59	1.78

N.A. : Not Available.

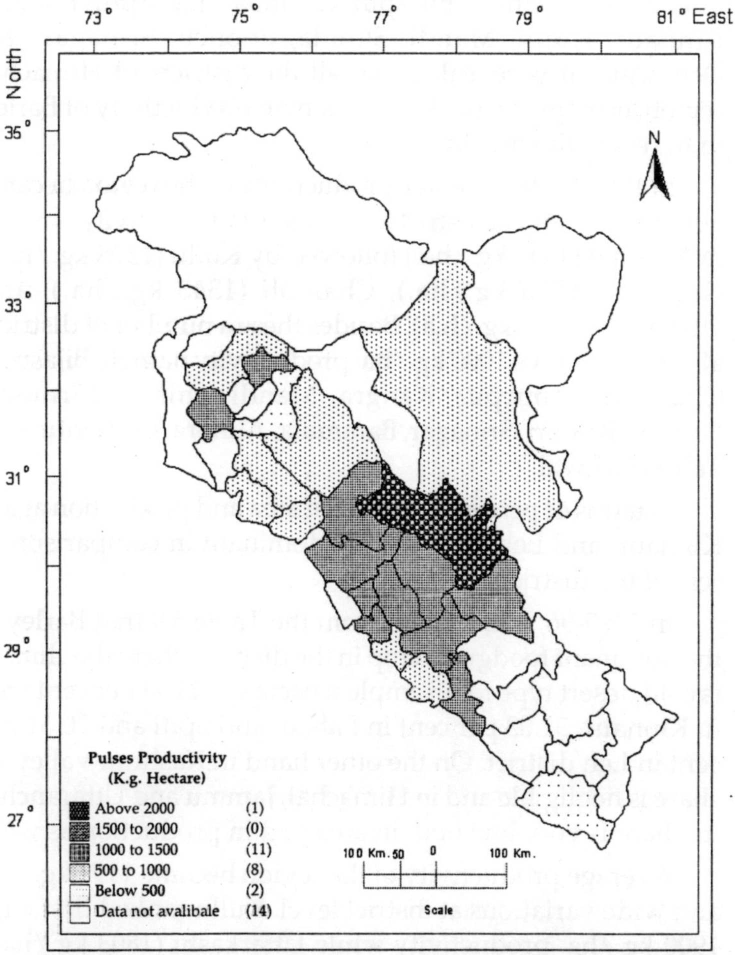

Fig. 5.23 : North-western Himalaya : Productivity of Barely (1974-75)

In 1974-75, highest productivity of Barley is noted in Kinnaur where area and production shares are 53.37 per cent and 55.36 per cent respectively. Leh maintained its dominance with 46.15 per cent area and 48.39 per cent production in total foodgrain. All other districts are very insignificant in barley with no one have even 15 per cent share either in area or in production. The number of districts above 1000 kg./ha. productivity is found in 11 districts namely Bilaspur, Chamba, Hamirpur, Kangra, Kinnaur, Kullu, Mandi, Shimla, Poonch, Srinagar and Dehradun. It is revealed that all the districts of Himachal region are comparatively dominant in productivity of Barley than rest of the region.

In 1985-86, the average productivity of the region became 1111 kg./ha. and at district level highest productivity is noted in Nainital (1443 kg./ha.) followed by Kullu (1398 kg./ha.), Garhwal (1390 kg./ha.), Chamoli (1343 kg./ha.), and Uttarkashi (1326 kg./h.a). Besides these a number of districts also found above 1000 kg./ha. productivity namely Bilaspur, Chamba, Hamirpur, Kangra, Mandi, Shimla, Sirmour, Poonch, Rajouri, Srinagar, Baramula, Almora, Dehradun and Tehri Garhwal.

In terms of percentage share of area and production again Kinnaur, and Leh are the most dominant in comparison to rest of the districts.

In 1995-96, it is evident from the Table 5.9 that Barley is the dominant foodgrain crop in the district where the climate is cold desert type for example it occupies 44.44 per cent area in Kinnaur, 31.03 per cent in Lahaul and Spiti and 70.61 per cent in Leh district. On the other hand in Kashmir valley its share is negligible and in Himachal, Jammu and Uttaranchal its share is very low both in area and in production as well.

Average productivity of the region became 1199 kg./ha. and wide variations at district level. Kullu ranked first with 1900 kg./ha. productivity while Uttarkashi (1803 kg./ha.), Dehradun (1741 kg./ha.), Chamoli (1621 kg./ha.), Kupwara (1523 kg./ha.) and Shimla (1505 kg./ha.) comes next in descending order.

Considerably high productivity is also found in Uttaranchal, Ladakh and Himachal region while Jammu region recorded low productivity and in Kashmir region either it is absent or in very poor condition.

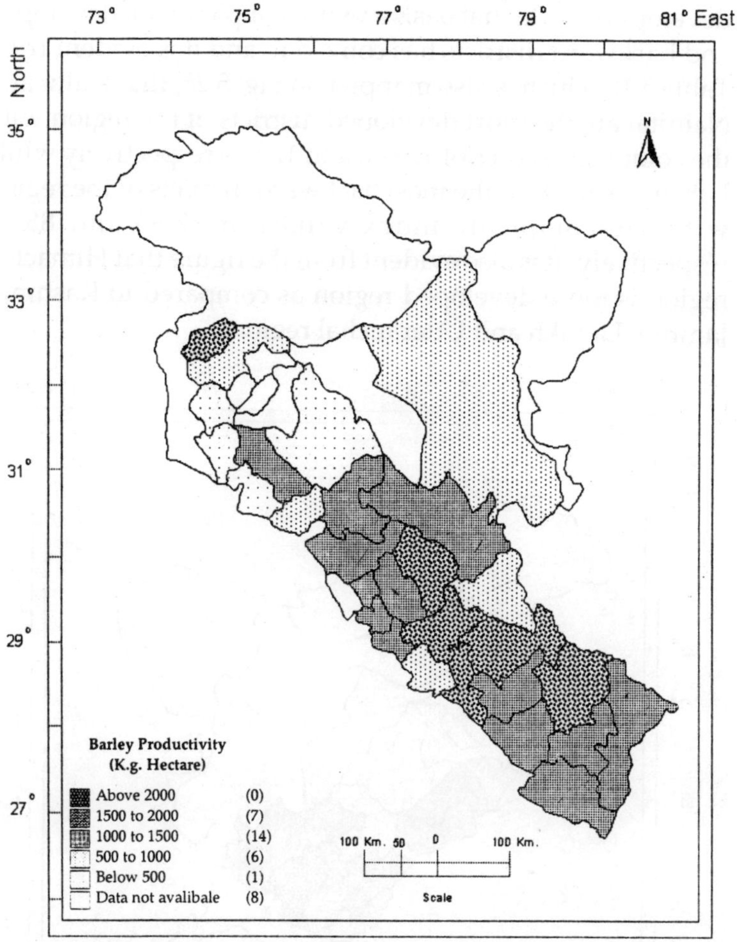

Fig. 5.24 : North-western Himalaya : Productivity of Barley (1995-96)

Inequalities based on Composite Development Index

Taking all those inputs of agricultural development which have been analysed above, along with the productivity of cereals and pulses and their composite development an index is development to find out the inequalities in agricultural development. On that basis a very clear picture of developed and backward districts has come out, and it is evident from Table 5.10 which is also mapped on Fig. 5.25, that Kullu and Nainital are the most developed districts of the region with the composite scores of 123.66 and 123.36 respectively while Leh and Kargil are the most backward districts of the region with the composite index values of 87.80 and 88.83 respectively. It is also evident from the figure that Himachal region is more developed region as compared to Kashmir, Jammu, Ladakh and Uttaranchal regions.

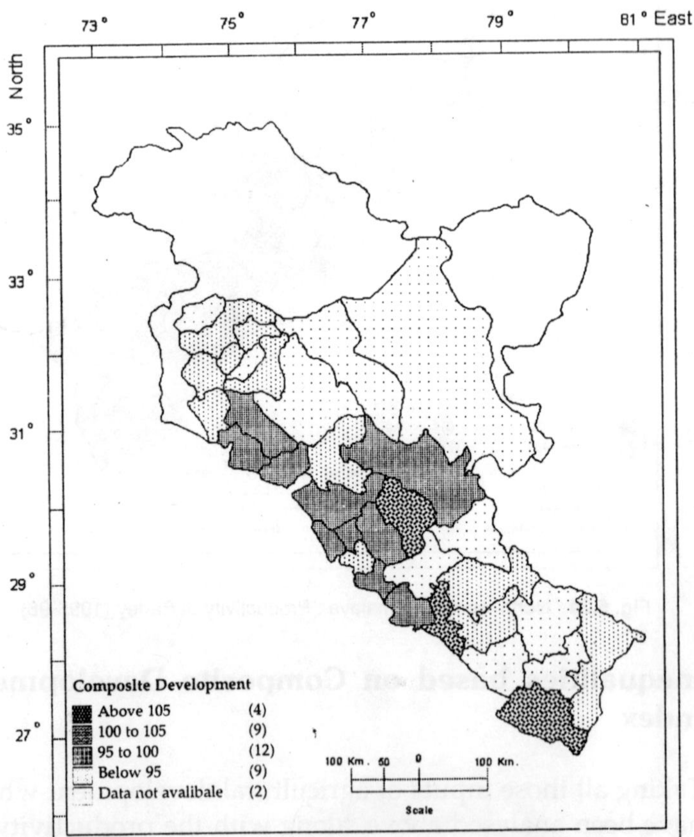

Fig. 5.25 : North-western Himalaya : Composite Development index

Table 5.10 : Composite Agriculture Developmental Status of North-western Himalayan Region

Districts	*Maize productivity*	*Rice productivity*	*Wheat productivity*	*Barley productivity*	*Pulses productivity*	*Foodgrains productivity*	*Cropping intensity*	*Forest cover*	*Cultivated land*	*Wooden Plough*
	1	2	3	4	5	6	7	8	9	10
CoK	N.A.	N.A.	N.A.	N.A.	N.A.	N.A.	N.A.	N.A.	N.A.	N.A.
PoK	N.A.	N.A.	N.A.	N.A.	N.A.	N.A.	N.A.	N.A.	N.A.	N.A.
Bilaspur	0.14	-0.88	-0.45	0.74	-0.76	-0.58	1.14	-0.77	0.70	-0.21
Chamba	1.62	-0.44	-0.10	0.43	-0.63	0.71	-0.14	0.40	-1.20	-0.07
Hamirpur	0.19	-0.36	0.31	0.91	-0.76	0	1.31	-0.45	1.22	-0.24
Kangra	-0.06	-0.07	0.47	0.34	-0.79	-0.15	0.92	0.38	0.03	-0.54
Kinnaur	0.64	-2.07	-0.71	0.05	1.93	-1.61	-0.66	-0.77	-1.41	-0.34
Kullu	0.81	-0.01	1.29	1.59	0.10	0.67	-3.46	-1.16	-1.05	5.28
Lahaul and Spiti	-1.85	-2.07	1.01	0.54	3.11	-0.29	-1.11	0.87	-1.51	-0.02
Mandi	1.77	0.01	0.22	0.17	-0.69	0.67	0.71	0.56	0.20	-0.21
Shimla	0.54	-0.24	-0.18	0.92	-0.58	-0.47	-0.17	-0.14	-0.55	-0.26
Sirmour	1.46	-0.79	0.54	-0.01	-0.81	0.61	1.03	-0.31	-0.08	-0.37
Solan	0.18	1.18	0.78	0.66	-0.90	0.17	0.62	-0.72	0.03	-0.46
Una	0.18	1.06	0.79	-1.66	-0.99	0.32	0.61	-0.43	0.41	-0.56
Jammu	0.09	0.86	1.13	-0.80	-1.57	0.03	0.95	-0.66	1.10	-0.66
Udhampur	1.44	0.22	0.45	0.56	-0.54	1.00	0.56	0.59	-0.44	-0.12
Kathua	1.15	0.43	0.83	-0.27	-1.13	0.12	1.55	-0.09	0.33	-0.54
Doda	0.47	-0.08	-0.19	-1.04	0.04	-0.11	-0.69	0.96	-0.82	0.43
Poonch	1.75	1.12	0.78	-0.80	-0.86	2.14	-0.39	0.01	-0.19	0.36
Rajouri	1.20	0.59	0.86	-0.80	-0.59	1.13	0.73	0.31	-0.11	-0.39
Srinagar	-0.52	1.44	-1.36	-1.66	0.23	0.96	-0.74	-1.03	0.75	0.14
Budgam	-0.98	0.48	-0.92	-1.66	1.73	1.28	-0.92	-1.13	1.66	-0.15
Anantnag	-0.85	0.24	-0.28	-1.66	1.08	-0.59	-0.51	-1.15	2.14	-0.11

(Contd.)

Table 5.10: (Contd.)

	1	2	3	4	5	6	7	8	9	10
Pulwama	-0.45	0.99	-0.95	-1.66	-0.40	0.46	0.06	-1.14	1.96	-0.20
Baramula	-0.58	0.58	-1.34	0.05	1.20	-0.29	-1.22	-1.14	1.00	0.43
Kupwara	-1.07	0.27	-1.09	0.95	-0.19	-1.19	-1.16	-1.12	1.38	1.07
Leh	-1.85	-2.07	-2.43	-0.12	-0.38	-2.02	-1.15	-1.17	-0.61	-0.17
Kargil	-1.85	-2.07	-2.43	-1.66	0.74	-2.13	-0.98	-1.15	0.58	0.21
Aimora	-0.37	-0.30	-0.43	0.12	0.19	-1.03	0.55	0.38	-0.82	-0.25
Chamoli	-0.39	-0.13	-0.03	1.10	-0.11	-0.41	0.18	1.30	-1.22	0.00
Dehradun	-0.55	0.43	0.62	1.32	0.22	-0.11	0.41	1.59	-0.38	-0.59
Garhwal	-0.53	-0.58	-0.24	0.20	-0.14	-0.86	0.21	1.20	-0.83	-0.21
Nainital	-0.33	2.13	2.11	0.81	1.44	2.30	0.55	1.120	0.50	-0.76
Pithoragarh	-0.33	-0.11	0.23	0.75	0.27	-0.32	0.78	0.9	-0.72	-0.19
Tehri Garhwal	-0.55	-0.05	0.03	0.20	0.45	-0.54	0.33	1.59	-0.72	-0.14
Uttarkashi	-0.49	0.30	0.70	1.43	0.09	0.14	0.11	2.38	-1.34	-0.16

(Contd...)

Table 5.10: (Contd)

Iron Plough	Pumping Sets Index	Sugarcane Crusher	Tractors	Wheat Thresher	Rice Thresher	Threshers and Shellers	Irrigated Area	Fertilizer Consumption	Composite Development Index
11	12	13	14	15	16	17	18	19	20
N.A.	N.A.	N.A.	N.A.	N.A.	N.A.	N.A.	N.A.	N.A.	N.A.
N.A.	N.A.	N.A.	N.A.	N.A.	N.A.	N.A.	N.A.	N.A.	N.A.
-0.48	-0.09	1.24	-0.17	0.47	0.14	0.44	-0.69	-0.72	99.21
-0.01	-0.36	-0.59	-0.16	-0.17	0.40	-0.19	-0.62	-0.68	98.19
-0.20	-0.36	1.78	-0.51	0.53	0.82	0.47	-0.85	-0.83	102.97
1.26	-0.16	0.56	0.04	0.08	2.75	0.17	-0.02	-0.26	104.96
-0.42	-0.36	-0.59	-0.51	-0.36	-0.60	-0.44	0.81	0.32	92.89
3.81	-0.08	-0.59	2.01	5.41	3.82	5.17	0.17	-0.12	123.66
0.30	-0.36	-0.59	0.14	0.34	-0.6	0.29	2.21	1.30	101.72
-0.39	-0.34	-0.53	-0.31	0.15	0.98	0.08	-0.46	-0.57	102.01
-0.69	-0.33	-0.59	-0.51	-0.14	-0.13	-0.24	-0.75	-0.77	94.72
-0.49	-0.23	2.23	0.34	-0.13	-0.09	0.04	0.08	-0.19	102.82
1.12	0.56	0.14	0.57	0.16	1.25	0.29	-0.18	-0.37	105.09
0.02	0.56	3.01	1.03	0.34	0.24	0.62	-0.50	-0.59	104.49
1.99	0.07	1.84	-0.02	-0.22	-0.16	-0.12	0.81	0.32	104.98
0.30	-0.35	0.80	-0.45	-0.34	-0.51	-0.35	-0.81	-0.81	101.18
0.76	-0.24	1.37	-0.29	-0.34	-0.29	-0.30	-0.11	-0.32	102.59
-0.77	-0.36	-0.59	-0.51	-0.36	-0.59	-0.44	-0.62	-0.68	94.04
-0.92	-0.35	0.46	-0.50	-0.36	-0.50	-0.44	-0.97	-0.92	99.41
0.20	-0.32	-0.59	-0.47	-0.35	-0.54	-0.42	-0.97	-0.92	98.54
0.72	-0.20	-0.59	-0.37	-0.36	-0.51	-0.42	1.45	0.77	98.69
0.42	0.71	-0.59	-0.40	-0.36	-0.26	-0.42	-0.97	-0.92	96.59
-0.31	0.26	-0.59	-0.27	-0.36	-0.47	-0.40	1.35	0.70	98.23

(Contd.)

Table 5.10: *(Contd.)*

11	12	13	14	15	16	17	18	19	20
-0.28	-0.35	-0.59	-0.11	-0.36	-0.60	-0.44	-0.97	-0.92	94.03
0.61	0.52	-0.59	-0.22	-0.36	-0.41	-0.44	0.65	0.21	98.65
0.87	-0.35	-0.59	-0.49	-0.36	-0.60	-0.44	-0.97	-0.92	93.97
-0.92	-0.31	-0.59	-0.51	-0.36	-0.60	-0.44	2.21	1.30	87.80
-0.92	-0.36	-0.59	-0.51	0.36	-0.60	-0.44	2.12	1.24	88.83
-0.88	-0.35	-0.59	-0.50	-0.36	-0.59	-0.44	-0.59	-0.65	93.07
-0.76	-0.36	-0.59	-0.51	-0.36	-0.60	-0.44	-0.77	-0.78	95.11
-0.37	-0.06	-0.37	1.63	-0.05	1.14	0.27	0.40	0.040	105.59
-0.80	-0.36	-0.59	-0.44	-0.36	-0.39	-0.42	-0.58	-0.65	93.61
-0.19	5.37	-0.14	4.49	0.44	-0.06	1.20	1.55	0.84	123.36
-0.86	-0.36	-0.54	-0.50	-0.36	-0.60	-0.44	-0.69	-0.72	96.19
-0.92	-0.36	-0.59	-0.51	-0.36	-0.60	-0.44	-0.44	-0.55	95.81
-0.84	-0.35	-0.59	-0.51	-0.36	-0.60	-0.44	-0.28	-0.44	98.73

N.A : Not Available

POSSIBILITIES OF AGRICULTURAL LAND EXTENSION

One of the most serious problems of the country today is its foodgrains demand of the growing population. In attempt to overcome this demand our agro-ecology had deteriorated to a considerable extent. It is therefore, imperative to plough the potential lands which are not under cultivation. The potential lands which is 23.74 per cent of the reporting area in the region during 1992-95, if put at par with cultivated land it can increase the production of the region by more than 20 per cent.

Identification and management of potential lands has significance in view of increasing pressure of population on agricultural land and resultant process of land degradation. Besides, there is no standard criteria for the identification of such lands because concept and meaning of potential lands vary place to place and from time to time, mainly due to variations in people's perception of their problems, needs and priorities. For some special reasons only those categories of land are included under potential lands which are brought under cultivation for raising the foodgrain output to meet the food demand of increasing population which includes pastures, land under miscellaneous trees and groves, culturable wastelands and fellow as well as current fellow lands. Although potential lands can be cultivated but at the cost of livestock population and a large number of people below poverty line which earn their livelihood from these cultivable lands.

Pastures are those lands which cover all grazing lands whether they are permanent pastures and meadows or not and village common grazing lands. The areas lying between agricultural land and reserved forests and which are not suitable for cultivation or from where forests have been cleared and the agricultural land which have been left fallow for several years have been categorized as pasture lands. These areas are left open for grazing for the village cattle.

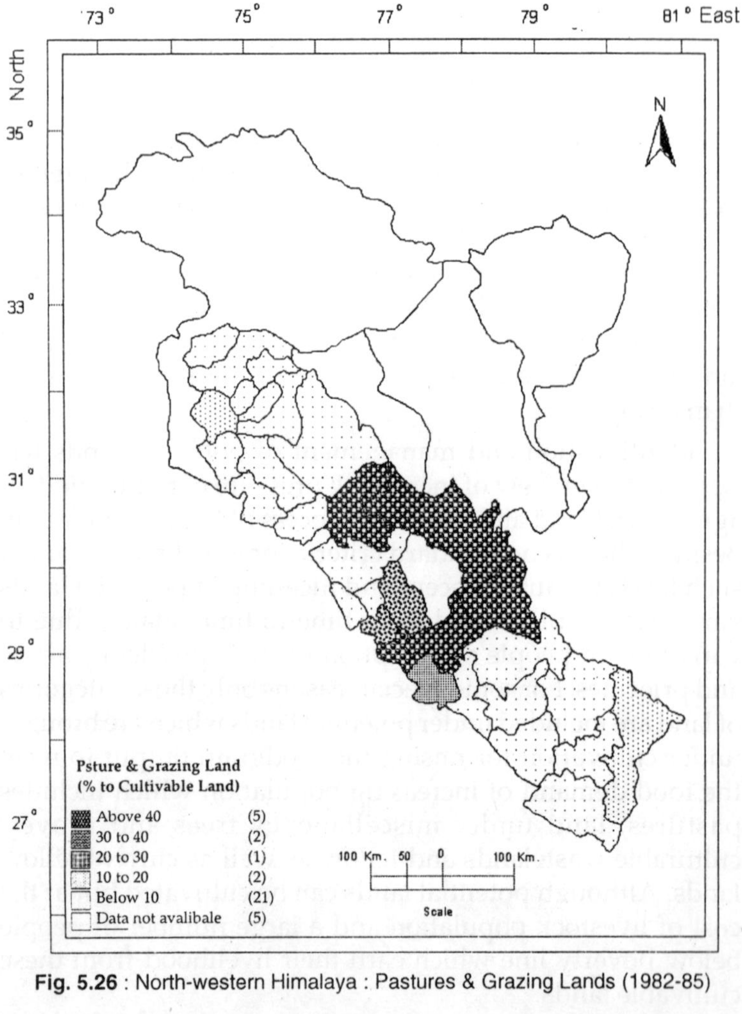

Fig. 5.26 : North-western Himalaya : Pastures & Grazing Lands (1982-85)

During 1992-95, it contributed more than half of total potential land i.e. out of total cultivable land of 23.74 per cent, pastures and grazing lands contribute 77.21 per cent in Himachal region, 14.64 per cent in U.P. Himalaya and about 5.11 per cent in Jammu region while 3.03 per cent in Kashmir region.

Culturable wastelands are those lands which area capable or have the potential for the development of vegetative cover and is not being used due to different constraints of varying degree are termed as culturable wastelands. It is seen that the culturable wastelands have a share of about 5 per cent in the landuse pattern. Out of this 5 per cent share of the culturable wasteland in the region as a whole 55 per cent is noted in the U.P. Himalaya which means a great scope for agricultural land extension in this region while the possibilities of its extension of Himachal region is 21 per cent, in Jammu region 17 per cent and in Kashmir region it is only 6.16 per cent, which is a indicator of good cultivation strength in the region.

Fellow lands which are only 2.22 per cent of the reporting area during 1992-95 contribute a share of 9.3 per cent of potential lands in the region as a whole and if considered 9.3 per cent as unity. 31 per cent of fellow land noted in Jammu region followed by 28.73 per cent in U.P. Himalaya, 29.42 per cent in Himachal region and 10.85 per cent in Kashmir region. So there is more or less same scope for agricultural land extension in all the region except the Kashmir region where the percentage of fellow land is less means high intensity of landuse.

But if we put these potential lands (Pastures and Grazing lands, culturable wastelands and Fellow lands) under plough how will our livestock population survive and where do get poor people their livelihood and how will ecology of the region maintain its balance. It must therefore, be necessary to reclaim the wastelands which is not an easy task and warrants careful approach with a defined methodology. The basic requirement for the development of wastelands is the recreation of a suitable vegetative cover which not only prevents further degradation but also improves the environment. In the lasts few decades there has been a rapid growth in population in Himalayan region. This growth has put more pressure on farming system to meet food, fodder and fuel needs. Crop, livestock and forestry have formed an integral part of the mountainous farming system. Varying degree of slope, soil infertility, harsh climate and water

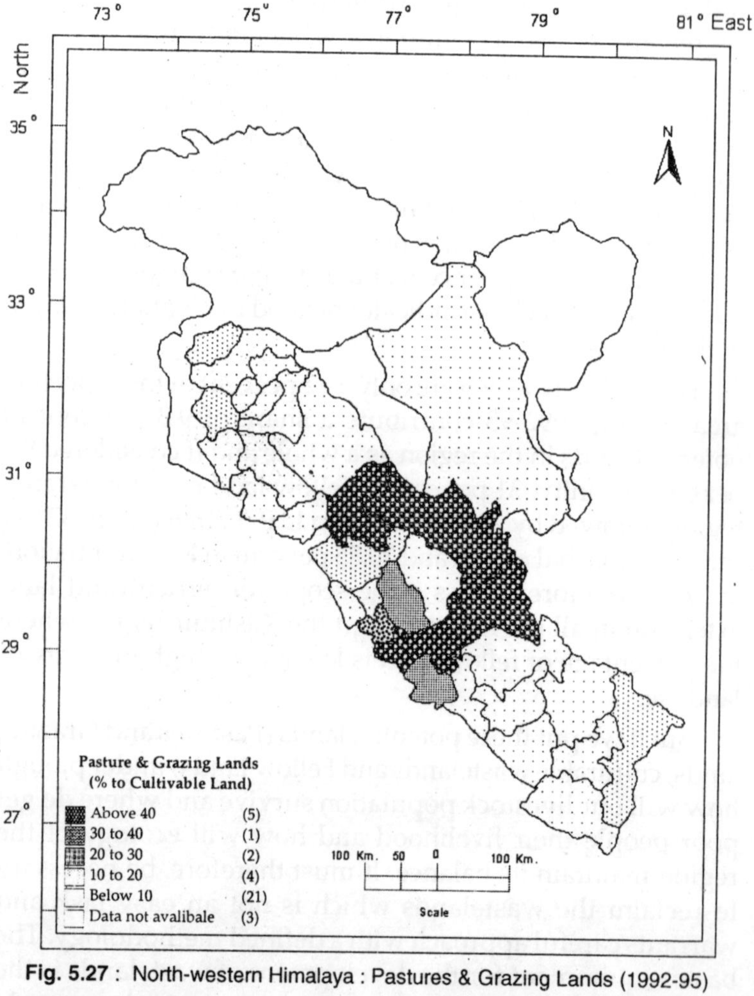

Fig. 5.27 : North-western Himalaya : Pastures & Grazing Lands (1992-95)

resource management limits the agricultural land extension in the highlands. A number of measures had been taken to manage these natural resources judiciously.

Maintenance of Soil Fertility

A number of different measures were employed by upland farmers to maintain soil fertility. The most important was the use of compost which involved using livestock manure, leaf litter from forests and crop residues from fields. Fellow land and shifting cultivation is also practised to recoup the soil fertility sometimes in this region.

Control of Soil Erosion

On sloping areas, it is almost impossible to completely halt soil erosion. However, in terms of the possibilities from the farmer's side, considerable efforts were made in the past to control soil erosion through different measures. Terracing is very significant in upland areas depending on soil characteristics, geology and the slope. This ethno-engineering knowledge has yet to be properly trapped. The use of different type of biomass has also been important for regulating run-off and controlling soil erosion. Shifting cultivation is another strategy for controlling soil erosion as well as improving soil fertility.

Management of Water Resources

The movement of water from one place to another on sloping terrain increases the risk of soil erosion and landslides, however, without water there can be no agriculture. In the mountainous areas the main challenge for water management has been to organize safe drinking water during a three-four month surplus period and frugal use of available water over a nine month dry period.

Climatic Preferences

Harsh climate of the high land areas also set limits for the agriculture in this mountainous belt. The growing season is too short for proper agricultural practices besides severe cold and accessibility of sun shine on the slope is a matter of concern and set limits before the farming communities.

Besides food crops, animal rearing, horticulture, floriculture and other form of agriculture diversification has tremendous scope to raise the economic status of the region if practiced judiciously.

POSSIBILITY OF AGRICULTURAL DIVERSIFICATION

Potential of Saffron Cultivation

Since the extension of agriculture in the new areas demands heavy investments especially in mountainous region. It is imperative to intensify agriculture by judicious utilization of land resources and by adopting suitable varieties of cereal and non-cereal crops which are eco-friendly for the agro-ecology of the region. The agro-ecology of the Kashmir valley especially of Karewas region of Pampur is most favourable for the saffron cultivation. Other traditional areas of saffron cultivation in the world are Iran and France.

Saffron is a leading cash crop of Kashmir valley. Its cultivation and allied operations provide employment to about 15 per cent of the rural workforce. More it constitute about 25 per cent of the total agricultural income of the Kashmir region. Kashmiri saffron has a very high reputation in the international market. It is used in more than one ways to give flavour and taste to the food. It is also needed as a raw material in the pharmaceutical industry for the manufacturing of invigorating, life saving and curative medicines.

Despite its high remunerative value and high consumption, saffron cultivation is confined mainly to the Karewas of Pampur. In recent past attempts have been made to diffuse it in the neighbouring Karewas and success has been achieved in some of the cases. The cultivation of saffron has now been diffused in many non-traditional areas. At present, it covers about 1.735 per cent of net sown area in the Kashmir valley which account for about 4.46 per cent of total area under commercial crops. The Saffron cultivation has spread over four out of six districts of Kashmir valley and covered an area of about 4406 Hectares.

The physical environment and socio-cultural milieu do not permit the farmers to diffuse new agricultural technology in their holdings. Consequently, the farmers are growing either paddy or maize and fodder crops. There are however, some lecustrine deposits locally known as Karewas are ideally suitable for the cultivation of saffron. It is very expensive condiment, which fetches handsome amount to its growers. It is noted that the productivity per hectare of saffron is declining in the traditional saffron growing areas of Pampur. The basic problem of saffron crop is its poor scientific base, as a result the growers could not adopt new scientific methods of cultivation. The traditional method of cultivation has kept it a labour intensive crop which ultimately effect on crop returns. The grower is not familiar with proper dosage of fertilizer and deficiency of nutrients in the Karewa soil. The crop is frequently damaged by various diseases and cultivators are helpless. Unfortunately, the high yielding varieties of saffron have not been developed. The inadequacies of the fund is however the main hindrance in the development of seeds of this precious crop. The seeds of saffron are quite expensive and their requirement per acre is very high. The prohibitory price is the single most important barrier in the spread of the this crop in unexploited areas of Karewas.

Potential of Dairy Farming

Animal husbandry is one of the important sub-sector of agricultural economy and plays an important role in the rural economy by providing gainful employment for the weaker sections i.e. small and marginal farmers and agricultural labourer. According to the National Sample Survey Organization (NSSO) an annual growth rate in employment in livestock sector in rural areas during 1972-88 was 4.15 per cent, which is much higher than the rate of growth in agriculture i.e. 1.10 per cent during the same time period.

The prospect for creating a regional balance between the rural and urban sectors can be achieved through the proper exploitation and management of local resources and development of various agro-based industries. The ventures should be made to develop the allied sectors i.e. agro-forestry, apiculture, pisciculture, sericulture, horticulture, floriculture, mushroom cultivation and dairy farming. These measures help in income generation besides being able to provide employment to a large number of rural population.

India has the largest cattle population in the world. Out of the world's total 1430 million, the total cattle and buffalo population in India is estimated at 271 million (Rajan, 1995). It means India maintains 20 per cent of total bovine population but produces hardly 10 per cent of the World's milk.

The per capita availability of milk in India is 185 grams per day or 68 kg./year which is much lower than the world average of about 100 kg./year. The Nutritional Advisory Committee of Indian Council of Medical Research has recommended a minimum quantity of 250 grams/day or 91 kg./year. This shows a wider gap between availability and requirement.

In this context of improving the nutritional standards of the people, we need for enhancing the production in various animal products particularly meat and milk. Besides providing nutritious human diet and protein through milk and meat, this sector also provide wool, skin, hides, and manures for the fields.

Potential of Poultry Farming

Poultry farming has great potential in the region. Agro-ecology and cold climate of the region sets demand of poultry products throughout the year. The region where terrain sets limits for extension in cultivated land, poultry farming help in producing calorific food requirement for a balanced diet.

Poultry rearing was practiced on a small scale for a long but commercial poultry production gained significance only over the last three decades. As a result, it is seen that India ranked sixth in the annual egg production and eighteenth in the broiler production in the world.

Potential of Horticulture

Horticultural production constitutes a highly specialized branch of agriculture and the intensive, market oriented nature of the industry is well expressed both by the British term Market-Gardening and North American equivalent Truck-Farming. With horticulture there is a particular incentive to make the optimum use of the climatic environment in order to produce crops of high yield and high quality. In areas which have short growing season like north-western Himalaya, great emphasis is also attached to early cropping in order to avoid the seasonal glut when supply may be in access of demand. The requirement means that horticulture is often concentrated where early growth is associated with a milder winter climate.

The promotion of horticulture in the mountainous areas falls on the national priority because undulating topography in mountainous region are more suitable for raising horticultural crops as it does not require the frequent tillage and large extent of levelled land. The development of horticulture is not only supplement of the national food requirement but also provides nutritious food in the form of fruits and vegetables and also plays a vital role in promoting environmental conservation in the mountainous region.

The increasing pressure of human population and wide extent of malnutrition calls to increase food production. This can be achieved only through the scientific knowledge and judicious use of natural resources. The failure of conventional agriculture to meet the nutritional needs of the growing population, the far-sighted agriculturists and horticulturists are turning their attention towards agriculture diversification like tree crops. Trees offer possibilities or higher quantity of food production per unit as compared to traditional agricultural crops.

The cold climate and well drained soils provide ideal agro-ecological conditions for the development and production of a wide range of fruit crops and vegetables in this mountainous region. The tree crops, plants of ornamental value, off-season vegetables and other horticulutral crops grown, have wide scope in markets of plain region. The temperature fruits like apple, peach, pear, plum, apricot, cherry, almond and walnut, and subtropical fruits like mango, litchi, guava, fig, pomegranate, olive and citrus fruits have per unit area is very high. There is a lot of variations in horticulture system and nature in its production according to altitudinal zones.

The sub-tropical zone ranging between 365 to 914 metres of height and average annual precipitation vary from 60 to 100 cm. The zone is potentially suitable for growing high quality mango, litchi, guava, fig and a variety of citrus fruits and varieties of peach, pear and plum are also grown here.

The sub-temperate zone ranging between 914 to 1523 metres has a moderate climate with an annual precipitation between 90 to 100 cm. This zone is favourable to grow peach, plum, apricot, pear and pomegranate. This agro-ecological zone has climatic similiarities with Mediterranean where rain occur in winter while summer are dry. Besides above crops olive and pista can also grown here.

The temperate zone ranging between 1523 to 2742 metres

and annual rain ranging between 90 to 100 cms. Snow in winter is a usual feature. Agro-ecology of this region is suitable for apple and pear besides cherry, walnut and chestnut.

The cold dry zone bordering with Tibet is extremely cold. The minimum temperature, on an average comes down to –15°C. The height ranges between 2742 to 3656 metres and rainfall varies from 25 to 40 cm annually. The ago-ecology of this region is suitable for growing quality dry fruits such as raisin, grapes, prunes, almonds, chilghoza, walnut and a varieties of apricots and apples. The apples grown here are of the best known quality in the world.

It is also noticed that tree crops are less affected by drought than annual crops and play important role in conservation of soil on sloppy grounds. Moreover, fruits crops offer better income prospects than any other agricultural activity as well as the capacity to generate supplementary sources of employment in rural areas.

The development of horticulture by all section of people is very much helpful for their economic upliftment. Horticulture not only contribute towards the achievement of broad national objectives but also increases the food production, productivity and employment. It is therefore, necessary to expand horticulture which is only possible through the improvement in existing infrastructure.

Potential of Floriculture

Floriculture is relatively new ancillary activity of horticulture, offers much potential in the mountains. Around 1700 species, of beautiful mountain flowers are available for diversification of farming into floriculture in the Himalayas are engaged in the flower trade. Cut flowers are sent to the cities in the plains during the off-season. The north-western Himalayas supply such flowers as roses, gladioli, lilies, narcissi, daffodils, carnations and chrysanthemum while the Kumaon Hills supply gladioli, orchids, gerbera, magnolia, camellias, irises, geraniums and other temperate species. Roses offer great scope for farming because of their tremendous genetic diversity in the Himalayan region. The existence of races with a good potential for rose oil brightens the chances of increasing yields to economically acceptable level.

After achieving the self-sufficiency on the food-front, the focus of Indian agriculture has shifted to other field of diversification and commercialization. One of the options emerged is in the field of floriculture. It is field of ornamental horticulture which consists of floriculture and landscape horticulture. Each is concerned with the associated activities of flower arrangement and landscape design respectively. According to *Encyclopedia Britannica*, 'Floriculture deals with the production of flowers and ornamental plants, generally cut flowers, pot plants and greenery'. The floriculture industry of India consists of florist trade, nursery plants, potted plants, seeds and bulbs production, micro-propagation and extraction and essential oils from the flowers. Floriculture has become a potential for the developing countries.

The prospects of floriculture in this region is very bright. This region endowed with diverse ecological region, varied climatic conditions favours the rich native flora which facilitate the availability and production of wide range of floriculture products throughout the year. Besides, it has fertile soils, abundant sunshine, sufficient water and cheap and skilled labour is also available easily. The agro-ecology of the region is like a mine for the varieties of ornamental plants such as Rhododendrons, Orchids, Primula, Lonicera, Aster, Iris, Begonia, Impassions, Hydechium, Pandanas, Gardenia, Lillium, Barleria, Ipomea, etc.

The world-wide consumption of floriculture product is estimated about 40 billion dollars, cut flowers contribute nearly 60 per cent. The U.S., Japan and Western Europe are the major markets for floriculture products. The global floriculture industry is growing at an annual growth rate of

10-15 per cent. There has been an increase in the per capita consumption of floricultural products in the developed and developing countries.

The Netherlands is the largest trader of floriculture product. But export from Netherlands to distant markets such as Japan and U.S. has been reduced. The East European markets are the targets of cut flowers and floricultural products in the U.S., Japan and South Asian countries. Hence, the prospects are bright for Indian floriculture which has unlimited opportunities because of its liberal policies towards horticulture and particularly floricultural trade.

India has immense potential of floriculture to exploit it commercially in the domestic market as well as for export. Estimates and survey shows tremendous growth in area, production and foreign exchange earnings from this sector but its share in the global market is very low i.e. less than one per cent. India's low share in global trade and unlimited potential reveals that still prospects of floriculture industry is very bright.

Potential of Aquaculture

There is wider scope for the fresh water aquaculture in the region in terms of ponds, tanks, canals and reservoir with plenty of water availability. The undulating topography of the region provides innumerable depression in the fold mountain ranges of Himalayas where fish culture can be practiced besides a number of small perennial streams can also be transformed into lakes by forming check-dams on it. Plenty of water throughout the year with natural depression in undulating topography provide the base for fresh water aquaculture in the region.

So it can be said that north-western Himalaya has unlimited unharnessed aquaculture potential which can be exploited on commercial scale if proper management and serious attention be given on it.

Potential of Apiculture

The region is a focal point in the origin and evolution of honey-bees. Modern intensive agriculture with its diversified cropping patterns and orchards is becoming increasingly popular in the mountain ecosystem. The current agricultural transformation, once linked to apicultural operations, offers much scope for income generation through bee-keeping.

Bee-keeping also increases production of fruit and vegetables, particularly cross-pollinated crops such as apples, pears, plums, and litchis and seed production for cabbages, cauliflowers, carrots, turnips, radishes, and other vegatables.

Traditionally, mountain women are rarely involved in bee-keeping perhaps due to ignorance. Bee-keeping requires light labour, no permanent infrastructure and exploits orchards flowers beneficially for extra income generation. Apicultural operations can easily be undertaken by women of mountain households without any extra drudgery.

Honey-bees perform several ecological and economic functions without competing for scarce land resources. The complementarity of bee-keeping with horticulture is scientifically proven, but so far it has not been widely adopted in mountain areas Bee-keeping is also an environmentally friendly activity and it helps to promote off-farm employment and high income opportunities.

Bee-keeping, besides bringing gains such as low inputs and high incomes, also has an important role in the promotion of crop yields through higher levels of pollination. In Himachal Pradesh in India, pollination by bees has been one of the key factors behind the horticultural revolution.

Angora Rabbits for Fibre and Meat Production

Angora rabbit keeping has been successful in the middle and high hills of Nepal and Himachal Pradesh, (India). Hill farmers have realized that rabbits are a potential source of cash income.

Angora rabbits survive and remain highly productive at altitudes ranging from 1,100 to 2,000 metres above sea level. They can also be reared on locally available resources, e.g., food, grass, and building materials (i.e., rabbit hutches can be made out of bamboo). A few years after they were introduced into the hills of Nepal and India, they began to gain popularity rapidly, mainly because of the high-value wool production from this breed of rabbits. A pair of Angora rabbits can produce wool worth thousands of rupees per year. As the wool is extremely light and of high value, rabbit keeping is particularly suitable and profitable for farmers in remote and inaccessible areas. Angora skin in also sold at high prices. The skin is used to make bags, hats, and gloves. The fibre is used to make sweaters. In addition, rabbit manure is valuable for crop production Rabbit meat is yet to gain popularity, but it is a rich source of protein.

Ghorlikharka (*Brassica campestris*) — A Mustard Variety

The *sarson* plant (*Brassica* spp.) is one of the major sources of vagetable oil in the hills of Nepal. The Ghorlikharka variety got its name from Ghorlikharka village, Dhankuta, Nepal, where it has been widely and successfully cultivated for a long time. This variety produces oilseeds at the rate of 1.48 tonnes per hectare with the use of only compost (at the rate of 15 tonnes per hectare) and under rainfed conditions, without using any chemical fertilizers. This option is suitable for resource-poor farmers of India who cannot afford external inputs such as chemical fertilizers and pesticides to increase crop production. According to the Overseas Development and Natural Resources Institute, U.K., Ghorlikharka has the highest oil content (57.1 %) of all the varieties of Sarson (mustard) in the world.

Sunflower Stalks for Crop Support

The use of sunflower stalks after harvest as support sticks for vegetable crops is an eco-friendly, low cost practice in vegetable growing areas. Vegetable farming, with better accessibility, is expanding rapidly in several mountain areas. Although a source of high income, most of the high-yielding vegetable crops, such as tomatoes, beans, cucumbers, and other creepers, have a high servicing cost in ecological terms. They need support sticks which are acquired by cutting trees/ branches, etc. The quantity and quality of the vegetables produced using this practice is also than that of vegetables produced using traditional plant support, thereby also lessening the rate of deforestation and should be adopted in India for eco-conservation.

Off-season Vegetable Production

Vegetable production in cold mountain deserts has traditionally been limited to the domestic scene. In recent years, development of infrastructure resulting from the tourism industry, etc. has created more demand and better marketing opportunities for locally produced vegetables. Vegetable cultivation, particularly, has a great potential for improving income generation in the cold deserts of Ladakh and Himachal Pradesh. In the Ladakh area, cabbage, cauliflower, turnip, knolkhal (*Brassica*), Chinese, radishes, carrot, spinach, methi, coriander, and potatoes are being grown successfully. In addition to local sales, these vegetables are marketed in neighbouring lowland areas as off-season vegetables and fetch good prices. Cauliflowers are especially produced as an off-season vegetable during July through to the end of September.

6

Management of Eco-Conservation and Sustainable Agriculture

CONSERVATION OF NATURAL RESOURCES

Environment can be defined as the totality of man's surrouding. It may be referred as "the sum of all social, biological and physical or chemical factors which constitute the surroundings of man." In the beginning, man was shaped by the environment which controlled and shaped its evolution and man remained in equilibrium within the ecosystem. But in the quest of its development, man broke this ecological equilibrium and exerting an ever increasing influence on environment. Due to growing urbanization, industrialization and in particular agricutural operation, the cultivation is now pushed to forested areas, marginal lands and up-slopes without taking into account the suitability of these lands for agriculture, badly affecting the agro-ecological balance and observed only when it exceeded the limits beyond which it became irreparable. All the developmental works either in agriculture or in any field may have short-term implications. The short-term implications may be negligible as compared to the benefits, but the long-term implications may be summarized in terms of vast areas of deforested lands, denudation of slope, soil erosion, silting of rivers, regular occurrence of floods and droughts, impoverishment of important fauna and flora, change in climate, polluted water unfit to drink and contaminated air unfit to breathe.

Sulphur-dioxide and nitrogen dioxide emitted into the air are converted into acids. At their deposition they have an acidifying effect on soil and water. Volatile organic substances might have ill effects on health. Most of them are carcinogenic. In combination with nitrogen oxides and in sunlight, some of them might form ozone and other photochemical oxidants. These are harmful to plants.

Table 6.1 : Some Major Pollutants and their Sources

Pollutants	*Sources*
Carbon dioxide	Incomplete fuel combustion (e.g., two stroke engine)
Sulphur dioxide	Burning of sulphur containing fuels like coal in power station and oil by vehicles.
Suspended particulate Matter	Smoke from domestic, industrial and vehicular sources.
Oxides of Nitrogen	Fuel combustion in motor vehicles, power stations and furnaces.
Volatile hydrocarbons	Partial combustion of carbonaceous fuels (two stroke engines, industrial processes, disposal of solid wastes).
Oxidants and Ozone	Emission from motor vehicles, photochemical, reactions of nitrogen oxides and reactive hydrocarbons.
Lead	Emission from motor vehicles.

The Greenhouse gases (Co_2 and CH_4 etc.) prevent some of the heat radiation from earth to escape into the space. So the concentration of green house gases is increasing which is presumed to raise the temperature of the earth in a long perspective.

Nutrients mainly nitrogen and phosphorus contribute to eutrophication of lakes, rivers and marine waters. Approximately, half of the nitrogen discharge is estimated to originate from agricultural land.

CONSERVATION OF WATER RESOURCES

Water is the most valuable natural resource of the developed, developing and underdeveloped world. It plays a very crucial role in domestic, agricultural and industrial sectors of activities. In a large measure, the spatial pattern of various activities particularly, agriculture and evolution and growth of human settlements is determined and regulated by the availability of surface and groundwater. The significance of water resources in the regional and economic development of an area hardly needs to be emphasized. Therefore, geohydrological investigations regarding the surface and groundwater potential of the region and planning and management of water resources are of paramount importance, keeping in view the ecologically sustainable economic development of a region. The ever increasing demand of water for agricultural and industrial development and domestic uses could be met by the intensive and extensive use and proper conservation and management of available water resources. Intensive use in this context is referred to as the local and village level planning and management of water resources, particularly to meet the domestic and agricultural requirements, whereas, the extensive use involves the regional and local level management of water resources and their conservation.

The Himalaya being geotechnically unstable, ecologically sensitive and the source region of all big rivers of northern India undoubtedly assumes added significances and critical importance in view of planning, management and conservation of water resources, at national as well as regional or local levels. The natural ecological system of the Himalaya is now passing through a process of anthropogenic transformation, and consequently, reckless deforestation and accelerated pace of soil erosion have considerably reduced the groundwater recharge and brought about major geohydrological disturbances in the entire region.

Precipitation in the form of snow which accumulates along the mountain slopes is the principal source of water in the high ranges of trans – and greater Himalaya. In the lesser Himalayan mountain ranges, rainwater maintains and consitutes the surface flow through overland system and forms the groundwater reserves through percolation and infiltration. Thus the rainwater accumulated through the process of percolation and infiltration when saturated on hard formation often appears along synclines-dip and fault lines, in the form of springs which in fact, are the most important source of water supply in these regions.

Water is a prime, natural, indispensable, finite and vulnerable resource. Covering more than two-third of the earth, water travels from sea to air, land and to sea again in an endless cycle. The tiny fraction suitable for drinking or irrigating is distributed unevenly. Worldwide agriculture uses 69 per cent of the total water available. About 23 per cent is used by industries and 8 per cent domestically. The pattern of use varies according to the economic development, climate and population.

Water problems arise from increasing demands generated by rapid population growth, urbanization, industrialization and irrigation need for additional food production. In many areas excessive pumping of groundwater not only brings down water quality, but also depletes it. This effects sustainability. The capacity of irrigation tanks numbering about five lakhs in the country is shrinking due to siltation and encroachment. Scarcity is noticed even in high rainfall areas. This is due to improper management and poor conservation of rainwater.

The demand of water for irrigation, drinking, industries, power-generation, fisheries, navigation and recreation is increasing day by day. The demands can be met by storing, diverting, conserving and managing useable water resources efficiently besides this construction of temporary and permanent check dams, gully control structure, percolation tanks, contour trenching, contour stone walls, vegetative barriers, catch pits, V-ditches on contour, contour bunds, compartmental bundings, micro-catchment, diversion of water of land areas and other *in situ* moisture conservation measures.

Land under irrigation increased from 22 million hectare in 1951 and touch the maximum feasible area of 113 million hectare at now i.e. 2001 A.D. The method of irrigation is more or less same, but now the farmers are aware of water scarcity and hence, efforts are to maximize the use of water.

A major quantity of water is utilized for irrigation in agriculture. Farmers prefer flood irrigation and the more recently introduced sprinkler and drip irrigation methods. Water is in constant demand for domestic use, animal culture, industries and agriculture.

Its judicious use is strongly recommended in all sectors including crop irrigation. Micro-irrigation or drip irrigation is the latest and most efficient method of water utilization for crop growth. It has high water use efficiency and hence should be adopted on a large scale for various crops.

Efficient Methods of Irrigation for Water Conservation

(i) ***Surface or Gravity Irrigation***: The most common way of irrigation from the olden days consists of allowing diverse streams to flow by gravity from the head of the field in furrows of borders. No basic changes have taken place in this surface or gravity irrigation methods through a long time. Under the gravity system water filters the soil traversing furrows or borders. The access irrigation water stands on the soil surface keeping the soil saturated for a long a time in this method more than 50 to 70 per cent of water is lost due to evaporation, transpiration and by other means. Hardly 30 to 50 per cent water is used by the crop. The irrigation efficiency of this method is lower than 30 to 50 per cent. In this method water distribution is not uniform and sometimes create the problem of salt accumulation on the soil.

The evapo-transpiration requirement for growing paddy is about 800 to 1000 mm, whereas in gravity or surface or flood irrigated areas, farmers used as much as 2000 to 2500 mm, which is not only wasteful but also affects the yield due to the drainage problem. Scientists have found that there is no need to flood the paddy field to a depth of 15 to 20 cm as practiced by farmers and it is enough to irrigate the field up to a depth of 3-5 cms. as soon as standing water disappers. This will reduce the water use by 30 per cent while productivity can be increased substantially. Therefore, it is not necessary to release water in the canal continuously, instead, the intermittent or warabandi system can be followed in all the command areas irrespective of water availability in the reservoirs.

Further, researchers have estimated the water needs of various crops based on the weather in each region of the country. Such information will help to know when to irrigate and how much to irrigate in order to save water and maximize production under surface irrigation for various crops. Orchard crops such as grapes and bananas, instead of covering the entire surface area and channel can be formed to irrigate the crop. Results have indicated that about 25 to 30 per cent of water can be saved without reducing the yield by adopting the above measures for irrigation.

(ii) ***Sprinkler Irrigation*** : It is an advance method of irrigation to increase the water use efficiency and reduces the water loss. In this method water is applied once in 4 to 7 days. This reduces the moisture stress of the crop area to some extent. However, the water application being controlled and only the needed water can be regulated in this system.

Sprinkler irrigation can be used for closely spaced crops such as millets, pulses and oilseeds. Large scale adoption of sprinkler in canal and tank irrigated areas is necessary to increase water use efficiently and to increase the productivity. Sprinkler irrigation can be used to prevent seepage losses and controlled irrigation. About 30 to 40 per cent of water can be saved in this period. This advance method of irrgation has an efficiency of up to 70 to 75 per cent.

(iii) ***Drip or Micro-irrigation*** : The recent advanced method of drip irrigation is the most efficient method of irrigating the crops. Water applied to the crop are near the root zone on a daily basis and as there is no water or moisture stress, the crop growth is not effected at all. The water is also controlled and the required quantity of water is given to each plant based on the evapo-transpiration requirement. This method has an irrigation efficiency of up to 92 to 95 per cent. Drip irrigation is also termed as micro-irrigation. This method can be introduced for all type of crops depending on soil, slope, water sources, farmer's capacity and crop spacing etc. Many new innovations have come in this system including fertigation i.e., applying fertilizer through drip irrigation. The above development will maximize farm productivity with available water.

To allocate water efficiently for crop production it is necessary to know the relationship between water supply and fertilizer use. Based on these results water can be regulated to the crops to obtain more yield from unit quantity of water and fertilizer.

The drip-irrigation is one of the most efficient economical method of irrigation for various crops. Irrigation is one of the limiting factors in crop production and hence while adopting any method of irrigation there are certain criteria to be observed in relation with agronomical practices of drip-irrigation. In drip-irrigation soil-water-crop relationship has to be considered and hence some points are of prime importance like, crop selection, soil selection, water quality, organic manuring and crop geometry, etc.

Problems and Prospects of Water Conservation

The main problem in using water economically is that no single government organization is responsible for this important input to maximize production. The irrigation department is responsible only to provide water to farmers through dams, reservoirs and tanks and does not bother about its optimal use. The agricultural department is only involved in increasing production through seeds, fertilizer and plant production measures. There is no subject matter specially on water management in either departments to advice farmers to operate the system according to needs.

There is enough water in the country to meet the entire demand but it is not equally distributed in all the areas. It is estimated by water experts that fresh water availability per capita per annum should be 1700 cubic metres. If it fall below 1000 cubic metres per person per year a region will experience chronic water scarcity. When it fall below 500 cubic metres per person per year, it is absolute scarcity. Fortunately, India's water availability is more than 2000 cubic metres per person per year. But the problem is that the transfer of water from surplus areas to deficit areas is not in a proper way.

Scientists have studied various options in water management practices both under surface irrigation like sprinkler and Micro-irrigation to reduce water loss and to increase production. But the problem is that the message has not reached to the farmer's level and still the age old methods are being practiced by farmers. This leads to water loss and its unequal distribution. The adverse effects are water-logging, salinity and low productivity. There is growing awareness that the farmers must be involved in all the phases of planning, design, implementation, operation and maintenance for good results. The Non-Governmental Organizations (NGOs) are creating the awareness and bringing pressure on the authorities concerned for transfer of water from surplus to deficit areas.

CONSERVATION OF LAND RESOURCES

The land capability map of India depicts the capability classes of different soils by various colour tints. The capability classification of land is very essential for the purpose of scientific and balanced landuse planning and agriculture. United States Department of Agriculture (USDA). Land capability classification has been modified by NATMO for application to Indian conditions. The grouping of soils into capability classes are based on the evaluation of the effects of one or more main variables mentioned hereunder (a) slope, (b) degree of erosion hazards, (c) danger of overflow and waterlogging, (d) soil depth, (e) soil's texture, (f) soil fertility, (g) drainage system, (h) water holding capacity of soil, (i) nutrient status of soils, (j) climatic condition, (k) land management practices, (l) stoniness of soil, (m) adverse effects of salinity, (n) physical obstacle such as rocks, sands, deep gullies and ravines, etc. The combination of these 14 variables indicate suitability of soils for the following four major uses of land:

(a) Crop Production
(b) Pasturage
(c) Forestry
(d) Preservation of wildlife.

According to the present capability classification, the soil of any region has been divided into two broad groups namely, the arable grades of cultivated land and land unsuitable for cultivation needing ecological improvements and soil management practices. The distribution along with a short description of each capability class is given below :

LAND SUITABLE FOR CULTIVATION

Very Good land

The lands are nearly lavel and erosion hazards (wind or water) are very low. The soil depth is very good which provides the plants sufficient rooting zones for their healthy growth. The water holding capacity of the soils is sufficient. Positive results can be raised by plant nutrients and fertilizers. As a result of all favourable conditions choice for cropping is immense. Such lands are found in pockets on the foothills of Jammu, Dehradun, Garhwal and Nainital.

Good Land

The soils classified and delineated as good land on the map require careful soil management and conservation practices to prevent degradation of the soil and also to maintain air and water relations when the same are put to cultivation. The soils have some limitations, but the practices to apply for removal of these limitations are also easily possible. So, the soils under this class could be used for growing various types of crops and plants without any damage and risk in their use and management. The limitations considered responsible in single or in combinations for differentiating the soils of this class from the Very Good Land are (a) gentle slopes, (b) moderate effect of past and present erosion hazards caused by wind or water, (c) slight to moderate salinity which can be neutralized easily but may occur,

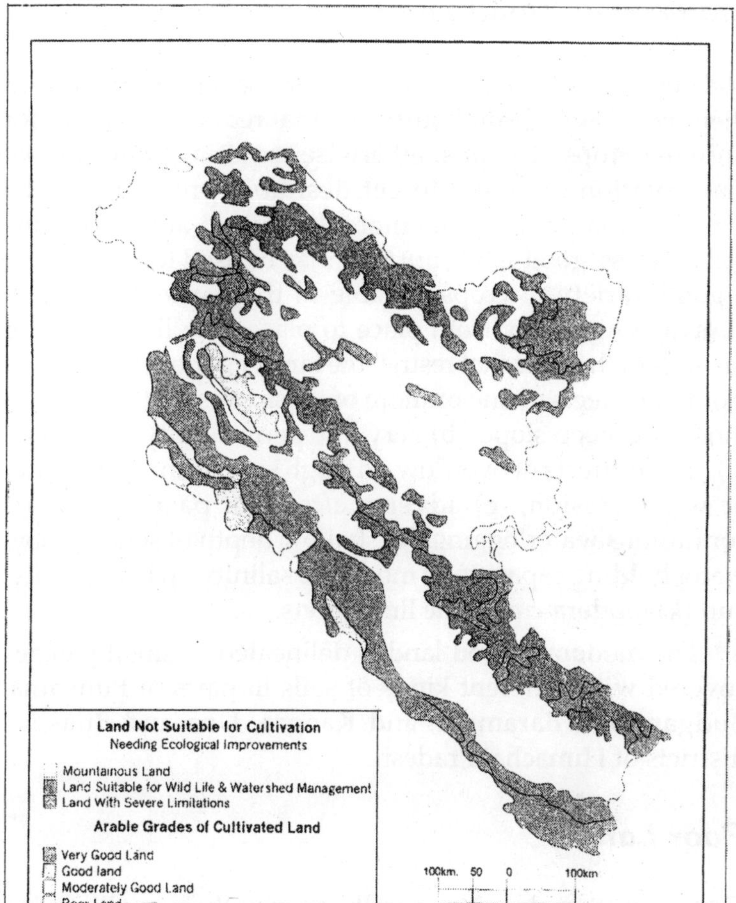

Fig. 6.1 : North-western Himalaya : Land Capability Classification

(d) less than ideal soil cover, (e) reclaimable wetness, (f) moderate permeability, (g) occassional overflow and, (h) adverse climatic conditions. These soils have been demarcated in part of Jammu district. The Kashmir valley is rich with soil of older alluvial soils with a moderate depth, silt and clay loam texture and slightly to neutral acidity, and found in parts of Anantnag, Pulwama, Srinagar and Baramula.

Moderately Good Land

Soils under this category of land have severe limitations in their use of agricultural purposes, that reduce the choice for growing crops or plants, otherwise requiring some special conservation practices to get desirable production. The conservation practices are more difficult to apply than the soils in class 'good-land', but they may be used for cultivating a good variety of crops suitable to the particular type of climate which vary from place to place. The limitations of the soils of the class that restrict the farmer's choice may result from the effects of one or more of these causes : (a) gentle to moderate steep slope, (b) very low permeability of top soil, (c) severe effects of overflow, (d) high susceptibility to water or wind erosion, (e) adverse effects of past erosion (f) continuous waterlogging, (g) shallow depth of soils, (h) low water holding capacity, (i) moderate salinity, (j) low fertility, and (k) moderate climatic limitations.

The moderate good land is delineated in small patches covered with different kinds of soils in parts of Pulwama, Budgam and Baramula, and Kangra, Una and Bilaspur districts of Himachal Pradesh.

Poor Land

The soil in this class is generally poor in their qualities due to low productivity. Utmost investments are needed to generate and upgrade the soil capability by choices of crops and farm management practices. Generally, the soil cover is shallow, waterlogging, salinity and low water holding capacity are some of the limitations of this group of soils. These are found in small patches in Jammu, Udhampur and Kathua districts of Jammu and Kashmir while some patches in Chamba and Kullu also delineated under this class. Traces of these soils also occur in Dehradun.

LAND UNSUITABLE FOR CULTIVATION

Land Suitable for Wildlife, Watershed Management

Areas as delineated in this major group of land capability need utmost attention for development in view of the gamut of wildlife and watershed management leading to have recreation and ecological balance. These lands are distributed in northern limits of Pithoragarh, Chamoli, Uttarkashi, Kinnaur and Lahaul and Spiti. Also found in patches on the heights of Chamba, Doda, Kargil, Anantnag and Srinagar while on the other extreme of valley floor in Budgam, Pulwama and Anantnag.

Land with Severe Limitations Suitable for Forestry and Grazing

Soils in this class have severe limitation that make them generally unsuitable for cultivation and limit their use largely for forestry, grazing, wildlife, food and fodder. A few of the soils in this class can be used for some cultivated crops by providing intensive management. Some of the soils in this class are also adapted for orchards, fruit cultivation and for some ornamental plants. Steep slopes, severe adverse effects of erosion, stoniness and rocks, shallow depth of soils, severe wetness and low-moisture holding capacity are the limitation. Such oil capability areas may be well or poorly suited to woodlands and grasslands depending on investment to suit land management. In most of the cases soils are almost occupied by these degraded forests and scrubs. These lands found scattered in small pockets in Pithoragarh, Chamoli, Lahaul and Spiti, Doda, Ladakh. But there is a continous stretch found from Budgam, Jammu, Kathua, Chamba, Kangra, Mandi, Solan, Sirmour, Dehradun, Garhwal, Almora to Pithoragarh without any discontinuity.

Mountainous Areas with Very Severe Limitations Suitable for Forestry and Grazing

These soils having limitations suitable for forestry as classified on the map also lie in various geographic areas. These have been classified for soil capabilities according to the terrain and climate. The degradation of forests are quite pronounced in such regions and hence suitability of forests are always recommendable for better man-land relationship. These soils are found extensively in the north-western Himalayan region and about two-third area come under this category from almost all the districts of the region.

CONSERVATION OF FOREST RESOURCES

The word forest is derived from the Latin 'foris' meaning outside, the reference being to village boundary or fence, and it must have included all uncultivated and uninhabited land. Today a forest is any land managed for the diverse purposes of forestry, whether covered with trees, shrubs, climbers, etc. or not. The Indian word 'jungle' has been adopted in the English language to describe a collection of trees, shrubs, etc. that are not grown in regular manner, contrasted with forests which is any vegetation under systematic management. The term forest in general may thus be defined as an area set aside for the production of timber and other forest produce or maintained under woody vegetation for certain indirect benefits such as climatic or protective. It is also ecologically a plant community predominantly of trees and other woody vegetation, usually with a closed canopy and legally it is an area of land proclaimed to be forest under forest law.

The role of forests as envisaged in the National Forest Policy of 1952, ensures the role of forests as productive, protective and bio-aesthetic. It entitles forests to an adequate share of land i.e. 66 per cent in the hilly and mountainous regions and 20 per cent in plains amounting 33 per cent of the total land area to promote public welfare and balanced economy.

In addition, the environment associated with them, is a source of attraction for tourists visiting the region. Considering the damage already done, immediate steps should be taken to increase the area under forests and develop their potential capacity. After determining the grazing capacity of the meadows and forest grasslands, the grazing policy should be redefined and enforced.

Discussing the problems of environment in developing countries, Gerasimov and Pulyarkin (1978), observe that at present danger to environment come from an intensive and uncontrolled techno-economic impact on the nature leading to violation of ecological balance. Hence a necessity arises to elaborate such a conception of technical, economic and social developments which would allow making effective use of scientfic and technical development to satisfy urgent material needed for the society. Not only ecological balance should be broken but environment of the society be enriched and improved. The developing countries suffer not so much from pollution as from degeneration of renewable natural resources, the burden which is increasing owing to various reasons. During the recent decades a serious threat to forest vegetation occurred in developing countries that they were considered quite rich in wood.

The entire Himalayan region is suffering due to reckless cutting, overgrazing, shifting cultivation, and extension of farming. Both in the interest of Himalayan agriculture and conservation of soil and water, integrated development of forests, soils and water, and scientific landuse is essential.

Usually forest management takes into consideration the multiple use of ecosystem (Ambasht, 1978). These are principally for :

☞ Timber and other economically important tree products

- Proper maintenance of the water cycle
- Use forests openings for grazing and forage
- Preservation and multiplication of wildlife
- Habitat for migratory birds in low lying areas
- Recreational areas

On the other hand man was compelled to clear-off the forests which resulted in the ecological havoc. The trees were cut down for the following reasons :

- To make room for agricultural crops
- To construct the houses for increasing population
- Wood for cooking food
- Making crude agricultural implements and carts
- Domestication of animals

The basic feature of all forest management methods are based on 'sustained yield concept.' This means forests have to be used for taking out necessary and desired products at a maximum possible level of harvest. Side by side adequate forest cover is to be maintained by the programme of afforestation or the regeneration. In managing the forests one has to constantly manipulate the growing stock, to achieve the set goals and keeping in mind the answer of the following questions :

(a) How much to cut?
(b) When to cut?
(c) Where to cut?
(d) How to cut?

Following practices are done for conservation and management :

(i) ***Selective cutting*** : The natural forests have diverse kind of trees, shrubs and herbs of different age. Due to the death of some old trees, or otherwise, the stunted seedlings become active and grow fast to cover the openings. Thus, for the management of such forests, a working plan is developed, the trees are marked according to time sequence and harvested. Then suitable steps are taken to fill-up the gaps.

(ii) ***Block Cuttings*** : The block cutting is practiced in the artificial raised plantation which are assumed to have the same age. The forest range is divided into some 30-40 blocks according to area covered. Then all trees, on maturity in a particular block are harvested. Thereafter, the cleared land is afforested with the same species. In the same manner next year another block is harvested. Every time the previously harvested block will be ready by the time one reaches to the last one.

(iii) ***Taungya System*** : For reducing the unnecessary expenditure, on plantation of trees, as system forest culture plus agriculture (forest trees + agriculture) has been evolved. This is designated as Taungya system. The landless farmers are plotted some cleared forest lands. In the first year, they cultivate the land with agricultural crops and take all the harvest. In the next year, the farmers prepare rows of beds for the raising tree seedlings and agricultural crops are cultivated in between them. The crops are grown for 2-3 years. When the seedlings grow to a metre or more the farmers are then shifted to some other cleared blocks. In this way, the forest department get the free land for cultivation of agricultural crops.

CONSERVATION OF BIODIVERSITY

India has rich heritage of species and genetic strains of flora. Over about six per cent of world species of flora are found in India. It is estimated that India is 10th among the plant rich countries of the world and sixth among the centres of diversity and origin of agrodiversity. Decrease of natural habitats results in the loss of biological diversity.

Biodiversity of north-western Himalayan region is at risk now-a-days because of developmental activities at the cost of natural habitats.

About 3500 plant species belong to 1500 genera reported only from the Kashmir Valley (Kaul and Dar, 1985), of which about 128 species distributed over 69 genera and 34 families of flowering plants were represented in the Pleistocene flora of Kashmir described by Puri and Mittre (1964). The physiographic changes and glaciation not only caused the devastation of these flora but also influenced significantly the climate of the Kashmir valley.

Plant exploration by the Botanical Survey of India has collected and identified a large number of useful plants in the western Himalayas which are summarized as follows :

Essential Oil Yielding Plants	:	27 species
Gum Yielding Plants	:	11 species
Resin Yielding Plants	:	3 species
Tannin Yielding Plants	:	10 species
Oil Yielding Plants	:	34 species
Timber Yielding Plants	:	30 species
Fibre Yielding Plants	:	31 species
Dye Yielding Plants	:	25 species
Alkaloid Yielding Plants	:	60 species
Medicinal Plants	:	212 species
Ornamental Plants	:	30 species

Dr. Promila Kapoor in her botanical exploration in Himachal Pradesh has identified 33 wild plants used as vegetables by local people, 10 species of edible seeds and yet another 28 species with edible fruits (Pal, 1982, p. 8).[1]

SCIENTIFIC AGRICULTURAL PRACTICES

Plastic Film Technology in Mountain Agriculture

For a long time in mankind's agricultural history, agricultural production was constrained by factors such as low temperatures, drought, short frost-free periods, overabundant precipitation, waterlogging, saline-alkali effects, and other adverse impacts. Traditionally, farmers used mulching methods to overcome these adverse conditions, for example, by spreading straw, fallen leaves ashes, and other agricultural residues on the ground to protect the roots of crops and by putting stones near the plants to increase the temperature of the soil and reducing evaporation, thus modifying the growing environment of food crops and other agricultural and horticultural crops. These methods are useful, but they have limitations which result in them not being very effective. These methods also cannot be used throughout all regions and on all crops. In more recent times, greenhouses have enabled farmers to advance and shorten the growing season of many crops.

In the 1950s, plastic film was introduced into agriculture making a modern version of traditional technology possible. Plastic Film Technology (or technology for the promotion of root cultivation) developed concomitantly with the petrochemical and high polymer chemical industries.

Plastic Film Technology (PFT) uses plastic film, usually polythene film (0.014 ± 0.003 mm) to cover the surface of the earth in order to increase temperature, retain moisture, promote seed germination and emergence, accelerate the growth and development of roots and of the plant as a whole, and achieve high yields and good crop qualities. PFT enables various crops to mature 5 to 29 days earlier than normally, and increases the yield by 30 to 50 per cent.

Remote Sensing Techniques

Efficient agricultural management involves appropriate application of production and conservation practices for the development of land and water resources, aimed at increasing the food production on a sustainable basis. Substantial increase in crop production are possible by bringing additional land under cultivation (horizontal approach) and improved crop management (vertical approach) technologies such as the use of high yielding, input responses, energy intensive and stress tolerant varieties, increased irrigation and integrated crop nutrition and protection.

Remote sensing may be defined as the science and art of acquiring information about objects from measurements made from a distance without coming into physical contact to the object. The human eye, cameras and scanners are some of the examples.

The remarkable development in space-born remote sensing technology and its application during the last decades have firmly established its immense potential for mapping and monitoring of various natural resources. This has opened new vistas in the area of application particularly in agricultural planning at the macro and micro watershed levels. The capabilities of the remote sensing technology in providing information about various agricultural resources, which influence agricultural production directly. A few examples are given below :

Soil Resources : Satellite data provides valuable information on physiography, which helps to map the special distribution of different soil unit with limited fieldwork. These maps also provide information on the inherent potential and limitation of the soil for agricultural production and thus aid in designing efficient cropping system and optimal landuse planning.

The non-adoption of proper soil and water conservation measures and overexploitation of available land resources has accelerated the process of land degradation. Remote sensed data is being regularly employed to map and monitor various degraded land viz., saline soils, eroded soils, water-logged areas, shifting cultivation areas, ravines and the like. Satellite data has been successfully used to map saline and alkaline soils. These maps depict the nature of problem, degree of salinity/alkalinity and spatial extent of the problem in each mapping unit.

Wastelands : Landuse information is important for spatial planning, management and utilization of land for various purposes like agriculture, forestry, environmental studies and to find out additional land resources that could be tilled. The information generated on landuse patterns, to convert single cropped areas to double cropped areas and allows cultivation of land to increase the food production.

Water resources : The availability of irrigation at critical period of crop growth is a major factor which determines the quantity of crop production. Only 42 million hectares of land has assured irrigation in the country as a whole, the rest of the agricultural land is dependant upon the vagaries of the monsoon. Hence, there is need to judiciously utilize the surplus and sub-surface water, to get maximum returns. In this context, remote sensing data proved effective inventorying, monitoring and managing both surface and groundwater resources to augment the water use efficiency.

Groundwater : Satellite data serves for extracting information on geology, geomorphology, drainage, landuse and soil which is essential in identifying not only the potential segment of water resources but also suitable site selection for recharge structure. The successful exploration of groundwater potential in rain-fed and upland situation can help to provide atleast one or two life saving irrigation during the critical stages of the crop growth thereby leading to the increased crop yields.

It is estimated that Indian farmers generally use 20 per cent more water than actually needed. Due to large scale dam and canal system and water seepage land get degraded very soon and allowing the good land to remain useless for a long time and then turned into wastelands. Excessive extraction of groundwater without adequate recharge leads to groundwater level going down sharply. One reason could be use of high power electric pumps and the government encouraging the use of such pumps by providing free power and loans at low rate of interest. While such measures give short-term relief, in the long run it is unsustainable.

Flood Monitoring : Remote sensing data is used for obtaining near real time information on areas effected by major flood and quantitative estimate of the damage to infrastructure and crops. These informations are provided to various departments for remedial measures. Flood prone zones can also be identified where appropriate action can be taken up, to reduce the losses due to inundation of crops.

Crops : Reliable and timely estimates of crop acreage and production are important for the formulation of marketing strategies and spectral reflectance data to identify and distinguish it, is helpful in deriving crop acreage and production estimates to monitor and assess the crop condition. The estimates were found to be in close agreement with the estimates of the Directorate of Economics and Statistics. The remote sensing based acreage estimations are available atleast one month before the harvest of the crop, which enable the administrator and planners to take strategic decisions on import-export policy matters and trade negotiations.

Production estimation : Crop yield is influenced by many factors such as genotype, soil characteristics, the cultural practices adopted, meteorological conditions and pest and diseases. The spectral data of a crop is the integration manifestation of the effects of all these factors.

Statistical, meteorological and spectral models are used for crop yield estimations. Remote sensing models adopt two approaches, namely, single-date data spectral index and multi-date data spectral index. The single date-data spectral index approach relies solely upon the data acquisition within a narrow critical period of maximum vegetative growth face while the multi-date approach depends on spectral data at different stages of crop growth within the season. The approach has the constraints obtaining the cloud free multi-temporal satellite data within the crop growth cycle.

Efforts are in progress to make the crop condition information more comprehensive through integration of information on cropping pattern, soil type, etc. In Geographic Information System (GIS) environment integrated mission for sustainable development of natural resources has to be based on maintaining the fragile balance between productivity function and conservation practices through constant monitoring and identification of problem areas. The integration of information on resources with socio-economic and meteorological inputs have a tremendous scope for optimum crop management strategies at the micro-level.

Weather Forecast

Recent natural calamities such as the tropical cyclone, land slides, floods and unseasonal rains during the post-monsoon season, points out how weather can influence the socio-economic aspects of a region. The yield in individual years continues to be significantly effected by weather conditions. The effect of weather on crop yield may be widespread or local, prolonged or shortlived. Vagaries of weather are perceived as a threat to food security. In order to reduce the impact of uncertainities of weather and for taking advance preventive measures against weather triggered crop production risk, there is a need to invest continuously in meteorological conditions, as it is vital to the sustainability of agriculture.

Prompt and well-planned actions and landuse policies based on detailed climatic information should lead productive and sustainable agriculture. It may influence different stages of crop growth. Forecast can help to optimize agricultural production. Weather forecast is generally divided into three categories according to their period of validity. These are (i) short range (two days), (ii) medium range (3 to 10 days) and (iii) long range (more than ten days). This unique service by the India Meteorological Department (IMD) intended to provide information to all those millions of farmers, who for a multiplicity of economic, social and crop related reasons, each day adopt some of their farm activities to the variation of weather. They require day to day forecast on a medium range scale. Even if farmers are warned three days in advance about vagaries of weather, agricultural losses can be minimized.

During April-May of 1983, the unusual and persistence cloudy weather in the north-west affected wheat cultivation. The then Prime Minister, Indira Gandhi asked meteorologists to study the problem and evolved ways to adjust cropping pattern to suit weather conditions. This lead to the establishment of the National Center for Medium Range Forecasting (NCMRFC) in 1988, whose principal purpose was providing agro-met advisories to the farmers. A super computer 'Cray X-MP' was commissioned in March 1989 to develop medium range weather forecast in quantitative terms. The forecast of six weather elements namely, precipitation, wind speed, wind direction, maximum temperature, minimum temperature and clouds is prepared. On the basis of these forecast and keeping in view the available crop information, the field units prepare the agromet advisory bulletins in the regional languages for the use of farmers.

Millions of farmers seek weather information each day which help them to make on-farm management decisions that may range from the seemingly trivial cutting of grass, to one that could lead to crop disaster, pesticides spray. All weather observations are complimentary and may be imagined as a part of the ever changing crop environment.

Newly emerging issues in sustainable developments include sustained agricultural production, security of food for the burgeoning population, environmental safety for clean air and water and the maintenance of an acceptable quality of the global atmosphere and weather forecast.

Biotechnology

Much is being accomplished using conventional, non-molecular tools of genetics for improving crops. However, crops are constantly under threat from frost, insect, diseases and drought. Recent advances in biotechnology promise support to plant breeders in evolving superior plants.

Biotechnology refers to a variety of techniques involving living organisms as a means of production, uses tissue culture or cell culture, cloning and fermentation, cell fusion, embryo transfer and recombinant DNA technology. Using tissue culture techniques plants can be produced from single cells obtained from roots, shoots, leaves, anthers, or any other plant organ. With the recombinant DNA technique, the desired piece of genetic material can be isolated, cut, cloned, transferred and expressed across the species among microbes, plants and animals. Even human genes can be transferred and expressed in plants. Thus, in principle genes available in the entire biodiversity, modified or synthesized can be transferred and expressed in plants. Biotechnology thus, provides unique opportunities for developing highly productive, sustainable agricultural systems not only to feed the growing population but also to enhance incomes in the rural sectors.

Genetically restructured plants that accommodate more crops per year (increased crop intensity) or more plants per hectare (increased plant population density) and permit an intercrop facility intensive farming. Many genes controlling desirable characteristics are available in related wild germ-plasm. However, through conventional hybridization procedures, transfer of these to cultivators is not possible due to sexual incompatibility barriers. In such cases *invitro*, entering of pollinated ovaries and hybrid embroys are used for obtaining hybrid plants. Following are the advantages of biotechnology in agricultural development :

Higher yield

First generation hybrid are characterized by greater growth vigour and productivity and have provided the material choice for intensive agriculture for a variety of crops.

Cost Reduction

In developing economies, reduced cost of production is

important for ensuring remunerative returns to the farmers and affordability for the consumer. Biotechnology reduces production cost by replacing part of the industrially produced inputs, such as plant nutrients and crop protection agro-chemicals by products of biological origin.

Eco-friendly

Environment protection is important in all sectors of production. Productivity increase have cause from intensive agricultural areas where High Yielding Varieties are cultivated under no input constraints situation. Since biotechnology offers viable and practicable approaches for meeting the needs of ecological preservation and environmental protection, a number of laboratories are trying to develop safer and easily degradable protection agrochemical schedules and integrated diseases and pest management and reduces dependence on chemicals by creating pest and disease resistant crops is an eco-friendly approach.

INTEGRATED APPROACHES FOR SUSTAINABLE AGRICULTURE

Integrated Nutrient Management

The integrated nutrient management philosophy is of ecological as well as economic of consumers. It measures a producer's endeavour to minimize ecological degradation and maximize economic value for the production cost to make it affordable for consumers.

The nutrient management scenario in the country has undergone three phases of development. The period 1950-1965 witnessed increased awareness of chemical fertilizer along with organic sources of nutrient and green manuring. The second phase started with the green revolution and initiated a sudden growth in the use of chemical fertilizers and the gradual decline in the use of organic resources and green manure. The third phase can be traced from late seventies onwards when the importance of organic resources and side effect of chemical fertilizers recognized.

In traditional agriculture, farmers use farm yard manure and cow dung as nutrient's source to the crops to supplement the natural supply available through soil and atmosphere. This system of low nutrient supply can only sustain low productivity of crop. Increasing needs for enhanced crop productivity due to ever increasing population necessitated the breeding of high yielding crops which require high amount of nutrition. With the availability of nutrient responses high yielding of crops, intensive nutrient application and improved farm management became an accepted practice to derive full benefits from such varieties of crops. The dramatic increase in fertilizer consumption and the increase in agricultural productivity recognized the crucial role of fertilizer.

A continuous increase in the chemical fertilizers and pesticides alone is not the best way to sustain agriculture production consistent with maintenance of soil fertility and the protection of the environment. A quick assessment of nutrient removal by the crops and quantities of chemical fertilizers added in the soil reveals that what is removed from the soil by the crop is much larger than what is added to it. This also calls for a further addition of nutrient in the soil to check the exhaustion of soil nutrients.

The answer to this problem is the development of integrated plant nutrient supply system involving an appropriate mix of organic, biological nitrogen, fixation, phosphate solubilizing microbes and need-based chemical fertilizer would be crucial for sustainability of production and the soil as the resource base for it.

The major sources of organic resources which need to be harnessed into the integrated plant nutrient supply system are crop residues, dung and urine from domesticated animals and wastes from slaughter houses, human excreta and sewage, bio-mass of weeds, organic wastes from fruit and vegetable production and household wastes, sugarcane trash, oil cakes, phosphogypsum and fly-ash from thermal power plants. There are several biological nitrogen fixation microbes such as blue-green algae and *azolla* for rice, *rhizobium* species for leguminous pulses and oil-seeds and trees, *azotobacter* and *azospirillum*. The phosphate solubilizing micro-organism available for increasing the availability of phosphate from unavailable soil forms and applied rock phosphate are the bacterial *Bacillus subtiles, Bacillus circulans* and the fungus, *Aspergillus niger*.

Green Manuring

The practice of green manuring is an important component of INM for improving soil fertility and supply of nutrient requirement for crops. The green manures are real nutrient integrators. They absorb mineral nutrients, convert them in to an organic form which then changed to mineral forms when the green manure decomposes in the field. The stem nodulating green manure plant *Seshania rostrata,* fix 100-285 kgs. of nitrogen per hectares in 45-55 days has great scope in rice culture. Depending on the crop grown, the nitrogen contribution by green manure crops varies from 60-280 kgs. per hectare.

Biofertilizers

Biofertilizers are an important component of integrated nutrient management but 90 per cent of their role is confined to nitrogen needs (biological nitrogen fixation). Biofertilizer such as *rhizobium* culture, is an effective source of nitrogen supply to leguminous crop, *Azotobacter* and *Azospirillum* help in nitrogen fixation and supply to crops like wheat, paddy, maize, cotton, sugarcane, fruit crops and vegetables. Phosphate Solubilizing Bacteria (PSB) help in making available soil phosphorus to the crops. Blue-green algae has shown promise in low-land rice. These bacterial fertilizers are renewable and eco-friendly supplementary source of nutrients. Also by fertilizer promoting balanced use of chemical fertilizers based on soil testing and correlation of secondary and micro nutrient deficiencies in soil and promotion of appropriate soil, water and nutrient management.

Integrated nutrient management is a flexible approach to minimize the use of chemical but maximize use efficiency and farmer's profit. Overuse and abuse harms the biological power of soil which must be prevented as all nutrient transformation are negotiated by soil micro-flora. As organic matter is the source of energy to soil micro-flora, ogranic carbon content is an index of soil health. Integrate nutrient management is based on three principles :

(i) Assessment of the basic soil fertility and climate.
(ii) Nature of crop in the cropping system and yield target.
(iii) Atleast 30 per cent of total nutrient level of NPK to inorganic form.

This help to estimate the fertilizer level and time of application to the crop. Soil fertility is not static but dynamic determined by many soil quality indicators (physical, chemical and biological) to regulate the function elaborated. Productivity is a function of fertility in interaction with the environment and management skill of the farmers.

There is wrong notion that sustainable, eco-friendly agriculture can be practiced using organic and bio-fertilizers and excluding chemical fertilizer. The nutrient need are so large that no single source can deliver the goods. Integrated nutrient management harmoniously integrate the use of minerals, fertilizers, organic, bio-fertilizer and recyclable waste. The integrated nutrient management packages should be technically sound, economically viable, practically feasible, socially acceptable and environmentally sustainable.

In short, different nutrient sources can be seen as different currencies which are used by farmers the world over to deposit nutrients into the soil bank so that latter on cheques can be written to meet the nutrient expense incurred in feeding the crops.

Integrated Pests Management

The Indian agriculture scenario has changed tremendously with the adoption of modern agricultural technology comprising High Yielding Varieties, chemical fertilizer, assured irrigation and improved agronomic practices. Under such an intensive cropping system certain pests such as brown hopper in rice, heliothis and whitefly in cotton, borers and pyrilla in sugarcane, aphids in mustards and phalaris weed in wheat emerged as a major problem. To overcome losses under such pest situation, prophylactic pest control measures mostly with chemical pesticides were adopted however, indiscriminate use of hazardous plant chemical has resulted in reduction of Biodiversity of natural enemies, out-break of secondary pest, development of resistance against pesticides, distribution of beneficial organism pesticides residue in food chain, contamination of food and ecosystem, environmental pollution and other pesticides associated with health hazards.

In order to overcome such adverse effects due to over-reliance on synthetic pesticides in the past, the current thrust in plant protection is on promoting integrated pest management means the conservation and augmentation of natural enemies of crop pest and adoption of a compatible, cultural, mechanical pesticides, tolerant varieties and other legal methods. Integrated pest management bring knowledge intensive and skill oriented to empower the farmer in decision-making of pest management tactics, the farmer further to be empowered to manage the overall crop production strategy on a sustainable basis.

Despite large scale use of pesticides which are basically toxic and do not increase the yield, pest's out-break like bollworms and whitefly in cotton, leaf folder in rice, pyrilla and borers in sugarcane, aphid in mustard and pod borer in pigeon-pea. Considering, the ill-effects of the indiscriminate use of chemical pesticides, bio-intensive integrated pest management is a more feasible and economical method of pest management and has become popular particularly among small and marginal farmers. Integrated pest management has several measures to check the pest damages on the crop. Some of them are as follows :

Host Plant Resistance : Resistance must be developed in the host plant against the natural enemies by the help of genetic engineering tools, the desired genes can be transferred from wild plant relatives or any source no matter whether it is bacteria, virus or animal.

The transfer of ***Bacillus thuringiensis*** insecticider toxin gene into crop plant have resulted in the creation of transgenic tobacco, tomato, cotton, potato and so on. Cow pea trypsin inhibitor gene effect the digestive process of attacking insect and thereby reduces the damage. Wheat resistance to rust, millets to powdery mildew, pulses to virus, rice to gallmidge and brown plant hopper, cotton to jassid and leave curled virus, potato to late blight and bacteria wilt, tomato and brinjal to bacterial wilt have also been evolved.

Biological Control

These are one of the most important components of integrated pest management. Natural enemies play a key role in managing the population of crop pests. Conservation and employing biological suppression agents such as frogs, birds, pathogens play an important role in pest suppression. There are 1005 species of plant exhibiting insecticiders properties' 384 with anti-feedant properties, 297 with repellant properties, 27 with attractant properties and 31 with growth inhibitor properties. The most commonly used botanical pesticides are Neem (*Azadirachta Indica*), Pongania (*Pongania glabra*), and Mahua (*Mahua indica*) has been found effective against several pests including rice cutworm, diamond backmoth, rice brown plant hopper, rice green leaf hopper and tobacco caterpillar. Mahua seed kernel extract (5 per cent) is effective against sawfly and others.

Bio-Pesticides

Future exports of commodities stand the risk of rejection, if they are not free of pesticide residue, bio-pesticide is an answer to this problem.

Bio-agents are an integral part of the integrated pest management. The integrated pest management has been developed for cotton-bollworm, sugarcane -inter-node borer, grapewine-mealy bug, coffee barrie borer, coconut rhinoceros beetle, wilt of oil seeds, pulses, banana, pepper, rhizome-rot of ginger, turmeric and damping of vegetables.

Bio-pesticide is a component of integrated pest management and are cheaper than pesticides by 50 per cent. They aid growth throughout and are easy to apply. They are eco-friendly having high cost-benefit ratio and do not pose risk of the pathogen developing resistance. They are compatible with bio-fertilizer and solublized phosphorus.

The advantage of bio-pesticide over-chemicals exist from the industrial point also. For a bio-pesticide the research and development cost in the United States is 0.8 to 1.6 million dollars as against 20 million dollars for a chemical pesticide (Jayarajan, 1997). The toxicology test costs only 0.5 million dollars against 10 million dollars. For chemical pesticide for getting one marketable pesticide, about 20 thousands candidates are to be tested while the number is much less for bio-pesticide.

Cultural Measures

Cultural measures are the cultural practices associated with crop production to make the environment less favourable for the survival, growth or reproduction of pest species. Cultural practices if followed provide relief from the pests, some of which are briefly enumerated:

(i) *Selection of site* : The selection of appropriate site for the cultivation of field crops and fruit trees can reduce future infestation from insect pests. The choice of cultivars should be such that it should be suitable for growing in the area and tolerant to pest and diseases of the area.

(ii) *Ploughing* : This exposes soil to inhibiting fauna to advere weather conditions and natural enemies. Ploughing exposes egg clusters of grasshopper, resting stages of girdle beetle, blue beetle, pupae of hairy caterpillars, tobacco caterpillars, white grubs, Plutella xylostello and Beetrocera species, rhizome and bulbs of perennial weeds and nematodes. Tractor operating cultivator open furrows and expose different stages of soil inhibiting pests which are eaten away by birds in the morning hours. Birds follow the track of the plough. Therefore, the time of tillage of operation is most important as in common, birds are not active at night.

(iii) *Planting* : Use of healthy seed, seedlings is one of the best integrated pest management practices.

(iv) *Planting date* : The date of planting takes advantages of the absence of the pest or avoid susceptibility stage of the crop. It should coincide with the most inactive period or lowest pest population, planting is done early and in a synchronized manner in a span of 10 to 15 days, it prevent carryover of pest from early sown crops to late sown crop and prevent building of damaging population. Timely harvesting can also minimize fruitfly damage to mango, guava, citrus, and many other fruit crops.

(v) *Destruction of off-type plants* : During the course of cultivation a new variety becomes improved and it poses some off-type plants. These off-type plants grow along with the standard variety and due to the difference in susceptibility they attract different types of insect pests. So the destruction of off-type plant is very important.

(vi) *Destruction of alternate host* : Many insects use a wide range of cultivated and uncultivated plants specially weeds as alternative host for off-season carryover of population. So the destruction of such alternative host plants is important as a cultural measure of integrated pest management.

(vii) *Pruning and destruction of crop refuse* : The pruning of infested parts of fruits and forest trees and defoliation in certain crops help to reduce the pest population. Clipping of effected leaves and twigs, pruning and fractured and effected branches, harbouring insects and pests.

(viii) *Crop rotation* : Continuous cropping encourages carryover of pest population from one planting season to another. Crop rotation or roto-cultivation is helpful in managing the pest population particularly of mono-phagus pests.

(ix) *Trap crops* : The pest attracted from the main crop to a preferred crop which is deliberately planted the attracted pests on a trap crop is more economically tackled in this restricted area. Growing of two rows of mustard after every ten rows of cabbage (one sown 15 days prior to planting while the other 25 days after planting cabbage) attract backmoth, leaf-webber and aphids which can be controlled in restricted areas easily.

REFERENCE

1. Pal, B.P. (1982), *Environmental Conservation and Development*, Natraj Publication, Dehradun.

7

Conclusion and Suggestions

In developing countries, agriculture is still the mainstay of approximately 70 per cent of the population; and in some countries it is as high as 80 to 90 per cent. The boundaries between ecology and agro-ecology are not clearly demarcated. Ecology and agro-ecology are like two sides of the same coin. The one with the national emblem represents ecology, because natural ecology is the heritage of human communities and societies, as a whole, and is not generally priced. The other side of the coin giving its value represents agro-ecology, since agro-ecology provides people with a living and its value can be assessed in terms of who depends on it and to what extent?

In the context of mountain areas, agro-ecology is the fundamental basis of agriculture and the overall economic system. It is the source of the resiliency and regeneration necessary for the sustainability of agricultural systems. The self-sufficiency of local mountain farming communities depends on agro-ecology. Considering the proportion of the earth occupied by farming systems, agro-ecology is a broad asset.

Today, mountain societies in this region are faced with the dilemma presented by increased populations and economic needs, rendering traditional methods of farming and use of agro-bio-resources economically unsustainable.

At the same time, the new technological options that can transform traditional farming may be ecologically unsustainable in the long run and cause loss of agro-bio-diversity from most mountainous areas.

In this region, agro-ecology and its management is under pressure. Transformation of agricultural systems resulting in land-use changes is having an adverse impact on native agro-ecology. The hunger, poverty, resource, scarcity, and land degradation faced by a large proportion of mountain households are forcing farmers to look for alternative occupations. Farmers are transforming their lands and farming methods to achieve better production and more benefits, and traditional agro-ecology is the loser in this process.

Inadequate technical understanding of mountain agro-ecology is another limitation to conservation and management. The main aspect is the absence of economic valuation of agro-ecology in the region – how much can it contribute to the economy of the community, nation, and global community, and on what terms? How can we assess the returns of investment value on saving agro-ecology and how can farming communities be encouraged to reorient their thinking along these lines? We also have little knowledge about the diversity of unique ethnic mountain communities and their micro-agro-ecosystems or the changes affecting these agro-ecosystems. People are less familiar with the slope and facing problems in developmental efforts.

Crops like other organisms, have distinct preferences of altitude, climate, soil, photoperiods, etc. at different period of their life cycle, but they can usually be grown over a wider range of environments with their wild relatives. This is particularly true for annual crops many of which can be grown under very diverse conditions from tropics to the temperate zones. But man often modify the environment by cultural and other methods to provide better conditions for growth and productivity.

It is therefore, imperative to integrate agricultural development policies with environmental management and planning in ecologically sensitive Himalayan region. The agro-ecological planning in this context is referred to as scientific and integrated framework for the ecologically sustainable and economically viable development of agriculture through ensuring optimum and most rational utilization of environmental resources.

Agro-ecological development planning needs information on various aspects like land capabilities, temperature, rainfall, relief, geo-hydrological balance, rate of soil erosion, extent of deforestation, extension of farming activities, magnitude of surface and sub-surface water pollution because of suggested biochemical techniques and existing pattern of agricultural resource utilization, etc.

On the basis of land resources status, it is revealed that the high productivity zone is either characterized by high irrigation-low rainfall or low irrigation-high rainfall situations. Here thrust will be on diversification of agriculture towards high value crops through the creation of relevant infrastructure and establishment of strong linkages with agro-processing industry and export.

There are some areas where productivity is found low but potentiality is high, it is characterized by low productivity despite abundant water availability and good soil. The strategy for this region focussed on flood control, drainage management, improvement of irrigation facilities and better inputs delivery system. The fertilizer application will have to be substantially stepped up and gross irrigated area in this region is to be increased to cover up to 50 per cent of gross sown area.

In low productivity zone of water scarcity the emphasis

will be on development of efficient water harvesting techniques and conservation methods, suitable irrigation packages on watershed approaches and appropriate farming system which economizes water use.

Most of the environmental problems are the result of non-development, backwardness and poverty at different levels. Age old practices of extractive resource utilization feasible under low pressure of population and low standards of living is gradually erasing the enrivonmental resource base of the area.

Himalayan agriculture is characterized by steep slopes, shallow fertile soils, nutrition deficiency and terraced fields. In such conditions, growing of quality forage is extremely difficult. Improved forages were introduced successfully in low lying valleys or relatively better-off areas where inputs and irrigation are available. However, they are not successful in marginal and fragile mountain areas, because of the high input dependency.

Pine looked as the greatest enemy of the Himalayan environment as its fallen needles are considered to be the prime cause of forest fires. Experiments are carried out to convert needles into fuel cakes at viable cost. As a result ecological degradation become ecological development by judicious use of natural resources.

Emerging towns in this ecologically fragile region are adding one more aspect to environmental degradation. Disposal of municipal wastes has taken significant dimension in the valley towns polluting the water of rivers in their downward course.

Traditional agricultural practices continued in the region are due to alien nature of fields, sparse infrastructure facilities, fragmentation of land and non-applicapability of modern equipment on account of difficult terrain conditions of hill area. The location of the agricultural fields is of paramount significance as it explains the surface conditions for adoptability of any farm inputs.

Female agricultural population is the backbone of the agricultural economy and take greater participation from sowing to post-harvest practices. A large number of farmers are not fully familiar with the modern agricultural equipments and integrated approaches of crop-rotation, fallow land and contour ploughing, because lack of knowledge and population pressure.

Man is the most important factor of change in most of the mountain ecosystems. Ecological balance is a sort of compromise between ecosystem and human material progress. The agricultural practices a that are governed by the principles of ecology and are within the ecological means have proved that, this form of natural farming is an excellent form of sustainable agriculture.

Increasing frequency of vehicular traffic is another cause of instability of hill slopes, heavy traffic causes vibrations which result in slipping and subsidence of weak strata. Unscientific exploitation of forest resources for timber, fuel-wood and fodder has caused depletion of forest resources.

The livestock population in the Himalayan region is still largely comprised of local species 95 per cent of the cattle, over 70 per cent of the sheep and most of the goats. This reveals the limited success of institutional policies geared towards promoting cross breeds.

Feed shortages has been a critical problem for livestock farmers throughout the Himalayas particularly in Uttaranchal where the current shortages of the feed and fodder is estimated to be 65 per cent.

The number of livestock per capita of pasture and grazing lands is increasing. Its number was 6.2 per hectare of pasture and grazing lands in 1972 which become 11.8 per hectare of pasture and grazing land in 1992. On the other hand, pastures and grazing lands available per livestock is decreasing i.e. 0.15 hectare in 1972 and 0.09 hectare in 1992.

Suggestions

To overcome the abovementioned problems of sustainability of agriculture, regional development and conservation of ecology following remedies are suggested:

There is a need to quantify the resource base of the Himalayan region. In order to harness the resources for economic development and environmental conservation immediate steps need to be taken by Central Government agencies and state governments to prepare a comprehensive data base assigning values to natural and socio-economic components of environmental resources. Data center should be set-up in every state as well as at district and furthermore at village level. Geographical Information System techniques should be adopted in natural resource management and creating data base for entire Himalayan region.

Regional development must take into consideration the ecological principles of environmental management and socio-economic needs of the society.

Environmentally benign farming technologies and knowledge intensive integrated farming system have been found as the best option to meet these challenges.

Tea is an important cash crop of north-eastern Himalaya from the ancient times. The climate of Palampur Tehsil of Kangra valley is also suitable for this crop up to 2000 m, so policies should be made to encourage it in the region for its economic gains.

Long-term loans should be given to farmers for establishement of new orchards and short-term credit is to be provided for rejuvenation of existing orchards. Production of vegetables should be encouraged, especially in neighbourhood of towns by supplying seeds and seedlings of quality on credit and making technical guidance available to the vegetable grower. Special attention is to be given for the organization of marketing cooperatives for fruits and vegetable growers. Integrated approaches of nutrient management and pests management should be adopted for sustainable agricultural development.

It is suggested that the construction activities should be banned or executed carefully so that natural beauty of landscape is not damaged, in particular it should be avoided on unfavourable and steep slope.

Afforestation programme of social forestry needs planting of such trees which have deeper and richer root system, it avoids soil erosion and accomplishes dewatering of the threatened area through evapo-transpiration. For example, populous alba consuming more water and deteriorating soil by incorporating some toxic material in the soil. On the other hand, Salix Celestia, locally known as Vir or Bissah is an air purifier which maintain the ecological balance disturbed by populous alba.

Tourism industry has vast economic potentials in this region. Different form of tourism related sports activities like skiing, river rafting, mountaineering and trekking are gradually taking shape. To percolate the goods of tourism into local community, it is necessary to encourage development of cottage industries, handicraft and create tourism attractions. To avoid traffic congestion bus stand should also be developed in peripheral areas. Disposal sites should also be developed away from the city center for the safeguard of environment.

Forest cover is an important factor of ecological balance. To encourage afforestation from the root level, its importance in our life should be taught from the primary level. It should make compulsory for every school administration that the tree plantation by every student make an annual event by conducting social forestry or afforestation programme on the eve of Republic Day, Independence Day, and Children's Day etc. Prizes and certificates should also be announced at school and local level for those who plant maximum number of trees.

It is considered to be the duty of every citizen that wherever he lives or built a home for himself must grow some trees in his vicinity.

Appendix - I

Log Y and Log Yt Values of Number of Wooden Plough in Selected Districts of North-Western Himalaya

Bilaspur			**Doda**		
Year	logY	logYt	*Year*	*logY*	*logYt*
1951			1951	4.45	4.46
1961	4.51	4.52	1961	4.64	4.58
1972	4.55	4.56	1972	4.68	4.70
1982	4.63	4.60	1982	4.74	4.82
1992	4.61	4.64	1992	4.99	4.94
Baramula			**Udhampur**		
Year	*logY*	*log Yt*	*Year*	*logY*	*logYt*
1951	5.18	5.12	1951	4.56	4.55
1961	4.99	5.10	1961	4.66	4.64
1972	5.17	5.08	1972	4.72	4.73
1982	5.02	5.07	1982	4.77	4.82
1992	5.07	5.05	1992	4.95	4.91
Srinagar			**Mandi**		
Year	*log Y*	*log Yt*	*Year*	*logY*	*logYt*
1951	4.69	4.71	1951	4.72	4.72
1961	4.60	4.65	1961	4.79	4.80
1972	4.74	4.59	1972	4.90	4.88
1982	4.47	4.54	1982	4.96	4.97
1992	4.47	4.48	1992	5.05	5.05
Chamba			**Jammu**		
Year	*logY*	*logYt*	*Year*	*logY*	*logYt*
1951	4.50	4.51	1951	4.54	4.62
1961	4.58	4.56	1961	4.68	4.65
1972	4.59	4.61	1972	4.75	4.68
1982	4.68	4.66	1982	4.80	4.70
1992	4.70	4.71	1992	4.62	4.73

Dehradun			**Leh**		
Year	*logY*	*logYt*	*Year*	*logY*	*logYt*
1951			1951	3.95	3.93
1961	4.42	4.45	1961	3.76	3.93
1972	4.45	4.44	1972	4.16	3.92
1982	4.48	4.43	1982	3.83	3.91
1992	4.38	4.42	1992	3.87	3.90
Shimla			**Pithoragarh**		
Year	*logY*	*logYt*	*Year*	*logY*	*logYt*
1951	4.75	4.77	1951		
1961	4.82	4.77	1961	4.64	4.68
1972	4.77	4.78	1972	4.83	4.77
1982	4.74	4.78	1982	4.87	4.87
1992	4.81	4.79	1992	4.94	4.96
Kathua			**Anantnag**		
Year	*logY*	*logYt*	*Year*	*logY*	*logYt*
1951	4.27	4.32	1951	4.86	4.95
1961	4.45	4.40	1961	4.99	4.96
1972	4.50	4.47	1972	5.08	4.97
1982	4.55	4.55	1982	5.02	4.98
1992	4.60	4.63	1992	4.91	5.00
Garhwal			**Poonch**		
Year	*logY*	*logYt*	*Year*	*logY*	*logYt*
1951			1951	4.16	4.22
1961	4.88	4.92	1961	4.51	4.32
1972	5.00	4.93	1972	4.30	4.42
1982	4.90	4.94	1982	4.45	4.52
1992	4.95	4.95	1992	4.68	4.62
Sirmour			**Total**		
Year	*logY*	*logYt*	*Year*	*logY*	*logYt*
1951	4.42	4.43	1951	5.76	5.85
1961	4.49	4.47	1961	6.04	5.97
1972	4.51	4.52	1972	6.16	6.09
1982	4.56	4.56	1982	6.21	6.21
1992	4.60	4.60	1992	6.27	6.33

Appendix-II

Log Y and Log Yt Values of Number of Iron Plough in Selected Districts of North-Western Himalaya

Bilaspur			Kathua		
Year	*logY*	*logYt*	*Year*	*logY*	*logYt*
1951			1951	3.49	3.64
1961	1.53	1.67	1961	4.01	3.90
1972	2.61	2.35	1972	4.31	4.15
1982	2.91	3.03	1982	4.37	4.40
1992	3.70	3.71	1992	4.57	4.65
Jammu			**Chamoli**		
Year	*logY*	*logYt*	*Year*	*logY*	*logYt*
1951	3.02	3.30	1951		
1961	3.97	3.76	1961	1.08	1.41
1972	4.54	4.23	1972	2.67	2.06
1982	4.56	4.69	1982	2.47	2.71
1992	5.03	5.15	1992	3.33	3.37
Almora			**Mandi**		
Year	*logY*	*logYt*	*Year*	*logY*	*logYt*
1951			1951	0.60	0.61
1961	1.76	1.82	1961	1.49	1.52
1972	2.26	2.31	1972	2.45	2.44
1982	3.09	2.80	1982	3.46	3.36
1992	3.12	3.29	1992	4.21	4.27
Chamba			**Ladakh**		
Year	*logY*	*logYt*	*Year*	*logY*	*logYt*
1951	0.48	2.09	1951		
1961	3.50	2.78	1961	3.58	3.26
1972	2.56	3.47	1972	3.09	3.19
1982	3.34	4.16	1982	2.89	3.12
1992	4.01	4.85	1992		

Dehradun

Year	logY	logYt
1951		
1961	3.03	3.21
1972	3.68	3.52
1982	4.07	3.84
1992	3.94	4.15

Srinagar

Year	logY	logYt
1951	1.83	1.45
1961	1.89	1.94
1972	1.92	2.42
1982	2.55	2.91
1992	3.93	3.40

Shimla

Year	logY	logYt
1951		
1961	1.28	1.25
1972	1.94	2.08
1982	3.09	2.90
1992	3.64	3.72

Pithoragarh

Year	logY	logYt
1951		
1961	2.30	2.47
1972	2.75	2.84
1982	3.92	3.22
1992	3.15	3.59

Poonch

Year	logY	logYt
1951	0.30	1.02
1961	2.45	1.96
1972	3.57	2.90
1982	3.93	3.84
1992	4.26	4.78

Anantnag

Year	logY	logYt
1951	1.18	0.35
1961	0.30	1.26
1972	1.68	2.18
1982	3.68	3.10
1992	4.07	4.01

Garhwal

Year	logY	logYt
1951		
1961	1.23	1.87
1972	3.33	2.57
1982	3.67	3.27
1992	3.45	3.97

Udhampur

Year	logY	logYt
1951	2.13	2.34
1961	2.98	2.91
1972	3.77	3.47
1982	4.08	4.04
1992	4.42	4.61

Sirmour

Year	logY	logYt
1951	0.60	0.81
1961	1.72	1.61
1972	2.59	2.41
1982	3.32	3.21
1992	3.80	4.00

Total

Year	logY	logYt
1951	3.64	3.94
1961	4.73	4.43
1972	5.06	4.91
1982	5.41	5.40
1992	5.72	5.88

Log Y and Log Yt Values of Pumping Sets in Selected Districts of North-Western Himalaya

Bilaspur			**Una**		
Year	*logY*	*logYt*	*Year*	*logY*	*logYt*
1951	0.30	0.26	1951		
1961	0.48	0.85	1961		
1972	1.72	1.43	1972	2.51	2.48
1982	2.35	2.01	1982	2.59	2.67
1992	2.28	2.60	1992	2.89	2.86
Solan			**Srinagar**		
Year	*logY*	*logYt*	*Year*	*logY*	*logYt*
1951			1951	3.42	2.258
1961			1961	0.90	1.878
1972	2.00	2.04	1972	0.95	1.498
1982	2.56	2.48	1982	0.48	1.118
1992	2.87	2.91	1992	1.73	0.738
Leh			**Mandi**		
Year	*logY*	*logYt*	*Year*	*logY*	*logYt*
1951			1951		
1961			1961	0.60	1.09
1972	0.30	0.32	1972	1.97	1.40
1982	0.60	0.56	1982	2.06	1.72
1992	0.78	0.80	1992	1.61	2.03
Kangra			**Baramula**		
Year	*logY*	*logYt*	*Year*	*logY*	*logYt*
1951			1951	3.53	2.16
1961			1961	1.15	2.01
1972	1.96	2.04	1972	0.85	1.87
1982	2.58	2.42	1982	0.85	1.72
1992	2.72	2.80	1992	2.96	1.58

Dehradun			**Sirmour**		
Year	*logY*	*logYt*	*Year*	*logY*	*logYt*
1951			1951	0.60	0.71
1961	0.70	0.84	1961	0.70	1.15
1972	1.57	1.44	1972	2.43	1.60
1982	2.23	2.04	1982	2.17	2.04
1992	2.48	2.64	1992	2.08	2.48
Shimla			**Kathua**		
Year	*logY*	*logYt*	*Year*	*logY*	*logYt*
1951			1951	2.75	1.66
1961	0.70	1.53	1961	0.30	1.74
1972	3.18	1.53	1972	1.65	1.83
1982	0.70	1.53	1982	2.22	1.91
1992	1.53	1.53	1992	2.21	1.99
Jammu			**Total**		
Year	*logY*	*logYt*	*Year*	*logY*	*logYt*
1951	2.99	2.18	1951	4.22	3.24
1961	1.18	1.35	1961	2.20	3.49
1972	2.59	2.51	1972	3.61	3.74
1982	2.79	2.68	1982	4.21	3.99
1992	3.01	2.84	1992	4.47	4.25
Almora					
Year	*logY*	*logYt*			
1951					
1961	0.48	0.48			
1972	0.70	0.70			
1982	0.90	0.90			
1992	1.15	1.14			

Appendix-IV

Log Y and Log Yt Values of Number of Tractors in Selected Districts of North-Western Himalaya

Bilaspur			Dehradun		
Year	*logY*	*logYt*	*Year*	*logY*	*logYt*
1961			1961	1.79	1.76
1972	0.78	0.78	1972	2.11	2.17
1982	1.41	1.42	1982	2.61	2.59
1992	2.06	2.06	1992	3.01	3.00
Una			**Mandi**		
Year	*logY*	*logYt*	*Year*	*logY*	*logYt*
1961			1961		
1972	1.72	1.72	1972	1.15	1.25
1982	2.26	2.26	1982	2.03	1.81
1992	2.79	2.79	1992	2.27	2.37
Poonch			**Baramula**		
Year	*logY*	*logYt*	*Year*	*logY*	*logYt*
1961	0.00	0.06	1961	0.00	0.02
1972	0.60	0.50	1972	0.78	0.71
1982	0.90	0.93	1982	1.32	1.41
1992	1.34	1.36	1992	2.15	2.11
Kullu			**Garhwal**		
Year	*logY*	*logYt*	*Year*	*logY*	*logYt*
1961			1961		
1972	1.04	1.14	1972	0.60	0.55
1982	1.81	1.59	1982	1.00	1.10
1992	1.93	2.04	1992	1.70	1.65
Anantnag			**Sirmour**		
Year	*logY*	*logYt*	*Year*	*logY*	*logYt*
1961	0.78	0.53	1961	0.60	0.78
1972	0.70	1.00	1972	1.61	1.44
1982	1.30	1.47	1982	2.28	2.09
1992	2.14	1.94	1992	2.57	2.75

Jammu			Kathua		
Year	*logY*	*logYt*	*Year*	*logY*	*logYt*
1961	1.64	1.85	1961	0.00	0.48
1972	2.43	2.22	1972	1.79	1.14
1982	2.78	2.58	1982	1.97	1.81
1992	2.73	2.94	1992	2.16	2.48
Nainital			**Total**		
Year	*logY*	*logYt*	*Year*	*logY*	*logYt*
1961	2.89	2.93	1961	2.99	2.80
1972	3.29	3.31	1972	2.86	3.26
1982	3.83	3.69	1982	3.95	3.72
1992	3.98	4.07	1992	4.16	4.18

Solan

Year	*logY*	*logYt*
1961		
1972	1.90	1.78
1982	1.90	2.14
1992	2.61	2.49

Appendix-V

Log Y and Log Yt Values of Number of Sugarcane Crushers in Selected Districts of North-Western Himalaya

Bilaspur			**Una**		
Year	*logY*	*logYt*	*Year*	*logY*	*logYt*
1951			1951		
1961	1.92	2.04	1961		
1972	2.24	2.11	1972	2.98	3.03
1982	2.25	2.18	1982	2.89	2.79
1992	2.15	2.25	1992	2.51	2.56
Solan			**Almora**		
Year	*logY*	*logYt*	*Year*	*logY*	*logYt*
1951			1951		
1961			1961	1.75	1.82
1972	2.32	2.38	1972	1.96	1.83
1982	2.23	2.12	1982	1.77	1.83
1992	1.80	1.86	1992		
Udhampur			**Kangra**		
Year	*logY*	*logYt*	*Year*	*logY*	*logYt*
1951			1951		
1961	2.16	2.17	1961		
1972	2.24	2.21	1972	2.82	2.88
1982	2.25	2.26	1982	2.83	2.72
1992	2.31	2.31	1992	2.50	2.56
Hamirpur			**Jammu**		
Year	*logY*	*logYt*	*Year*	*logY*	*logYt*
1951			1951	2.63	2.83
1961			1961	2.89	2.85
1972	2.67	2.77	1972	3.29	2.86
1982	2.80	2.60	1982	2.71	2.88
1992	2.34	2.44	1992	2.79	2.89

Dehradun		
Year	*logY*	*logYt*
1951		
1961	2.09	2.19
1972	2.05	1.97
1982	1.90	1.75
1992	1.40	1.53

Sirmour		
Year	*logY*	*logYt*
1951	2.36	2.59
1961	2.73	2.60
1972	2.83	2.62
1982	2.71	2.63
1992	2.45	2.65

Mandi		
Year	*logY*	*logYt*
1951	2.08	2.02
1961	1.65	1.79
1972	1.68	1.56
1982	1.30	1.34
1992	1.11	1.11

Poonch		
Year	*logY*	*logYt*
1951		
1961	1.08	1.33
1972	1.89	1.49
1982	1.62	1.65
1992	1.70	1.81

Kathua		
Year	*logY*	*logYt*
1951	2.25	2.44
1961	2.61	2.47
1972	2.72	2.51
1982	2.49	2.54
1992	2.47	2.58

Total		
Year	*logY*	*logYt*
1951	2.98	3.28
1961	3.54	3.38
1972	3.79	3.49
1982	3.69	3.60
1992	3.44	3.70

Pithoragarh		
Year	*logY*	*logYt*
1951		
1961	2.29	2.30
1972	1.72	1.89
1982	1.81	1.48
1992	0.90	1.07

Appendix-VI

Log Y and Log Yt Values of Number of Wheat Threshers in Selected Districts of North-Western Himalaya

Bilaspur			**Kathua**		
Year	*logY*	*logYt*	*Year*	*logY*	*logYt*
1972	0.78	1.08	1972	1.08	1.36
1982	2.87	2.26	1982	2.32	1.76
1992	3.14	3.44	1992	1.89	2.17
Mandi			**Hamirpur**		
Year	*logY*	*logYt*	*Year*	*logY*	*logYt*
1972	1.73	1.88	1972	0.30	0.70
1982	2.98	2.69	1982	2.98	2.17
1992	3.35	3.50	1992	3.24	3.64
Jammu			**Sirmour**		
Year	*logY*	*logYt*	*Year*	*logY*	*logYt*
1972	2.02	2.15	1972	0.70	1.02
1982	2.86	2.59	1982	2.64	2.01
1992	2.89	3.03	1992	2.69	3.01
Chamba			**Dehradun**		
Year	*logY*	*logYt*	*Year*	*logY*	*logYt*
1972	0.30	0.61	1972	1.82	1.92
1982	2.26	1.69	1982	2.65	2.44
1992	2.49	2.77	1992	2.86	2.96
Shimla			**Kangra**		
Year	*logY*	*logYt*	*Year*	*logY*	*logYt*
1972	0.90	1.01	1972	2.03	2.25
1982	2.16	1.95	1982	3.39	2.94
1992	2.79	2.90	1992	3.41	3.63

Solan			Una		
Year	*logY*	*logYt*	*Year*	*logY*	*logYt*
1972	1.79	1.89	1972	1.92	2.10
1982	2.67	2.48	1982	3.04	2.70
1992	2.98	3.08	1992	3.13	3.31
Garhwal			**Total**		
Year	*logY*	*logYt*	*Year*	*logY*	*logYt*
1972	0.30	023	1972	2.74	2.94
1982	0.48	0.62	1982	4.15	3.74
1992	1.08	1.01	1992	4.34	4.54
Kullu					
Year	*logY*	*logYt*			
1972	1.11	1.30			
1982	2.60	2.23			
1992	2.98	3.16			

Appendix-VII

Log Y and Log Yt Values of Number of Rice Threshers in Selected Districts of North-Western Himalaya

Bilaspur			Jammu		
Year	*logY*	*logYt*	*Year*	*logY*	*logYt*
1972	0.90	0.87	1972	0.60	0.78
1982	1.11	1.19	1982	1.75	1.40
1992	1.54	1.51	1992	1.84	2.02
Mandi			**Hamirpur**		
Year	*logY*	*logYt*	*Year*	*logY*	*logYt*
1972	1.30	1.32	1972	0.30	0.40
1982	1.87	1.82	1982	1.40	1.20
1992	2.30	2.32	1992	1.90	2.00
Baramula			**Sirmour**		
Year	*logY*	*logYt*	*Year*	*logY*	*logYt*
1972	1.08	1.19	1972	0.00	0.57
1982	1.43	1.21	1982	2.44	1.31
1992	1.11	1.22	1992	1.49	2.06
Chamba			**Almora**		
Year	*logY*	*logYt*	*Year*	*logY*	*logYt*
1972	0.60	0.79	1972	0.85	0.89
1982	1.70	1.32	1982	0.70	0.62
1992	1.67	1.86	1992	0.30	0.35
Shimla			**Kangra**		
Year	*logY*	*logYt*	*Year*	*logY*	*logYt*
1972	0.70	0.60	1972	1.28	1.39
1982	0.85	1.04	1982	2.35	2.13
1992	1.58	1.48	1992	2.75	2.86

Solan		
Year	*logY*	*logYt*
1972	0.95	0.97
1982	1.51	1.48
1992	1.99	2.00

Una		
Year	*logY*	*logYt*
1972	1.23	1.32
1982	1.72	1.54
1992	1.67	1.76

Dehradun		
Year	*logY*	*logYt*
1972	0.00	0.20
1982	1.64	1.24
1992	2.07	2.27

Total		
Year	*logY*	*logYt*
1972	2.48	2.54
1982	3.00	2.90
1992	3.20	3.26

Kullu		
Year	*logY*	*logYt*
1972	0.85	0.96
1982	1.43	1.20
1992	1.32	1.44

Appendix-VIII

Log Yand Log Yt Values of Total Number of Threshers and Shellers in Selected Districts of North-Western Himalaya

	Bilaspur			Rajouri	
Year	*logY*	*logYt*	*Year*	*logY*	*logYt*
1972	2.97	2.96	1972	1.40	1.40
1982	3.24	3.25	1982	1.68	1.69
1992	3.54	3.53	1992	1.98	1.98
	Mandi			**Hamirpur**	
Year	*logY*	*logYt*	*Year*	*logY*	*logYt*
1972	2.62	2.61	1972	3.10	3.10
1982	2.85	2.87	1982	3.39	3.39
1992	3.15	3.14	1992	3.69	3.68
	Kathua			**Sirmour**	
Year	*logY*	*logYt*	*Year*	*logY*	*logYt*
1972	0.78	0.79	1972	2.35	2.35
1982	1.08	1.07	1982	2.63	2.64
1992	1.34	1.35	1992	2.92	2.92
	Chamba			**Almora**	
Year	*logY*	*logYt*	*Year*	*logY*	*logYt*
1972	1.85	1.84	1972	1.26	1.26
1982	2.08	2.10	1982	1.56	1.56
1992	2.38	2.37	1992	1.86	1.86
	Shimla			**Kangra**	
Year	*logY*	*logYt*	*Year*	*logY*	*logYt*
1972	2.23	2.19	1972	2.91	2.90
1982	2.29	2.37	1982	3.13	3.16
1992	2.59	2.55	1992	3.43	3.42

Solan			Nainital		
Year	*logY*	*logYt*	*Year*	*logY*	*logYt*
1972	2.87	2.87	1972	3.08	3.06
1982	3.16	3.16	1982	3.29	3.32
1992	3.45	3.45	1992	3.59	3.57
Dehradun			**Lahaul & Spiti**		
Year	*logY*	*logYt*	*Year*	*logY*	*logYt*
1972	2.50	2.49	1972	1.38	1.33
1982	2.80	2.79	1982	1.38	1.48
1992	3.09	3.09	1992	1.68	1.63
Kullu			**Jammu**		
Year	*logY*	*logYt*	*Year*	*logY*	*logYt*
1972	1.34	1.34	1972	2.21	2.20
1982	1.57	1.58	1982	2.50	2.50
1992	1.82	1.82	1992	2.80	2.80
Una			**Total**		
Year	*logY*	*logYt*	*Year*	*logY*	*logYt*
1972	2.94	2.94	1972	3.86	3.86
1982	3.21	3.22	1982	4.12	4.13
1992	3.51	3.51	1992	4.42	4.41

Appendix-IX

Log Y & Log Yt Values per cent Irrigated area in Selected Districts of North-Western Hinalaya

Bilaspur			Jammu		
Year	logY	logYt	Year	logY	logYt
1961-63	0.83	0.83	1961-63	1.52	1.52
1971-74	0.88	0.87	1971-74	1.60	1.60
1982-85	0.92	0.93	1982-85	1.66	1.67
1992-95	0.97	0.97	1992-95	1.75	1.75
Shimla			**Kangra**		
Year	logY	logYt	Year	logY	logYt
1961-63	1.00	1.00	1961-63		
1971-74	0.95	0.94	1971-74	1.45	1.45
1982-85	0.85	0.88	1982-85	1.46	1.46
1992-95	0.85	0.82	1992-95	1.48	1.48
Doda			**Solan**		
Year	logY	logYt	Year	logY	logYt
1961-63	1.15	1.13	1961-63		
1971-74	1.08	1.10	1971-74	1.20	1.21
1982-85	1.04	1.06	1982-85	1.30	1.30
1992-95	1.04	1.03	1992-95	1.40	1.40
Chamba			**Kathua**		
Year	logY	logYt	Year	logY	logYt
1961-63			1961-63	1.18	1.22
1971-74	1.08	1.07	1971-74	1.34	1.30
1982-85	1.04	1.05	1982-85	1.43	1.39
1992-95	1.04	1.03	1992-95	1.43	1.47
Sirmour			**Kinnaur**		
Year	logY	logYt	Year	logY	logYt
1961-63	1.38	1.37	1961-63	1.72	1.68
1971-74	1.40	1.42	1971-74	1.72	1.65
1982-85	1.46	1.46	1982-85	1.34	1.62
1992-95	1.52	1.51	1992-95	1.75	1.59

Una			**Udhampur**		
Year	*logY*	*logYt*	*Year*	*logY*	*logYt*
1961-63			1961-63	0.85	0.87
1971-74	0.60	0.57	1971-74	0.85	0.83
1982-85	0.78	0.85	1982-85	0.85	0.79
1992-95	1.18	1.14	1992-95	0.70	0.74
Srinagar			**Mandi**		
Year	*logY*	*logYt*	*Year*	*logY*	*logYt*
1961-63	1.80	1.79	1961-63	1.23	1.24
1971-74	1.81	1.82	1971-74	1.26	1.23
1982-85	1.87	1.86	1982-85	1.20	1.22
1992-95	1.88	1.89	1992-95	1.20	1.20
Kullu			**Baramula**		
Year	*logY*	*logYt*	*Year*	*logY*	*logYt*
1961-63			1961-63	1.66	1.67
1971-74	0.70	0.58	1971-74	1.68	1.68
1982-85	0.78	1.01	1982-85	1.70	1.70
1992-95	1.56	1.44	1992-95	1.71	1.72
Anantnag			**Total**		
Year	*logY*	*logYt*	*Year*	*logY*	*logYt*
1961-63	1.79	1.79	1961-63	1.20	1.20
1971-74	1.82	1.82	1971-74	1.26	1.30
1982-85	1.84	1.84	1982-85	1.46	1.40
1992-95	1.86	1.87	1992-95	1.48	1.50

Appendix-X

Log Y & Log Yt Values of Intensity of Landuse in Selected Districts of North-Western Himalaya

Bilaspur			Dehradun		
Year	*logY*	*logYt*	*Year*	*logY*	*logYt*
1961-63	2.25	2.24	1961-63	2.13	2.17
1971-74	2.23	2.25	1971-74	2.25	2.19
1982-85	2.28	2.26	1982-85	2.19	2.20
1992-95	2.28	2.28	1992-95	2.21	2.22
Doda			**Mandi**		
Year	*logY*	*logYt*	*Year*	*logY*	*logYt*
1961-63	2.07	2.07	1961-63	2.21	2.21
1971-74	2.09	2.08	1971-74	2.22	2.22
1982-85	2.08	2.08	1982-85	2.23	2.23
1992-95	2.08	2.08	1992-95	2.24	2.24
Chamoli			**Kathua**		
Year	*logY*	*logYt*	*Year*	*logY*	*logYt*
1961-63	2.05	2.08	1961-63	2.12	2.11
1971-74	2.16	2.13	1971-74	2.15	2.17
1982-85	2.18	2.17	1982-85	2.26	2.24
1992-95	2.19	2.21	1992-95	2.29	2.30
Chamba			**Nainital**		
Year	*logY*	*logYt*	*Year*	*logY*	*logYt*
1961-63	2.18	2.19	1961-63	2.14	2.17
1971-74	2.18	2.18	1971-74	2.23	2.19
1982-85	2.18	2.17	1982-85	2.23	2.22
1992-95	2.15	2.16	1992-95	2.23	2.24
Jammu			**Shimla**		
Year	*logY*	*logYt*	*Year*	*logY*	*logYt*
1961-63	2.20	2.22	1961-63	2.19	2.19
1971-74	2.27	2.24	1971-74	2.19	2.18
1982-85	2.29	2.27	1982-85	2.17	2.17
1992-95	2.26	2.29	1992-95	2.15	2.15

Leh			Tehri Garhwal		
Year	*logY*	*logYt*	*Year*	*logY*	*logYt*
1961-63	2.03	2.03	1961-63	2.20	2.21
1971-74	2.03	2.03	1971-74	2.22	2.21
1982-85	2.01	2.02	1982-85	2.20	2.20
1992-95	2.02	2.01	1992-95	2.20	2.20
Pithoragarh			**Anantnag**		
Year	*logY*	*logYt*	*Year*	*logY*	*logYt*
1961-63	2.00	2.07	1961-63	2.01	2.02
1971-74	2.23	2.14	1971-74	2.06	2.05
1982-85	2.25	2.22	1982-85	2.10	2.09
1992-95	2.25	2.30	1992-95	2.11	2.12
Sirmour			**Almora**		
Year	*logY*	*logYt*	*Year*	*logY*	*logYt*
1961-63	2.23	2.22	1961-63	2.00	2.07
1971-74	2.24	2.24	1971-74	2.25	2.14
1982-85	2.25	2.25	1982-85	2.21	2.20
1992-95	2.27	2.27	1992-95	2.23	2.27
Srinagar			**Total**		
Year	*logY*	*logYt*	*Year*	*logY*	*logYt*
1961-63	2.03	2.02	1961-63	2.10	2.11
1971-74	2.02	2.04	1971-74	2.15	2.14
1982-85	2.06	2.05	1982-85	2.18	2.17
1992-95	2.08	2.07	1992-95	2.19	2.20

Appendix-XI

Log Y & Log Yt Values of Area, Production and Productivity of Cereals and Pulses in North-Western Himalaya

Maize	Area	
	logY	*logYt*
1954-55	5.38	5.46
1964-65	5.61	5.56
1974-75	5.75	5.67
1985-86	5.79	5.77
1995-96	5.81	5.87
Maize	**Production**	
	logY	*logYt*
1954-55	5.21	5.32
1964-65	5.55	5.54
1974-75	5.94	5.76
1985-86	6.03	5.98
1995-96	6.08	6.20
Maize	**Productivity**	
	logY	*logYt*
1954-55	2.83	2.86
1964-65	2.95	2.98
1974-75	3.20	3.10
1985-86	3.23	3.21
1995-96	3.27	3.33
Rice	**Area**	
	logY	*logYt*
1954-55	5.46	5.46
1964-65	5.55	5.56
1974-75	5.64	5.65
1985-86	5.79	5.74
1995-96	5.80	5.83

Rice	Production	
	logY	*logYt*
1954-55	5.46	5.44
1964-65	5.52	5.62
1974-75	5.85	5.79
1985-86	6.07	5.97
1995-96	6.06	6.14
Rice	**Productivity**	
	logY	*logYt*
1954-55	3.00	2.95
1964-65	2.97	3.06
1974-75	3.21	3.17
1985-86	3.27	3.28
1995-96	3.40	3.39
Wheat	**Area**	
	logY	*logYt*
1954-55	5.64	5.58
1964-65	5.61	5.69
1974-75	5.78	5.80
1985-86	5.98	5.91
1995-96	5.99	6.02
Wheat	**Production**	
	logY	*logYt*
1954-55	5.48	5.39
1964-65	5.45	5.60
1974-75	5.82	5.82
1985-86	6.09	6.03
1995-96	6.24	6.25

Wheal	Productivity		Pulses	Production	
	logY	*logYt*		*logY*	*logYt*
1954-55	2.85	2.80	1954-55	4.49	4.58
1964-65	2.84	2.91	1964-65	4.63	4.64
1974-75	3.04	3.02	1974-75	4.90	4.69
1985-86	3.11	3.12	1985-86	4.73	4.74
1995-96	3.25	3.23	1995-96	4.70	4.79
Barley	**Area**		**Pulses**	**Productivity**	
	logY	*logYt*		*logY*	*logYt*
1954-55	4.87	4.79	1954-55	2.82	2.77
1964-65	4.69	4.79	1964-65	2.67	2.76
1974-75	4.76	4.79	1974-75	2.75	2.75
1985-86	4.86	4.80	1985-86	2.80	2.74
1995-96	4.80	4.80	1995-96	2.71	2.73
Barley	**Production**		**F'grain**	**Area**	
	logY	*logYt*		*logY*	*logYt*
1954-55	4.63	4.61	1954-55	6.01	6.03
1964-65	4.64	4.69	1964-65	6.14	6.14
1974-75	4.77	4.76	1974-75	6.28	6.25
1985-86	4.90	4.84	1985-86	6.42	6.37
1995-96	4.87	4.91	1995-96	6.43	6.48
Barley	**Productivity**		**F'grain**	**Production**	
	logY	*logYt*		*logY*	*logYt*
1954-55	2.34	2.57	1954-55	5.89	5.90
1964-65	2.95	2.73	1964-65	6.04	6.10
1974-75	3.01	2.88	1974-75	6.38	6.31
1985-86	3.05	3.04	1985-86	6.57	6.51
1995-96	3.08	3.20	1995-96	6.65	6.72
Pulses	**Area**		**F'grain**	**Productivity**	
	logY	*logYt*		*logY*	*logYt*
1954-55	4.66	4.82	1954-55	2.87	2.86
1964-65	4.97	4.88	1964-65	2.91	2.96
1974-75	5.15	4.94	1974-75	3.10	3.05
1985-86	4.92	5.00	1985-86	3.15	3.15
1995-96	4.99	5.06	1995-96	3.22	3.24

Appendix-XII

Log Y & Log Yt Values of Area, Production and Productivity of Cereals and Pulses in Nainital

Maize	Area	
	logY	logYt
1954-55	4.38	4.46
1964-65	4.36	4.32
1974-75	4.28	4.17
1985-86	4.00	4.03
1995-96	3.83	3.88

Maize	Production	
	logY	logYt
1954-55	4.25	4.32
1964-65	4.29	4.23
1974-75	4.21	4.13
1985-86	4.01	4.04
1995-96	3.92	3.95

Maize	Productivity	
	logY	logYt
1954-55	2.86	2.86
1964-65	2.93	2.91
1974-75	2.93	2.96
1985-86	3.01	3.01
1995-96	3.09	3.07

Rice	Area	
	logY	logYt
1954-55	4.72	4.75
1964-65	4.82	4.84
1974-75	4.98	4.92
1985-86	5.06	5.01
1995-96	5.03	5.09

Rice	Production	
	logY	logYt
1954-55	4.55	4.59
1964-65	4.80	4.85
1974-75	5.22	5.10
1985-86	5.44	5.35
1995-96	5.49	5.61

Rice	Productivity	
	logY	logYt
1954-55	2.83	2.83
1964-65	2.97	3.00
1974-75	3.17	3.16
1985-86	3.38	3.33
1995-96	3.46	3.50

Wheat	Area	
	logY	logYt
1954-55	4.69	4.64
1964-65	4.66	4.76
1974-75	4.91	4.88
1985-86	5.08	5.00
1995-96	5.07	5.12

Wheat	Production	
	logY	logYt
1954-55	4.59	4.52
1964-65	4.65	4.78
1974-75	5.02	5.04
1985-86	5.46	5.30
1995-96	5.49	5.56

Wheal	Productivity	
	logY	*logYt*
1954-55	2.90	2.87
1964-65	2.98	3.02
1974-75	3.11	3.16
1985-86	3.37	3.30
1995-96	3.42	3.44

Barley	Area	
	logY	*logYt*
1954-55	3.83	3.64
1964-65	3.43	3.50
1974-75	3.04	3.36
1985-86	3.29	3.21
1995-96	3.19	3.07

Barley	Production	
	logY	*logYt*
1954-55	3.73	3.43
1964-65	3.29	3.37
1974-75	2.70	3.31
1985-86	3.45	3.24
1995-96	3.35	3.18

Barley	Productivity	
	logY	*logYt*
1954-55	2.91	2.79
1964-65	2.86	2.87
1974-75	2.66	2.95
1985-86	3.16	3.03
1995-96	3.16	3.11

Pulses	Area	
	logY	*logYt*
1954-55	4.46	4.58
1964-65	4.52	4.40
1974-75	4.29	4.21
1985-86	3.97	4.03
1995-96	3.81	3.84

Pulses	Production	
	logY	*logYt*
1954-55	4.30	4.25
1964-65	4.10	4.13
1974-75	3.95	4.00
1985-86	3.86	3.88
1995-96	3.80	3.75

Pulses	Productivity	
	logY	*logYt*
1954-55	2.84	2.67
1964-65	2.58	2.73
1974-75	2.66	2.79
1985-86	2.89	2.85
1995-96	2.99	2.91

F'grain	Area	
	logY	*logYt*
1954-55	5.24	4.92
1964-65	5.23	5.12
1974-75	5.33	5.32
1985-86	5.42	5.52
1995-96	5.39	5.72

F'grain	Production	
	logY	*logYt*
1954-55	5.09	5.05
1964-65	5.17	5.26
1974-75	5.47	5.46
1985-86	5.77	5.67
1995-96	5.81	5.87

F'grain	Productivity	
	logY	*logYt*
1954-55	2.86	2.83
1964-65	2.94	2.99
1974-75	3.14	3.14
1985-86	3.35	3.29
1995-96	3.42	3.45

Appendix-XIII

Log Y & Log Yt Values of Area, Production and Productivity of Cereals and Pulses in Dehradun

Maize	Area		Rice	Production	
	logY	*logYt*		*logY*	*logYt*
1954-55	3.98	3.96	1954-55	3.96	3.94
1964-65	4.01	3.99	1964-65	4.11	4.06
1974-75	3.90	4.02	1974-75	4.05	4.18
1985-86	4.12	4.06	1985-86	4.33	4.30
1995-96	4.10	4.09	1995-96	4.45	4.42
Maize	**Production**		**Rice**	**Productivity**	
	logY	*logYt*		*logY*	*logYt*
1954-55	3.86	3.67	1954-55	2.81	2.82
1964-65	3.55	3.81	1964-65	3.01	2.92
1974-75	3.83	3.94	1974-75	2.89	3.02
1985-86	4.34	4.07	1985-86	3.14	3.12
1995-96	4.12	4.21	1995-96	3.24	3.21
Maize	**Productivity**		**Wheat**	**Area**	
	logY	*logYt*		*logY*	*logYt*
1954-55	2.87	2.72	1954-55	4.38	4.35
1964-65	2.54	2.82	1964-65	4.33	4.37
1974-75	2.93	2.92	1974-75	4.36	4.38
1985-86	3.22	3.01	1985-86	4.46	4.40
1995-96	3.02	3.11	1995-96	4.39	4.41
Rice	**Area**		**Wheat**	**Production**	
	logY	*logYt*		*logY*	*logYt*
1954-55	4.14	4.11	1954-55	4.15	3.98
1964-65	4.10	4.13	1964-65	4.00	4.15
1974-75	4.14	4.16	1974-75	4.11	4.31
1985-86	4.19	4.18	1985-86	4.66	4.47
1995-96	4.21	4.20	1995-96	4.64	4.64

Wheat	Productivity		Pulses	Production	
	logY	*logYt*		*logY*	*logYt*
1954-55	2.77	2.63	1954-55	3.49	3.38
1964-65	2.67	2.78	1964-65	3.16	3.33
1974-75	2.75	2.93	1974-75	3.25	3.28
1985-86	3.20	3.07	1985-86	3.33	3.23
1995-96	3.25	3.22	1995-96	3.15	3.18
Barley	**Area**		**Pulses**	**Productivity**	
	logY	*logYt*		*logY*	*logYt*
1954-55	3.63	3.60	1954-55	2.78	2.64
1964-65	3.51	3.59	1964-65	2.52	2.68
1974-75	3.48	3.58	1974-75	2.65	2.72
1985-86	3.90	3.57	1985-86	2.86	2.76
1995-96	3.39	3.59	1995-96	2.79	2.79
Barley	**Production**		**F'grain**	**Area**	
	logY	*logYt*		*logY*	*logYt*
1954-55	3.56	3.50	1954-55	4.82	4.79
1964-65	3.47	3.56	1964-65	4.77	4.80
1974-75	3.50	3.61	1974-75	4.77	4.80
1985-86	3.91	3.67	1985-86	4.82	4.80
1995-96	3.63	3.73	1995-96	4.80	4.80
Barley	**Productivity**		**F'grain**	**Production**	
	logY	*logYt*		*logY*	*logYt*
1954-55	2.93	2.90	1954-55	4.63	4.52
1964-65	2.96	2.96	1964-65	4.55	4.64
1974-75	3.02	3.03	1974-75	4.60	4.75
1985-86	3.00	3.10	1985-86	5.00	4.87
1995-96	3.24	3.16	1995-96	4.99	4.99
Pulses	**Area**		**F'grain**	**Productivity**	
	logY	*logYt*		*logY*	*logYt*
1954-55	3.71	3.73	1954-55	2.81	1.68
1964-65	3.65	3.65	1964-65	2.78	2.32
1974-75	3.61	3.56	1974-75	2.83	2.96
1985-86	3.47	3.47	1985-86	3.18	3.60
1995-96	3.36	3.38	1995-96	3.19	4.23

Appendix-XIV

Log Y & Log Yt Values of Area, Production and Productivity of Cereals and Pulses in Shimla

Maize	Area	
	logY	*logYt*
1954-55	4.36	4.38
1964-65	4.40	4.36
1974-75	4.33	4.34
1985-86	4.34	4.32
1995-96	4.29	4.30

Maize	Production	
	logY	*logYt*
1954-55	4.24	4.32
1964-65	4.40	4.40
1974-75	4.63	4.48
1985-86	4.58	4.57
1995-96	4.57	4.65

Maize	Productivity	
	logY	*logYt*
1954-55	2.88	2.93
1964-65	3.00	3.04
1974-75	3.30	3.14
1985-86	3.24	3.25
1995-96	3.28	3.35

Rice	Area	
	logY	*logYt*
1954-55	3.92	3.96
1964-65	3.92	3.86
1974-75	3.72	3.75
1985-86	3.68	3.65
1995-96	3.52	3.54

Rice	Production	
	logY	*logYt*
1954-55	3.99	3.93
1964-65	3.90	3.85
1974-75	3.50	3.77
1985-86	3.85	3.70
1995-96	3.62	3.62

Rice	Productivity	
	logY	*logYt*
1954-55	3.08	2.97
1964-65	2.98	3.00
1974-75	2.78	3.02
1985-86	3.18	3.05
1995-96	3.10	3.07

Wheat	Area	
	logY	*logYt*
1954-55	4.55	4.60
1964-65	4.63	4.57
1974-75	4.53	4.53
1985-86	4.51	4.49
1995-96	4.41	4.45

Wheat	Production	
	logY	*logYt*
1954-55	4.38	4.48
1964-65	4.58	4.52
1974-75	4.63	4.55
1985-86	4.63	4.59
1995-96	4.53	5.62

Wheat	**Productivity**		**Pulses**	**Production**	
	logY	*logYt*		*logY*	*logYt*
1954-55	2.83	2.88	1954-55	2.05	2.28
1964-65	2.95	2.95	1964-65	2.79	2.31
1974-75	3.10	3.03	1974-75	3.02	2.34
1985-86	3.12	3.10	1985-86	0.48	2.38
1995-96	3.12	3.17	1995-96	3.38	2.41
Barley	**Area**		**Pulses**	**Productivity**	
	logY	*logYt*		*logY*	*logYt*
1954-55	4.15	4.14	1954-55	2.47	2.65
1964-65	4.00	4.03	1964-65	3.04	2.58
1974-75	3.97	3.92	1974-75	2.35	2.51
1985-86	3.81	3.82	1985-86	2.11	2.44
1995-96	3.70	3.71	1995-96	2.59	2.37
Barley	**Production**		**F'grain**	**Area**	
	logY	*logYt*		*logY*	*logYt*
1954-55	3.88	3.98	1954-55	5.02	4.87
1964-65	4.03	3.96	1964-65	4.96	4.87
1974-75	4.09	3.94	1974-75	4.95	4.87
1985-86	3.82	3.92	1985-86	4.83	4.87
1995-96	3.88	3.90	1995-96	4.83	4.87
Barley	**Productivity**		**F'grain**	**Production**	
	logY	*logYt*		*logY*	*logYt*
1954-55	2.73	2.84	1954-55	4.86	4.91
1964-65	3.03	2.93	1964-65	4.94	4.94
1974-75	3.12	3.02	1974-75	5.05	4.96
1985-86	3.02	3.10	1985-86	4.98	4.99
1995-96	3.18	3.19	1995-96	4.97	5.01
Pulses	**Area**		**F'grain**	**Productivity**	
	logY	*logYt*		*logY*	*logYt*
1954-55	2.58	2.63	1954-55	2.84	2.89
1964-65	2.75	2.73	1964-65	2.98	2.96
1974-75	3.67	2.83	1974-75	3.10	3.04
1985-86	1.36	2.94	1985-86	3.15	3.12
1995-96	3.79	3.04	1995-96	3.14	3.19

Appendix-XV

Log Y & Log Yt Values of Area, Production and Productivity of Cereals and Pulses in Sirmour

Maize	Area		Rice	Production	
	logY	logYt		logY	logYt
1954-55	4.29	4.29	1954-55	3.57	3.56
1964-65	4.31	4.33	1964-65	3.64	3.60
1974-75	4.39	4.36	1974-75	3.48	3.63
1985-86	4.41	4.39	1985-86	3.78	3.67
1995-96	4.40	4.43	1995-96	3.68	3.70
Maize	**Production**		**Rice**	**Productivity**	
	logY	logYt		logY	logYt
1954-55	4.08	4.16	1954-55	2.84	2.82
1964-65	4.36	4.34	1964-65	2.88	2.86
1974-75	4.67	4.52	1974-75	2.78	2.90
1985-86	4.64	4.69	1985-86	3.08	2.95
1995-96	4.83	4.87	1995-96	2.95	2.99
Maize	**Productivity**		**Wheat**	**Area**	
	logY	logYt		logY	logYt
1954-55	2.80	2.87	1954-55	4.36	4.36
1964-65	3.04	3.01	1964-65	4.39	4.39
1974-75	3.29	3.16	1974-75	4.44	4.42
1985-86	3.23	3.30	1985-86	4.47	4.45
1995-96	3.42	3.45	1995-96	4.47	4.48
Rice	**Area**		**Wheat**	**Production**	
	logY	logYt		logY	logYt
1954-55	3.74	3.74	1954-55	4.19	4.22
1964-65	3.76	3,74	1964-65	4.35	4.34
1974-75	3.70	3.73	1974-75	4.53	4.47
1985-86	3.70	3.72	1985-86	4.56	4.59
1995-96	3.73	3.71	1995-96	4.70	4.72

Wheat	Productivity		Pulses	Production	
	logY	*logYt*		*logY*	*logYt*
1954-55	2.83	2.86	1954-55	2.91	3.49
1964-65	2.96	2.95	1964-65	3.31	3.28
1974-75	3.09	3.04	1974-75	3.36	3.07
1985-86	3.10	3.14	1985-86	2.62	2.86
1995-96	3.24	3.23	1995-96	3.15	2.65

Barley	Area		Pulses	Productivity	
	logY	*logYt*		*logY*	*logYt*
1954-55	3.53	3.56	1954-55	2.56	2.79
1964-65	3.55	3.53	1964-65	3.08	2.76
1974-75	3.55	3.51	1974-75	2.63	2.73
1985-86	3.47	3.49	1985-86	2.87	2.70
1995-96	3.46	3.47	1995-96	2.50	2.67

Barley	Production		F'grain	Area	
	logY	*logYt*		*logY*	*logYt*
1954-55	3.26	3.37	1954-55	4.75	4.75
1964-65	3.53	3.41	1964-65	4.76	4.78
1974-75	3.47	3.45	1974-75	4.83	4.80
1985-86	3.54	3.49	1985-86	4.81	4.82
1995-96	3.45	3.53	1995-96	4.83	4.84

Barley	Productivity		F'grain	Production	
	logY	*logYt*		*logY*	*logYt*
1954-55	3.73	2.82	1954-55	4.55	4.60
1964-65	2.98	2.88	1964-65	4.75	4.74
1974-75	2.92	2.94	1974-75	4.96	4.87
1985-86	3.07	3.00	1985-86	4.96	5.00
1995-96	2.98	3.06	1995-96	5.11	5.13

Pulses	Area		F'grain	Productivity	
	logY	*logYt*		*logY*	*logYt*
1954-55	3.35	3.32	1954-55	2.80	2.85
1964-65	3.23	3.33	1964-65	2.99	2.96
1974-75	3.73	3.34	1974-75	3.12	3.07
1985-86	2.75	3.35	1985-86	3.15	3.18
1995-96	3.64	3.36	1995-96	3.27	3.29

Appendix-XVI

Log Y & Log Yt Values of Area, Production and Productivity of Cereals and Pulses in Mandi

Maize	Area	
	logY	*logYt*
1954-55	4.43	4.43
1964-65	4.49	4.49
1974-75	4.56	4.56
1985-86	4.65	4.63
1995-96	4.68	4.69

Maize	Production	
	logY	*logYt*
1954-55	4.17	4.23
1964-65	4.49	4.47
1974-75	4.82	4.71
1985-86	4.94	4.95
1995-96	5.14	5.19

Maize	Productivity	
	logY	*logYt*
1954-55	2.74	2.80
1964-65	3.00	2.98
1974-75	3.26	3.15
1985-86	3.29	3.33
1995-96	3.46	3.50

Rice	Area	
	logY	*logYt*
1954-55	4.35	4.38
1964-65	4.40	4.37
1974-75	4.37	4.37
1985-86	4.38	4.36
1995-96	4.34	4.36

Rice	Production	
	logY	*logYt*
1954-55	4.22	4.27
1964-65	4.43	4.34
1974-75	4.35	4.40
1985-86	4.51	4.47
1995-96	4.50	4.53

Rice	Productivity	
	logY	*logYt*
1954-55	2.86	2.89
1964-65	3.04	2.96
1974-75	2.93	3.02
1985-86	3.13	3.09
1995-96	3.16	3.16

Wheat	Area	
	logY	*logYt*
1954-55	4.61	4.64
1964-65	4.75	4.70
1974-75	4.73	4.75
1985-86	4.84	4.81
1995-96	4.84	4.86

Wheat	Production	
	logY	*logYt*
1954-55	4.44	4.49
1964-65	4.69	4.64
1974-75	4.78	4.78
1985-86	4.97	4.93
1995-96	5.03	5.07

Wheat	Productivity	
	logY	*logYt*
1954-55	2.83	2.85
1964-65	2.94	2.94
1974-75	3.05	3.03
1985-86	3.13	3.12
1995-96	3.19	3.21
Barley	**Area**	
	logY	*logYt*
1954-55	3.77	3.70
1964-65	3.78	3.73
1974-75	3.73	3.75
1985-86	3.80	3.77
1995-96	3.65	3.79
Barley	**Production**	
	logY	*logYt*
1954-55	3.48	3.62
1964-65	3.78	3.67
1974-75	3.76	3.71
1985-86	3.86	3.76
1995-96	3.68	3.81
Barley	**Productivity**	
	logY	*logYt*
1954-55	2.71	2.83
1964-65	3.00	2.90
1974-75	3.03	2.97
1985-86	3.06	3.04
1995-96	3.03	3.10
Pulses	**Area**	
	logY	*logYt*
1954-55	2.73	2.76
1964-65	2.78	2.90
1974-75	3.75	3.03
1985-86	2.23	3.16
1995-96	3.65	3.29

Pulses	Production	
	logY	*logYt*
1954-55	2.15	2.50
1964-65	2.78	2.63
1974-75	3.71	2.76
1985-86	1.96	2.89
1995-96	3.20	3.02
Pulses	**Productivity**	
	logY	*logYt*
1954-55	2.42	2.74
1964-65	3.01	2.74
1974-75	2.97	2.73
1985-86	2.73	2.73
1995-96	2.55	2.73
F'grain	**Area**	
	logY	*logYt*
1954-55	5.04	5.05
1964-65	5.10	5.09
1974-75	5.12	5.12
1985-86	2.17	5.16
1995-96	5.18	5.19
F'grain	**Production**	
	logY	*logYt*
1954-55	4.83	4.88
1964-65	5.09	5.03
1974-75	5.22	5.19
1985-86	5.35	5.34
1995-96	5.45	5.49
F'grain	**Productivity**	
	logY	*logYt*
1954-55	2.79	2.83
1964-65	2.98	2.95
1974-75	3.09	3.07
1985-86	3.18	3.18
1995-96	3.28	3.30

Appendix-XVII

Log Y & Log Yt Values of Area, Production and Productivity of Cereals and Pulses in Bilaspur

Maize	**Area**	
	logY	*logYt*
1954-55	4.27	4.26
1964-65	4.29	4.30
1974-75	4.33	4.34
1985-86	4.37	4.38
1995-96	4.43	4.42

Maize	**Production**	
	logY	*logYt*
1954-55	4.00	4.13
1964-65	4.29	4.28
1974-75	4.68	4.44
1985-86	4.57	4.59
1995-96	4.63	4.74

Maize	**Productivity**	
	logY	*logYt*
1954-55	2.73	3.17
1964-65	3.00	3.28
1974-75	3.34	3.40
1985-86	3.21	3.51
1995-96	3.20	3.63

Rice	**Area**	
	logY	*logYt*
1954-55	3.66	3.69
1964-65	3.64	3.62
1974-75	3.57	3.54
1985-86	3.49	3.47
1995-96	3.36	3.39

Rice	**Production**	
	logY	*logYt*
1954-55	3.37	3.50
1964-65	3.58	3.48
1974-75	3.49	3.47
1985-86	3.62	3.45
1995-96	3.28	3.44

Rice	**Productivity**	
	logY	*logYt*
1954-55	2.71	2.80
1964-65	2.94	2.86
1974-75	2.92	2.92
1985-86	3.13	2.98
1995-96	2.92	3.04

Wheat	**Area**	
	logY	*logYt*
1954-55	4.22	4.23
1964-65	4.30	4.28
1974-75	4.30	4.34
1985-86	4.41	4.39
1995-96	4.44	4.44

Wheat	**Production**	
	logY	*logYt*
1954-55	4.00	4.09
1964-65	4.29	4.21
1974-75	4.39	4.32
1985-86	4.40	4.43
1995-96	4.50	4.54

Wheat	Productivity		Pulses	Production	
	logY	logYt		logY	logYt
1954-55	2.78	2.87	1954-55	3.62	3.73
1964-65	2.98	2.92	1964-65	3.52	3.45
1974-75	3.08	2.98	1974-75	3.52	3.16
1985-86	2.99	3.04	1985-86	2.38	2.88
1995-96	3.06	3.09	1995-96	2.78	2.60
Barley	**Area**		**Pulses**	**Productivity**	
	logY	logYt		logY	logYt
1954-55	2.35	2.25	1954-55	2.91	2.96
1964-65	2.24	2.36	1964-65	2.97	2.83
1974-75	2.45	2.46	1974-75	2.65	2.70
1985-86	2.55	2.56	1985-86	2.46	2.57
1995-96	2.70	2.66	1995-96	2.52	2.45
Barley	**Production**		**F'grain**	**Area**	
	logY	logYt		logY	logYt
1954-55	2.25	2.18	1954-55	4.66	4.66
1964-65	2.26	2.34	1964-65	4.68	4.69
1974-75	2.49	2.49	1974-75	4.73	4.71
1985-86	2.60	2.64	1985-86	4.73	4.74
1995-96	2.85	2.80	1995-96	4.77	4.77
Barley	**Productivity**		**F'grain**	**Production**	
	logY	logYt		logY	logYt
1954-55	2.90	2.93	1954-55	4.43	4.52
1964-65	3.02	2.98	1964-65	4.66	4.63
1974-75	3.04	3.03	1974-75	4.90	4.74
1985-86	3.05	3.08	1985-86	4.83	4.85
1995-96	3.15	3.13	1995-96	4.89	4.96
Pulses	**Area**		**F'grain**	**Productivity**	
	logY	logYt		logY	logYt
1954-55	3.71	3.77	1954-55	2.77	2.88
1964-65	3.55	3.62	1964-65	3.01	2.96
1974-75	3.87	3.46	1974-75	3.17	3.04
1985-86	2.92	3.31	1985-86	3.10	3.12
1995-96	3.26	3.15	1995-96	3.12	3.19

Appendix-XVIII

Log Y & Log Yt Values of Area, Production and Productivity of Cereals and Pulses in Chamba

Maize	Area		Rice	Production	
	logY	*logYt*		*logY*	*logYt*
1954-55	4.38	4.37	1954-55	3.62	3.62
1964-65	4.37	4.39	1964-65	3.58	3.61
1974-75	4.41	4.40	1974-75	3.62	3.60
1985-86	4.41	4.42	1985-86	3.62	3.58
1995-96	4.45	4.44	1995-96	3.54	3.57
Maize	**Production**		**Rice**	**Productivity**	
	logY	*logYt*		*logY*	*logYt*
1954-55	4.11	4.18	1954-55	3.07	3.08
1964-65	4.37	4.36	1964-65	3.05	3.08
1974-75	4.70	4.55	1974-75	3.12	3.08
1985-86	4.67	4.73	1985-86	3.12	3.09
1995-96	4.89	4.92	1995-96	3.05	3.09
Maize	**Productivity**		**Wheat**	**Area**	
	logY	*logYt*		*logY*	*logYt*
1954-55	2.73	2.80	1954-55	4.18	4.17
1964-65	2.99	2.97	1964-65	4.18	4.20
1974-75	3.29	3.14	1974-75	4.21	4.23
1985-86	3.27	3.31	1985-86	4.27	4.26
1995-96	3.44	3.48	1995-96	4.29	4.29
Rice	**Area**		**Wheat**	**Production**	
	logY	*logYt*		*logY*	*logYt*
1954-55	3.55	3.54	1954-55	4.01	3.99
1964-65	3.53	3.53	1964-65	4.13	4.10
1974-75	3.50	3.51	1974-75	4.07	4.20
1985-86	3.49	3.50	1985-86	4.38	4.31
1995-96	3.49	3.48	1995-96	4.42	4.42

Wheat	Productivity	
	logY	*logYt*
1954-55	2.83	2.87
1964-65	3.06	2.93
1974-75	2.86	3.00
1985-86	3.10	3.06
1995-96	3.13	3.13

Barley	Area	
	logY	*logYt*
1954-55	3.77	3.82
1964-65	3.75	3.77
1974-75	3.78	3.71
1985-86	3.77	3.66
1995-96	3.51	3.61

Barley	Production	
	logY	*logYt*
1954-55	3.32	3.54
1964-65	3.77	3.60
1974-75	3.78	3.66
1985-86	3.85	3.72
1995-96	3.59	3.79

Barley	Productivity	
	logY	*logYt*
1954-55	2.56	2.73
1964-65	3.02	2.84
1974-75	3.01	2.95
1985-86	3.08	3.06
1995-96	3.09	3.17

Pulses	Area	
	logY	*logYt*
1954-55	A	A
1964-65	A	A
1974-75	3.52	1.71
1985-86	0.30	2.45
1995-96	3.54	3.19

Pulses	Production	
	logY	*logYt*
1954-55	A	A
1964-65	A	A
1974-75	2.95	1.94
1985-86	0.00	2.02
1995-96	3.11	2.10

Pulses	Productivity	
	logY	*logYt*
1954-55	A	A
1964-65	A	A
1974-75	2.43	2.03
1985-86	2.70	3.11
1995-96	2.57	3.65

F'grain	Area	
	logY	*logYt*
1954-55	4.72	4.70
1964-65	4.68	4.72
1974-75	4.77	4.73
1985-86	4.73	4.75
1995-96	4.77	4.77

F'grain	Production	
	logY	*logYt*
1954-55	4.48	4.52
1964-65	4.67	4.66
1974-75	4.87	4.80
1985-86	4.92	4.94
1995-96	5.05	5.08

F'grain	Productivity	
	logY	*logYt*
1954-55	2.76	2.82
1964-65	2.99	2.94
1974-75	3.10	3.06
1985-86	3.19	3.19
1995-96	3.28	3.31

Appendix-XIX

Log Y & Log Yt Values of Area, Production and Productivity of Cereals and Pulses in Kathua

Maize	**Area**	
	logY	*logYt*
1954-55	3.92	4.03
1964-65	4.16	4.08
1974-75	4.20	4.13
1985-86	4.25	4.19
1995-96	4.14	4.24

Maize	**Production**	
	logY	*logYt*
1954-55	3.79	3.87
1964-65	4.06	4.05
1974-75	4.35	4.24
1985-86	4.48	4.43
1995-96	4.52	4.62

Maize	**Productivity**	
	logY	*logYt*
1954-55	2.87	2.84
1964-65	2.91	2.97
1974-75	3.15	3.11
1985-86	3.23	3.24
1995-96	3.38	3.38

Rice	**Area**	
	logY	*logYt*
1954-55	4.12	4.16
1964-65	4.28	4.25
1974-75	4.34	4.33
1985-86	4.44	4.41
1995-96	4.46	4.49

Rice	**Production**	
	logY	*logYt*
1954-55	4.11	4.02
1964-65	4.06	4.19
1974-75	4.34	4.37
1985-86	4.63	4.54
1995-96	4.70	4.72

Rice	**Productivity**	
	logY	*logYt*
1954-55	2.99	2.86
1964-65	2.78	2.95
1974-75	3.00	3.04
1985-86	3.19	3.13
1995-96	3.24	3.22

Wheat	**Area**	
	logY	*logYt*
1954-55	4.22	4.28
1964-65	4.45	4.40
1974-75	4.55	4.51
1985-86	4.64	4.63
1995-96	4.71	4.75

Wheat	**Production**	
	logY	*logYt*
1954-55	4.03	4.01
1964-65	4.20	4.23
1974-75	4.56	4.45
1985-86	4.48	4.67
1995-96	4.99	4.89

Wheat	**Productivity**		**Pulses**	**Production**	
	logY	*logYt*		*logY*	*logYt*
1954-55	2.81	2.73	1954-55	2.89	3.10
1964-65	2.75	2.83	1964-65	3.33	3.13
1974-75	3.01	2.94	1974-75	3.30	3.15
1985-86	2.84	3.04	1985-86	3.14	3.18
1995-96	3.28	3.14	1995-96	3.10	3.20
Barley	**Area**		**Pulses**	**Productivity**	
	logY	*logYt*		*logY*	*logYt*
1954-55	3.44	3.48	1954-55	2.81	2.72
1964-65	3.39	3.42	1964-65	2.54	2.61
1974-75	3.48	3.36	1974-75	2.45	2.50
1985-86	3.22	3.30	1985-86	2.33	2.39
1995-96	3.22	3.24	1995-96	2.35	2.27
Barley	**Production**		**F'grain**	**Area**	
	logY	*logYt*		*logY*	*logYt*
1954-55	3.31	3.29	1954-55	4.65	4.72
1964-65	3.14	3.26	1964-65	4.87	4.81
1974-75	3.37	3.23	1974-75	4.93	4.89
1985-86	3.20	3.20	1985-86	5.01	4.98
1995-96	3.13	3.17	1995-96	5.01	5.06
Barley	**Productivity**		**F'grain**	**Production**	
	logY	*logYt*		*logY*	*logYt*
1954-55	2.87	2.81	1954-55	4.53	4.52
1964-65	2.75	2.85	1964-65	4.65	4.70
1974-75	2.89	2.88	1974-75	4.95	4.88
1985-86	2.97	2.91	1985-86	5.06	5.06
1995-96	2.91	2.94	1995-96	5.22	5.24
Pulses	**Area**		**F'grain**	**Productivity**	
	logY	*logYt*		*logY*	*logYt*
1954-55	3.08	3.38	1954-55	2.88	2.80
1964-65	3.78	3.52	1964-65	2.79	2.90
1974-75	3.85	3.65	1974-75	3.02	2.99
1985-86	3.81	3.79	1985-86	3.05	3.08
1995-96	3.75	3.93	1995-96	3.21	3.18

Appendix-XX

Log Y & Log Yt Values of Area, Production and Productivity of Cereals and Pulses in Jammu

Maize	Area	
	logY	logYt
1954-55	3.92	3.95
1964-65	4.10	4.04
1974-75	4.14	4.13
1985-86	4.18	4.22
1995-96	4.32	4.31

Maize	Production	
	logY	logYt
1954-55	3.78	3.74
1964-65	3.82	3.94
1974-75	4.22	4.15
1985-86	4.38	4.35
1995-96	4.51	4.55

Maize	Productivity	
	logY	logYt
1954-55	2.87	2.79
1964-65	2.72	2.90
1974-75	3.08	3.01
1985-86	3.21	3.13
1995-96	3.19	3.24

Rice	Area	
	logY	logYt
1954-55	4.21	4.25
1964-65	4.43	4.38
1974-75	4.50	4.52
1985-86	4.72	4.65
1995-96	4.74	4.79

Rice	Production	
	logY	logYt
1954-55	4.20	4.13
1964-65	4.26	4.37
1974-75	4.60	4.60
1985-86	4.88	4.83
1995-96	5.05	5.06

Rice	Productivity	
	logY	logYt
1954-55	2.99	2.89
1964-65	2.82	2.99
1974-75	3.10	3.08
1985-86	3.17	3.18
1995-96	3.31	3.28

Wheat	Area	
	logY	logYt
1954-55	4.50	4.62
1964-65	4.84	4.73
1974-75	4.89	4.84
1985-86	4.99	4.95
1995-96	4.97	5.06

Wheat	Production	
	logY	logYt
1954-55	4.31	4.22
1964-65	4.24	4.47
1974-75	4.97	4.72
1985-86	4.80	4.97
1995-96	5.29	5.22

Wheat	Productivity	
	logY	*logYt*
1954-55	2.81	2.60
1964-65	2.40	2.74
1974-75	3.08	2.88
1985-86	2.81	3.03
1995-96	3.32	3.17
Barley	**Area**	
	logY	*logYt*
1954-55	3.67	3.73
1964-65	3.45	3.41
1974-75	3.20	3.08
1985-86	2.64	2.76
1995-96	2.44	2.43
Barley	**Production**	
	logY	*logYt*
1954-55	2.48	2.95
1964-65	3.22	2.81
1974-75	3.02	2.67
1985-86	2.48	2.53
1995-96	2.14	2.39
Barley	**Productivity**	
	logY	*logYt*
1954-55	1.81	2.22
1964-65	2.78	2.41
1974-75	2.82	2.59
1985-86	2.83	2.77
1995-96	2.70	2.95
Pulses	**Area**	
	logY	*logYt*
1954-55	3.36	3.70
1964-65	4.10	3.81
1974-75	4.11	3.92
1985-86	4.12	4.03
1995-96	3.90	4.14

Pulses	Production	
	logY	*logYt*
1954-55	3.17	3.52
1964-65	3.66	3.47
1974-75	3.67	3.42
1985-86	3.68	3.37
1995-96	2.90	3.31
Pulses	**Productivity**	
	logY	*logYt*
1954-55	2.81	2.82
1964-65	2.56	2.66
1974-75	2.56	2.50
1985-86	2.56	2.34
1995-96	2.00	2.17
F'grain	**Area**	
	logY	*logYt*
1954-55	4.87	4.97
1964-65	5.18	5.06
1974-75	5.18	5.16
1985-86	5.27	5.25
1995-96	5.28	5.34
F'grain	**Production**	
	logY	*logYt*
1954-55	4.71	4.69
1964-65	4.81	4.90
1974-75	5.17	5.11
1985-86	5.37	5.32
1995-96	5.48	5.53
F'grain	**Productivity**	
	logY	*logYt*
1954-55	2.84	2.71
1964-65	2.63	2.83
1974-75	2.99	2.95
1985-86	3.09	3.07
1995-96	3.20	3.19

Appendix-XXI

Log Y & Log Yt Values of Area, Production and Productivity of Cereals and Pulses in Doda

Maize	Area	
	logY	logYt
1954-55	3.97	4.13
1964-65	4.47	4.29
1974-75	4.51	4.45
1985-86	4.63	4.62
1995-96	4.70	4.78
Maize	**Production**	
	logY	logYt
1954-55	3.83	4.01
1964-65	4.39	4.29
1974-75	4.74	4.57
1985-86	4.91	4.85
1995-96	4.97	5.13
Maize	**Productivity**	
	logY	logYt
1954-55	2.87	2.88
1964-65	2.91	3.00
1974-75	3.23	3.11
1985-86	3.28	3.23
1995-96	3.27	3.35
Rice	**Area**	
	logY	logYt
1954-55	3.51	3.54
1964-65	3.63	3.56
1974-75	3.53	3.58
1985-86	3.60	3.59
1995-96	3.61	3.61

Rice	Production	
	logY	logYt
1954-55	3.50	3.49
1964-65	3.50	3.57
1974-75	3.70	3.65
1985-86	3.81	3.74
1995-96	3.75	3.82
Rice	**Productivity**	
	logY	logYt
1954-55	2.99	2.95
1964-65	2.87	3.02
1974-75	3.17	3.08
1985-86	3.21	3.14
1995-96	3.14	3.20
Wheat	**Area**	
	logY	logYt
1954-55	3.69	3.64
1964-65	4.11	3.63
1974-75	2.70	3.62
1985-86	3.84	3.61
1995-96	3.76	3.59
Wheat	**Production**	
	logY	logYt
1954-55	3.43	3.59
1964-65	4.13	3.64
1974-75	3.28	3.70
1985-86	3.78	3.75
1995-96	3.88	3.81

Wheat	Productivity	
	logY	*logYt*
1954-55	2.74	2.75
1964-65	3.02	2.81
1974-75	2.58	2.88
1985-86	2.94	2.95
1995-96	3.12	3.01
Barley	**Area**	
	logY	*logYt*
1954-55	3.73	3.78
1964-65	3.83	3.74
1974-75	3.69	3.70
1985-86	3.65	3.66
1995-96	3.61	3.62
Barley	**Production**	
	logY	*logYt*
1954-55	3.54	3.65
1964-65	3.66	3.55
1974-75	3.49	3.46
1985-86	3.41	3.36
1995-96	3.17	3.26
Barley	**Productivity**	
	logY	*logYt*
1954-55	2.81	2.87
1964-65	2.84	2.81
1974-75	2.81	2.75
1985-86	2.76	2.70
1995-96	2.56	2.64
Pulses	**Area**	
	logY	*logYt*
1954-55		
1964-65	3.57	3.52
1974-75	3.46	3.52
1985-86	3.51	3.52
1995-96	3.54	3.52

Pulses	Production	
	logY	*logYt*
1954-55		
1964-65	3.18	3.05
1974-75	2.97	3.07
1985-86	2.88	3.10
1995-96	3.30	3.11
Pulses	**Productivity**	
	logY	*logYt*
1954-55		
1964-65	2.61	2.53
1974-75	2.51	2.55
1985-86	2.37	2.58
1995-96	2.75	2.59
F'grain	**Area**	
	logY	*logYt*
1954-55	4.40	4.52
1964-65	4.76	4.62
1974-75	4.76	4.72
1985-86	4.81	4.82
1995-96	4.86	4.92
F'grain	**Production**	
	logY	*logYt*
1954-55	4.25	4.37
1964-65	4.67	4.54
1974-75	4.79	4.71
1985-86	4.80	4.89
1995-96	5.05	5.06
F'grain	**Productivity**	
	logY	*logYt*
1954-55	2.85	2.84
1964-65	2.92	2.92
1974-75	3.03	2.99
1985-86	2.98	3.07
1995-96	3.19	3.14

Appendix-XXII

Log Y & Log Yt Values of Area, Production and Productivity of Cereals and Pulses in Poonch

Maize	Area	
	logY	*logYt*
1954-55	4.43	4.55
1964-65	4.74	4.50
1974-75	4.32	4.44
1985-86	4.35	4.39
1995-96	4.37	4.34

Maize	Production	
	logY	*logYt*
1954-55	4.30	4.42
1964-65	4.71	4.54
1974-75	4.62	4.65
1985-86	4.79	4.76
1995-96	4.83	4.88

Maize	Productivity	
	logY	*logYt*
1954-55	3.77	3.81
1964-65	3.92	3.74
1974-75	3.49	3.67
1985-86	3.61	3.60
1995-96	3.56	3.53

Rice	Area	
	logY	*logYt*
1954-55	3.77	3.81
1964-65	3.92	3.74
1974-75	3.49	3.67
1985-86	3.61	3.60
1995-96	3.56	3.53

Rice	Production	
	logY	*logYt*
1954-55	3.51	3.54
1964-65	3.68	3.64
1974-75	3.74	3.74
1985-86	3.85	3.84
1995-96	3.91	3.93

Rice	Productivity	
	logY	*logYt*
1954-55	2.74	2.73
1964-65	2.75	2.90
1974-75	3.25	3.07
1985-86	3.24	3.24
1995-96	3.35	3.40

Wheat	Area	
	logY	*logYt*
1954-55	3.93	4.06
1964-65	4.36	4.04
1974-75	3.84	4.02
1985-86	3.97	4.01
1995-96	4.04	3.99

Wheat	Production	
	logY	*logYt*
1954-55	3.74	3.63
1964-65	3.66	3.78
1974-75	3.91	3.93
1985-86	4.05	4.09
1995-96	4.31	4.24

Wheat	**Productivity**	
	logY	*logYt*
1954-55	2.81	2.57
1964-65	2.31	2.74
1974-75	3.07	2.91
1985-86	3.08	3.08
1995-96	3.27	3.25

Barley	**Area**	
	logY	*logYt*
1954-55	2.20	2.15
1964-65	1.94	1.81
1974-75	1.36	1.47
1985-86	0.78	1.13
1995-96	1.08	0.79

Barley	**Production**	
	logY	*logYt*
1954-55	2.07	2.13
1964-65	1.91	1.77
1974-75	1.40	1.40
1985-86	0.85	1.04
1995-96	0.78	0.67

Barley	**Productivity**	
	logY	*logYt*
1954-55	2.87	2.98
1964-65	2.97	2.95
1974-75	3.04	2.93
1985-86	3.07	2.90
1995-96	2.70	2.88

Pulses	**Area**	
	logY	*logYt*
1954-55	1.00	1.90
1964-65	3.24	1.98
1974-75	1.90	2.05
1985-86	2.26	2.13
1995-96	1.86	2.20

Pulses	**Production**	
	logY	*logYt*
1954-55	0.70	1.54
1964-65	2.80	1.56
1974-75	1.36	1.57
1985-86	1.64	1.58
1995-96	1.34	1.60

Pulses	**Productivity**	
	logY	*logYt*
1954-55	2.70	2.64
1964-65	2.56	2.58
1974-75	2.46	2.52
1985-86	2.38	2.46
1995-96	2.48	2.40

F'grain	**Area**	
	logY	*logYt*
1954-55	4.62	4.73
1964-65	4.95	4.69
1974-75	4.50	4.65
1985-86	4.58	4.60
1995-96	4.58	4.56

F'grain	**Production**	
	logY	*logYt*
1954-55	4.46	4.51
1964-65	4.79	4.60
1974-75	4.58	4.70
1985-86	4.69	4.80
1995-96	4.99	4.89

F'grain	**Productivity**	
	logY	*logYt*
1954-55	2.84	2.77
1964-65	2.84	2.92
1974-75	3.08	3.06
1985-86	3.11	3.20
1995-96	3.41	3.34

Log Y & Log Yt Values of Area, Production and Productivity of Cereals and Pulses in Srinagar

Maize	Area		Rice	Production	
	logY	logYt		logY	logYt
1954-55	3.78	4.22	1954-55	4.60	4.69
1964-65	4.55	4.11	1964-65	4.64	4.65
1974-75	4.33	4.01	1974-75	4.83	4.62
1985-86	3.71	3.90	1985-86	4.53	4.59
1995-96	3.66	3.79	1995-96	4.49	4.55
Maize	**Production**		**Rice**	**Productivity**	
	logY	logYt		logY	logYt
1954-55	3.65	4.04	1954-55	3.09	3.06
1964-65	4.37	3.99	1964-65	3.08	3.16
1974-75	4.21	3.94	1974-75	3.24	3.25
1985-86	3.76	3.89	1985-86	3.44	3.35
1995-96	3.69	3.83	1995-96	3.39	3.44
Maize	**Productivity**		**Wheat**	**Area**	
	logY	logYt		logY	logYt
1954-55	2.87	2.82	1954-55	3.99	3.08
1964-65	2.82	2.88	1964-65	3.84	3.01
1974-75	2.88	2.93	1974-75	3.75	2.93
1985-86	3.05	2.99	1985-86	1.88	2.85
1995-96	3.03	3.04	1995-96	1.20	2.78
Rice	**Area**		**Wheat**	**Production**	
	logY	logYt		logY	logYt
1954-55	4.51	4.63	1954-55	3.86	4.33
1964-65	4.56	4.50	1964-65	3.69	3.56
1974-75	4.58	4.37	1974-75	3.72	2.79
1985-86	4.09	4.24	1985-86	1.70	2.02
1995-96	4.11	4.11	1995-96	1.00	1.25

Wheat	Productivity	
	logY	*logYt*
1954-55	2.87	2.90
1964-65	2.85	2.88
1974-75	2.97	2.86
1985-86	2.82	2.84
1995-96	2.80	2.83

Barley	Area	
	logY	*logYt*
1954-55	2.84	2.39
1964-65	1.74	1.70
1974-75	1.63	1.02
1985-86	0.60	0.34
1995-96		

Barley	Production	
	logY	*logYt*
1954-55	2.71	2.70
1964-65	1.85	2.05
1974-75	1.69	1.40
1985-86	0.60	0.75
1995-96		

Barley	Productivity	
	logY	*logYt*
1954-55	2.87	2.95
1964-65	3.10	2.99
1974-75	3.06	3.02
1985-86	3.00	3.06
1995-96		

Pulses	Area	
	logY	*logYt*
1954-55		
1964-65	3.41	3.60
1974-75	3.71	3.39
1985-86	2.80	2.97
1995-96	2.81	2.76

Pulses	Production	
	logY	*logYt*
1954-55		
1964-65	3.57	3.82
1974-75	3.86	3.53
1985-86	2.94	2.96
1995-96	2.61	2.68

Pulses	Productivity	
	logY	*logYt*
1954-55		
1964-65	3.16	2.90
1974-75	3.16	3.01
1985-86	3.15	3.12
1995-96	2.79	3.23

F'grain	Area	
	logY	*logYt*
1954-55	4.64	4.83
1964-65	4.83	4.70
1974-75	4.85	4.57
1985-86	4.27	4.44
1995-96	4.26	4.30

F'grain	Production	
	logY	*logYt*
1954-55	4.68	4.84
1964-65	4.81	4.80
1974-75	5.11	4.75
1985-86	4.59	4.71
1995-96	4.56	4.66

F'grain	Productivity	
	logY	*logYt*
1954-55	3.04	3.01
1964-65	2.98	3.10
1974-75	3.26	3.18
1985-86	3.32	3.27
1995-96	3.31	3.35

Log Y & Log Yt Values of Area, Production and Productivity of Cereals and Pulses in Anantnag

Maize	Area	
	logY	logYt
1954-55	4.01	4.18
1964-65	4.33	4.25
1974-75	4.54	4.31
1985-86	4.36	4.38
1995-96	4.32	4.44

Maize	Production	
	logY	logYt
1954-55	3.88	4.02
1964-65	4.07	4.13
1974-75	4.55	4.24
1985-86	4.46	4.35
1995-96	4.23	4.46

Maize	Productivity	
	logY	logYt
1954-55	2.87	2.84
1964-65	2.74	2.88
1974-75	3.01	2.92
1985-86	3.10	2.97
1995-96	2.90	3.01

Rice	Area	
	logY	logYt
1954-55	4.81	4.88
1964-65	4.86	4.81
1974-75	4.86	4.75
1985-86	4.60	4.69
1995-96	4.62	4.62

Rice	Production	
	logY	logYt
1954-55	4.90	4.97
1964-65	4.86	4.98
1974-75	5.24	4.99
1985-86	5.10	5.00
1995-96	4.83	5.00

Rice	Productivity	
	logY	logYt
1954-55	3.09	3.09
1964-65	3.00	3.16
1974-75	3.38	3.24
1985-86	3.50	3.31
1995-96	3.21	3.38

Wheat	Area	
	logY	logYt
1954-55	4.09	4.50
1964-65	3.80	3.61
1974-75	3.53	2.72
1985-86	1.28	1.83
1995-96	0.90	0.94

Wheat	Production	
	logY	logYt
1954-55	3.96	2.82
1964-65	3.71	2.73
1974-75	3.58	2.65
1985-86	1.00	2.56
1995-96	1.00	2.48

Wheat	Productivity		Pulses	Production	
	logY	logYt		logY	logYt
1954-55	2.87	2.87	1954-55		
1964-65	2.91	2.90	1964-65	3.52	3.82
1974-75	3.05	2.93	1974-75	3.46	3.43
1985-86	2.72	2.96	1985-86	3.35	3.05
1995-96	3.10	2.98	1995-96	3.13	2.67
Barley	**Area**		**Pulses**	**Productivity**	
	logY	logYt		logY	logYt
1954-55	2.84	1.83	1954-55		
1964-65	1.76	1.09	1964-65	2.78	2.77
1974-75	0.85	0.35	1974-75	2.82	2.81
1985-86			1985-86	2.85	2.87
1995-96			1995-96	2.94	2.92
Barley	**Production**		**F'grain**	**Area**	
	logY	logYt		logY	logYt
1954-55	2.76	2.21	1954-55	4.95	5.05
1964-65	1.32	1.52	1964-65	5.08	5.00
1974-75	0.48	0.84	1974-75	5.06	4.94
1985-86			1985-86	4.82	4.89
1995-96			1995-96	4.81	4.84
Barley	**Productivity**		**F'grain**	**Production**	
	logY	logYt		logY	logYt
1954-55	2.92	3.13	1954-55	4.99	4.08
1964-65	2.57	2.71	1964-65	5.02	5.08
1974-75	2.63	2.29	1974-75	5.31	5.08
1985-86			1985-86	5.14	5.08
1995-96			1995-96	4.93	5.08
Pulses	**Area**		**F'grain**	**Productivity**	
	logY	logYt		logY	logYt
1954-55			1954-55	3.04	3.03
1964-65	3.74	3.40	1964-65	2.94	3.08
1974-75	3.65	3.40	1974-75	3.25	3.13
1985-86	3.50	3.40	1985-86	3.32	3.19
1995-96	3.20	3.40	1995-96	3.12	3.24

Log Y & Log Yt Values of Area, Production and Productivity of Cereals and Pulses in Baramula

Maize	Area	
	logY	logYt
1954-55	4.30	4.50
1964-65	4.65	4.50
1974-75	4.66	4.50
1985-86	4.48	4.50
1995-96	4.38	4.49

Maize	Production	
	logY	logYt
1954-55	4.17	4.43
1964-65	4.70	4.45
1974-75	4.61	4.47
1985-86	4.48	4.49
1995-96	4.39	4.51

Maize	Productivity	
	logY	logYt
1954-55	2.87	2.94
1964-65	3.07	2.97
1974-75	2.99	2.99
1985-86	3.00	3.01
1995-96	3.01	3.03

Rice	Area	
	logY	logYt
1954-55	4.62	4.69
1964-65	4.70	4.65
1974-75	4.71	4.62
1985-86	4.52	4.59
1995-96	4.55	4.56

Rice	Production	
	logY	logYt
1954-55	4.71	4.74
1964-65	4.68	4.78
1974-75	5.01	4.83
1985-86	4.91	4.87
1995-96	4.82	4.91

Rice	Productivity	
	logY	logYt
1954-55	3.09	3.05
1964-65	2.98	3.13
1974-75	3.29	3.20
1985-86	3.39	3.28
1995-96	3.27	3.35

Wheat	Area	
	logY	logYt
1954-55	4.07	4.07
1964-65	3.79	3.71
1974-75	3.31	3.34
1985-86	2.80	2.98
1995-96	2.74	2.61

Wheat	Production	
	logY	logYt
1954-55	3.74	3.80
1964-65	3.63	3.45
1974-75	3.08	3.11
1985-86	2.54	2.76
1995-96	2.54	2.41

Wheat	**Productivity**		**Pulses**	**Production**	
	logY	*logYt*		*logY*	*logYt*
1954-55	2.66	2.73	1954-55		
1964-65	2.84	2.74	1964-65	3.64	3.09
1974-75	2.77	2.76	1974-75	4.12	3.85
1985-86	2.74	2.78	1985-86	4.00	4.62
1995-96	2.80	2.80	1995-96	3.65	5.39
Barley	**Area**		**Pulses**	**Productivity**	
	logY	*logYt*		*logY*	*logYt*
1954-55	3.03	2.79	1954-55		
1964-65	2.19	2.16	1964-65	2.64	2.81
1974-75	1.32	1.54	1974-75	3.05	2.89
1985-86	0.30	0.91	1985-86	3.26	3.06
1995-96	0.85	0.29	1995-96	2.95	3.14
Barley	**Production**		**F'grain**	**Area**	
	logY	*logYt*		*logY*	*logYt*
1954-55	2.84	1.50	1954-55	4.88	5.00
1964-65	2.05	1.45	1964-65	5.06	4.97
1974-75	0.90	1.39	1974-75	5.06	4.93
1985-86	0.30	1.33	1985-86	4.85	4.90
1995-96	0.85	1.27	1995-96	4.82	4.87
Barley	**Productivity**		**F'grain**	**Production**	
	logY	*logYt*		*logY*	*logYt*
1954-55	2.81	2.82	1954-55	4.86	4.97
1964-65	2.86	2.87	1964-65	5.02	5.00
1974-75	2.93	2.92	1974-75	5.22	5.03
1985-86	3.00	2.97	1985-86	5.07	5.06
1995-96	3.00	3.02	1995-96	4.98	5.09
Pulses	**Area**		**F'grain**	**Productivity**	
	logY	*logYt*		*logY*	*logYt*
1954-55			1954-55	2.92	2.94
1964-65	4.00	4.06	1964-65	2.97	3.01
1974-75	4.07	3.97	1974-75	3.16	3.09
1985-86	3.74	3.78	1985-86	3.22	3.16
1995-96	3.70	3.69	1995-96	3.16	3.24

Bibliography

Books

Agarwala, V.P. (1985), *Forest in India*, Oxford and IBH Publishing Co., New Delhi.

Ahmad, A.J. and Muller, F.G. (1982), *Integrated Physical, Socio-Economic and Environmental Planning*, T. Cooly International Publishing Limited, Dublin.

Alan Wild (1993), *Soils and the Environment*, Cambridge University Press, London.

Allee, W.C. *et al.* (1949), *Principle of Animal Ecology*, W.B. Saunders, Philadelphia, pp. i-xii, 1-873.

Amani, K.Z. (1976), *Agricultural Land Use in Aligarh District*, Kumar Publications, Aligarh.

Anderson, M.S. (1968), *Geography of Living things*, The English University Press Ltd., London.

Andrewartha, H.G. and L.C. Birch (1954), *The Distribution and Abundance of Animals*, Chicago : University of Chicago Press, pp. i-xv, 1-782.

Ashish, M. (1983), Agricultural Economy of Kumaon Hills : Threat to Ecological Disaster, in O.P. Singh (ed.) *The Himalaya : Nature, Man and Culture*, New Delhi.

Atkinson, E.T. (1973), *Himalayan Gezetteer*, Cosmos Publications. New Delhi.

Azzi, G. (1956), *Agricultural Ecology*, Constable and Company, Ltd. London.

Baker, O.E. (1973), *A Geographic Summary of Physical Features and Land Utilization in USA*, U.S. Deptt. of Agri., Miscellaneous Publications, Washington DC.

Bamzai, P.N.K. (1978), *Geography of Jammu and Kashmir State,* Light and Life Publishers, New Delhi.

Bandyopadhyay, J. Dayal, N.D., Schoettli, U. and C. Singh (1985), *India's Environment : Crises and Responses,* Natraj Publishers, Dehradun.

Banett, C.F. (1975), *Man and Earth's Ecosystem,* John Wiley, New York.

Barnes, C.P. (1936), *Land Classification: Objectives and Requirements,* U.S. Resettlement Administrative Land Utilization Division, Washington.

Basu, D.N. and G.S. Guha (1996), *Agro-climatic Regional Planning in India,* Concept Publishing Co. Pvt. Ltd., New Delhi.

Basu, D.N. and S.P. Kashyap (1996), *Agro-climatic Regional Planning in India,* Concept Publishing Co. Pvt. Ltd., New Delhi.

Bates, C.E. (1980) *Gazetteer of Kashmir,* Light and Life Publishers, New Delhi.

Bennet, Hugh H. (1955), *Element of Soils Conservation,* McGraw-Hill, 12.

Bensin, B.M. (1930), *Possibilities for International Cooperation in Agroecological Investigations,* Internatl. Rev. Agr. Mo. Bull. Agr. Sci. and Pract., Rome.

Bharghouti, S., Garbus, L. and D. Umali (1992), *Trend in Agricultural Diversification,* The World Bank, Washington D.C.

Bose, S.C. (1968), *Land and People of the Himalaya,* Indian Publications, Calcutta.

Braun-Blanquet, J. (1964), *Pflanzensoziologie Grundzuge der Vegetationskunde,* 3rd ed., Springer, Vienna.

Briggs, D.J. and F.M. Courtney (1985), *Agriculture and Environment,* Longman, London.

Chadha, S.K. (1988), *Himalayas : Ecology and Environment,* Mittal Publications, Delhi.

Chadha, S.K. (1989), *Environmental Holocaust in Himalaya,* Ashish Publishing House, New Delhi.

Chadha, S.K. (1989), *Himalayan Ecology,* Ashish Publishing House, New Delhi.

Chakravarty, A. (1988) *Post Harvest Technology of Cereals, Pulses and Oil Seeds,* Oxford and IBH Publishing Co. Pvt. Ltd., New Delhi.

Charak, S.S. (1978), *History and Culture of Himalayan States : Himachal Pradesh (part One),* Light and Life Publications, New Delhi.

Chaudhary, B.D.N. (1983), *An Introduction to Environmental Management,* Interprint, New Delhi.

Chaudhuri, B. and A.K. Maiti (1989), *Forests and Forest Development in India,* Inter-India Publications, New Delhi.

Chauhan, I.S. and A. Chauhan (1998), *Environmental Degradation,* Rawal Publication, New Delhi.

Chorley, R.J. (1969), *Water, Earth and Man,* Methuen Pub., London.

Chute, R.M. (1971), *Environmental Insight,* Harper and Row Pub., New York.

Clements, F.E. (1905), *Research Methodology in Ecology,* Univ. Publ. Co. Lincoln, Nebaraska.

Colinvaux, P.A. (1973), *Introduction to Ecology,* John Wiley and Sons, Inc., New York.

Cox, G.W. (1969), *Reading in Conservation Ecology,* Appleton Century Pub. New York.

Cox, G.W. and M.D. Atkins, (1979), *Agricultural Ecology : An Analysis of World Productions Systems,* W.H. Freeman and Co., San Francisco.

Cox, G.W. and M.D. Atkins, (1964), *Agricultural Ecology,* W.H. Freeman and Company, San Francisco.

Cox, R.R. (1972), *Man, Location and Behaviour,* John Wiley and Sons, Inc., New York.

Crocker, R.L. (1952), Soil Genesis and Pedogenic Factors, Quart, *Rev. Biol.* 27, 139-168.

Dale, T. and V.G. Carter (1955), *Top Soil and Civilization,* University of Oklahama Press, New York.

Darwin, C. (1859), *The Origin of Species by Natural Selection,* Modern Library, New York.

Dasman, R.F., (1968) *Environmental Conservation,* John Wiley and Sons, New York.

Dasman, R.F., Milton, J.P. and P.H. Freeman (1973), *Ecological Principles for Economic Development,* John Wiley and Sons, London.

Daubenmire, R.F. (1974), *Plant and Environment,* Wiley Eastern Pvt. Ltd., New Delhi.

Davenport, C.B. (1903), *The Animal Ecology of the Cold Spring Harbour Sand Split with Remark on the Theory of Adaptation,* Decennial Publication, University of Chicago, 10 : 157-176.

Dhaliwal, G.S. and D.S. Kler, (1995), *Principles of Agricultural Ecology,* Himalayan Publishing House, New Delhi.

Dhar, S.N. (1976), *Jammu and Kashmir,* National Book Trust, New Delhi.

Dice, L.R. (1952), *Natural Communities,* Ann Arbor, University of Michigan Press, Michigan, pp. i-x, 1-547.

Dlij, De H.G. (1973), *Man Shapes the Earth : A Tropical Geography,* Santa Barbara, California.

Dombois, D.M. and Ellenberg, H. (1977), *Aims and Methods of Vegetation Ecology,* John Wiley and Sons, New York.

Douglas Pearce (1987), *Tourism Today : A geographical analysis,* John Wiley and Sons, New York.

Dubey, Y.P. and B.P. Kaistha (1998), Production Potential of Lahaul Valley–Constraints and Remedies, *Indian Farming,* vol. 48, No. 9, p. 21.

Ehrlick, P.R., Ehrlick, A.H. and J.P. Hodren (1977), *Ecoscience : Population, Resources and Environment,* Freeman, San Francisco.

Eric Duffey (1970), *Conservation of Nature,* Collins Publishers, McGraw-Hill Book Company, New York.

Evans, F.C. (1956), Ecosystem as the Basic unit in Ecology, *Science* 123, 1127-1128.

Evans, L.T. (1979), *Crop Physiology,* Blackie and Sons Pvt. Ltd., Bombay.

Eyre, S.R. and G.R.T.John (1961), *Geography and Human Ecology,* Edward Arnold, London.

Fedoseyev, P.N. (1983), *Environmental Protection and Soceity,* Academy of Sciences, Moscow.

Forbes, S.A. (1907), On the Local Distribution of Certain Iliinois

Fishes : An Essay in Statistical Ecology, *Bull. Ill. Lab. Nat. Hist. Survey*, 15: 537-550.

Garner, W.W. and H.A. Allard (1920), Effect of the Relative Length of Day and Night and other factors of the Environment on Growth and Production in Plants, *Journal of Agricultural Resource*, 18:553-606, New York.

Gause, G.F. (1935), *The Struggle for Existence*, McMillon Hafner Press, New York.

Gerasimov, T.P. (1983), *Geography and Ecology*, Progress Publishers, Moscow.

Ghosh, T.K. and S.P. Kashyap (1996), District Level Planning in India : An Overview of Agroclimatic Regional Planning Approach, G.S. Guha and S.P. Kashyap, (eds.) *Agroclimatic Regional Planning in India*, vol. 1, pp. 214-251, Concept Publishing Company Pvt. Ltd., New Delhi.

Goldsmith, E. (1992), *The Great U-Turn : Essays on Ecology*, Ashish Publishing House, New Delhi.

Gopal, B. and N. Bhardwaj (1979), *Elements of Ecology*, Vikas Publishing House Pvt. Ltd., New Delhi.

Gray, L.C. (1923), *Utilization of our Lands for Crops, Pastures and Forests*, U.S. Department of Agriculture Publication, New York.

Guderian, R. (1977), *Air Pollution : Ecological Studies*, Springer Verlag Berlin, Heidelberg, New York.

Guha, R. (1989), *The Unquiet Woods : Ecological Change and Peasant in the Himalaya*, Oxford University Press, New Delhi.

Haeckel, E. (1866), *Generelle Mophologie der Organismen*, vol. 2, Reimer, Berlin.

Harcourt, A.F.P. (1972), *The Himalayan Districts of Kooloo, Lahaul and Spiti*, Vivek Publishing House, New Delhi.

Hart, R.A.J. (1984), *Ecosociety : A Historical Study of Ecological Man*, Natraj Pub., Dehradun.

Holdren, J.P. and Ehrlick, P.R. (1971), *Global Ecology*, Harcourt Brack Jovanovick Inc., New York.

Holling, C.S. (1978), *Adaptive Environmental Assessment and Management*, John Wiley and Sons, New York.

Hugel, B.C. (1972), *Kashmir and Punjab*, Light and Life Publishers, Jammu.

Hussain, M. (1985), *Geography of Jammu and Kashmir*, Ariana Publishing House, New Delhi.

Hussain, M. (1994), Ecology and Environment, Anmol Publications Pvt. Ltd., New Delhi.

Hussain, M. and S.S.H. Zaidi (1996), *Environment Management in India*, Manak Publications Pvt. Ltd., New Delhi.

Hussain, Majid (1979), *Agricultural Geography*, Inter-India Publications, New Delhi.

Hussain, Majid (1982), *Crop Combinations in India*, Concept Publishing Company Pvt. Ltd., New Delhi.

Hussain, Majid (1994), *Agricultural Geography*, Anmol Publications Pvt. Ltd., New Delhi.

Hutchinson, J.B. (1965), *Crop Plant Evolution : A General Discussion*, Cambridge University Press, New York.

Jhina, P.S. (1962), *Agriculture in Hill Regions of North India*, Ministry of Food and Agriculture, New Delhi.

Jordan, C.F. (1975), What is Ecology, *Bull, Eco. Amer.* 56, 2-3.

Joshi, B.H. (1992), *Problems of Indian Agriculture*, B.R. Publishing Corporation, New Delhi.

Joshi, S.C. (1896), *Nepal Himalaya : Geo-Ecological Perspectives*, Himalayan Research Group, Nainital.

Joshi, S.C., Joshi, D.R. and D.D. Dani (1983), *Kumaon Himalaya: A Geographical Perspective on Resource Development*, Gyanodaya Prakashan, Nainital.

Joshi, S.N. (1995), Yield Gap Analysis in Agro-climatic Sub-region, D.N. Basu and S.P. Kashyap, (Eds.) *Agroclimatic Regional Planning in India*, vol. 2, Concept Publishing Company Pvt. Ltd., New Delhi.

Joy, Tivy and Greg O'Hare (1986), *Human Impact on the Environment*, Oliver and Boyed, New York.

Kapoor, A. (1995), *Paradise in Peril : An Ecological Profile of the Kashmir Valley*, Allied Publishers Limited, New Delhi.

Kataria, M.S. (1982), *Geography of Indian Livestock and White Revolution*, Vishal Publications, New Delhi.

Kaul, V. and U. Dar (1985), *Changing Vegetation of Kashmir in Geography of Jammu and Kashmir*, M. Hussain, *et al.*, Ariana Publishing House, New Delhi, pp. 44-59.

Kellogg, C.E. (1936), *Development and Significance of the Great Soil Groups of United States*, U.S. Department of Agriculture, Miscellaneous Publications, New York.

Klages, K.H.W. (1942), *Ecological Crop Geography*, The MacMillan Co., New York.

Knight, S. (1984), *Kashmir and Tibet*, Mittal Pub., New Delhi.

Kochhar, P.L. (1984), *Plant Ecology*, Ratan Prakashan Mandir, Delhi.

Kormondy, E.J. (1986), *Concept of Ecology*, Prentice Hall of India Pvt. Ltd. New Delhi.

Krebs, C.J. (1985), *Ecology : The Experimental Analysis of Distribution and Abundance*, Harper and Row Publishers, New York.

Kumar, A. (1988), *Foodgrain Production in India : Impact of Techno-Environmental factor*, Criterion Publications, New Delhi.

Kumar, H.D. (1977), *Modern Concept of Ecology*, Vikas Publishing House Pvt. Ltd., New Delhi.

Lawrence, W.R. (1956), *Valley of Kashmir*, Kesar Pub., Srinagar, Kashmir.

Lemay, J. and Harrison, E. (1974), *Environmental Landuse Problems*, Marcel Dekker, Inc., New York.

Liebig, J. (1840), *Chemistry in its Application to Agriculture and Physiology*, Taylor and Walton, London.

Love, G.A. and Love R.M. (1970), *Ecological Crises : Reading for Survival*, Harcourt Brace, New York.

M.L. Dewan, M.C. Nautiyal and V.K. Sah (1997), *Nut Fruits for the Himalayas*, Concept Publishing Co. Pvt. Ltd., New Delhi.

MacDougal, D.T. (1901), *Practical Text Book of Plant Physiology*, Longman, New York.

Maithani, B.P. (1992), *Towards Sustainable Hill Area Development*, Himalaya: Man, Nature and Culture, vol. 16, No. 2, 4, 7.

Mann, K.H. (1982), *Ecology of Coastal waters*, Blackwell Scientific Publication, England.

Mathur, R.S. (1988), National Resources : A Plea for their Conservation, H.S. Mathur, eds. *Environmental Resources: The Crisis of Development*, R.B.S.A. Publishers, Jaipur.

McHale, J. (1971), *Ecological Context*, Studio Vista, Ltd., London.

Memoria, C.B. (1953), *Agricultural Problems of India*, Kitab Mahal, Allahabad.

Miller, E.C. (1938), *Plant Physiology with reference to Green Plants*, McGraw Hill, New York.

Mishra, P.L. (1994), *Agro-Industrial Development in India*, Mohit Publications, New Delhi.

Misra, R. and G.S. Puri (1954), *Indian Manual of Plant Ecology*, The English Book Depot, Dehradun.

Misra, R.P. (1989), *Research Methodology*, Concept Publishing Company Pvt. Ltd., New Delhi.

Mitchell, R. (1939), *Analysis of Indian Agro-Ecosystems*, Interprint, New Delhi.

Mittre, V. (1964), Floristic and Ecological Consideration of the Pliestocene Plants Impression from Kashmir, *Palaeobotanists* 13(3): 308-327.

Mobius, K. (1877), *Die Auster and die Austerwirtschaft wiegundt*, Hempel and Parey, Berlin.

Mohammad, A. (1989), *Food Production and Food Problems in India*, Concept Publishing Company Pvt. Ltd., New Delhi.

Mohammad, Ali (1978), *Studies in Agricultural Geography*, Rajesh Pub., New Delhi.

Mohammad, Noor (1981), *Perspectives in Agricultural Geography*, Concept Publishing Company Pvt. Ltd., New Delhi.

Molish, H. (1896), Das Esfrieren Von Pflanzer bei Temperaturen Uber den Eispunkt, *Sitzber Akad. Wiss. Matt. Naturev. Kl.* 105 (1) : 85-95.

Moore, P.G. and R. Seed (1986), *The Ecology of Rocky Coasts*, Columbia University Press, New York.

Morgan, T.H. (1938), *Evolution and Adaptation*, MacMillan, New York.

Moutia, L.A. and R. Mamet (1946), A Review of 25 Years of Economic Entomology in the Islands of Mauritius, *Bull. Ent. Res.* 36:439-472.

Muneer, A. (1992), *Agricultural Productivity and Regional Development*, Manak Publications Pvt. Ltd., New Delhi.

Nag, P. and S. Sengupta, (1992), *Geography of India*, Concept Publishing Company Pvt. Ltd., New Delhi.

Nag, P., Kumar, V.K. and J. Singh (1997), *Geography and Environment*, Concept Publishing Co. Pvt. Ltd., New Delhi.

Nagarajan, S. (1997), *Wheat : Promise of Value-added Products*, The Hindu Survey of Indian Agriculture, Kasturi and Sons Ltd., Chennai.

Nagarajan, S. and G. Singh (1998), Sustaining wheat revolution : Retrospect and Prospect, *Indian Journal of Agricultural Sciences*, vol. 68, No. 8, pp. 396-404.

Nair, B.K., Goswami, U.N. and K. Kunhikrishan (1972), *Cropping Pattern in India*, Indian Council of Agricultural Research, New Delhi.

Nicholson, M. (1970), *The Environmental Revolution*, Hodder and Stoughton, London.

Odum, E.P. (1959), *Fundamentals of Ecology*, 2nd ed. Saunders, Philadelphia, Pennsylvania.

Odum, E.P. (1963), *Ecology*, Holt, New York.

Odum, E.P. (1971) *Fundamental of Ecology*, W.B. Saunders, Philadelphia.

Pal, B.P. (1982), *Environmental Conservation and Development*, Natraj Publication, Dehradun.

Paliwal, G.S. (1984), *The Vegetational Wealth of the Himalayas*, Puja Publishers, New Delhi.

Pandit, S. (1983), *Critical Study of Agriculture Productivity in U.P.– 1951-75*, Concept Publishing Company Pvt. Ltd., New Delhi.

Pangtey, Y.P.S. and S.C. Joshi (1987), *Western Himalaya*, Gyanopadaya Prakashan, Nainital.

Park, S. Nobel (1983), *Biophysical Plant Physiology and Ecology*, W.H. Freeman and Company, New York.

Patrick, J. Webber, (1979), *High Altitude Geoecology*, Westview Press, U.S.A.

Pearsall, W.H. (1950), *Mountains and Moorlands*, New Naturalist, 11, Collins, London.

Peesi, M. and Probald, F. (1974), *Man and Environment*, Academiai Kiado, Budapest.

Peter Farb (1968), *Ecology : Life Nature Library*, Time-Life International, Neatherland.

Peter Laut (1968), *Agricultural Geography*, Thomus Nelson Limited, Melbourn, Australia.

Pokhriyal, H.C. (1994), *Human Environment and Socio-economic Development in the Himayalas*, B.R. Publishing Corporation, New Delhi.

Puri, G.S. (1948), *A Preliminary Note on the Pliestocene Flora of the Karewa formation of Kashmir*, Journal of Mining Metal Society, India, 20 : 61-66.

Purohit, V.B. (1988), *Recent Advances in Ecology and Environment*, Proceedings of Conference on Ecology and Environment, Today and Tomorrow's Printers and Publishers, New Delhi.

Purseglove, J.W. (1968), *Tropical Crops : Dicotyledons*, Longman Group Limited, London.

Putman, R.J. and S.D. Wratten (1986), *Principles of Ecology*, Croom Helm Ltd., London.

Rajan, K. (1995), *Integrated Part of Farming Systems*, The Hindu, Survey of Indian Agrculture, pp. 48-57.

Ramaley, F. (1940), *The Growth of the Science*, University of Colorado, 26 : 3-14.

Randhawa, M.S. (1961), *Agriculture and Animal Husbandry in India*, I.C.A.R., New Delhi.

Raychaudhuri, *et al.* (1963), *Soils of India*, Indian Council of Agricultural Research, New Delhi.

Raza, M. (1987), *India : Diversities and Disparities, Contemporary Affairs*, vol. 1, No. 1, pp. 36-41.

Raza, M. (1992), *Development and Ecology*, Rawat Publications, New Delhi.

Redclift, M. (1984), *Development and the Environmental Crises*, Methuen Pub., London.

Riddle, R. (1981), *Ecodevelopment*, Gower Pub. Co., Hampshire, England.

Robbins, W.W. (1927), *Principles of Plant Growth,* John Wiley and Sons Ltd., New York.

Sapru, R.K. (1987), *Environment Management in India,* Ashish Publishing House, New Delhi.

Scamoni, A. (1966), *Biogeozonose-Phytozonose,* Biosoziologie, Bericht uber das internationale symposium in Stolzenau/ Weser, Junk Publication, The Haque, pp. 14-22.

Schimid-Nielson, K. (1964), *Desert Animals : Physiological Problems of Heat and Water,* Oxford University Press, England.

Sejwar, C. (1996), *The Geography of Nutrition in India,* Indian Board of International Literature, New Delhi.

Semple, E.C. (1911), *Influences of Geographical Environment,* Russell and Russell, London.

Shafi, M. (1994), Sustainable Agriculture and Environmental Security in India, in *Geography of Environment* edited by Shafi, M. and M. Raza, Rawat Publications, New Delhi, pp. 63-70.

Sharma, B.L. and P. Bhardwaj, (1998), Sustainable Agricultural Development, H.S. Sharma, S. Chatopadhyay, (eds.) *Sustainable Development,* Concept Publishing Company Pvt. Ltd., New Delhi.

Sharma, R.C. (1981), *Environmental Education,* Metropolitan Book, Co., New Delhi.

Shelford, V.E. (1907), Preliminary notes on the distribution of the tiger beetles (Cicindela), and its relation to plan succession, *Biol. Bull;* 14:9-14.

Shelford, V.E. (1913), *Animal Communities in Temperate America,* University of Chicago Press, Chicago.

Shinda, S.D. (1980), *Agriculture in an Underdeveloped Region : A Geographical Survey,* Himalayan Publishing House, Bombay.

Shirley, H.L. (1929), *The Influence of Light Intensity and Light Quality upon the Growth of the Plants,* American Journal of Botany, New York.

Simmons, I.G. (1980), *Biogeography : Natural and Cultural,* Arnold Heinemann Publishers Pvt. Ltd. New Delhi.

Singh, R.B. and S. Mishra (1996), *Environmental Law in India*, Concept Publishing Co. Pvt. Ltd., New Delhi.

Singh R.L. (1971), *Regional Geography of India*, National Geography Society, Varanasi.

Singh, A.L. (1997), *Land Resource Management*, B.R. Publishing Corporation, New Delhi.

Singh, C. (1983), *Modern Techniques of Raising Field Crops*, Oxford and IBH Publishing Co. Pvt. Ltd. Delhi.

Singh, H., Dewan, M.L. and V.R. Suri (1997), *Horticulture Guide for Uttarakhand*, Concept Publishing Co. Pvt. Ltd., New Delhi.

Singh, H.H. *et al.* (eds.), *Geography and Environment, Issues and Challenges*, Concept Publishing Co. Pvt. Ltd., New Delhi.

Singh, J. (1974), *An Agricultural Atlas of India : A Geographical Analysis*, Vishal Publications, Haryana.

Singh, J.S. (1985), *Environmental Regeneration in Himalaya*, The Central Himalayan Association and Gyanopadaya, Nainital.

Singh, O., Nag, P., Kumar, V.K. and J. Singh (1993), *Frontier in Environmental Geography*, Concept Publishing Co. Pvt. Ltd., New Delhi.

Singh, O.P. (1983), *The Himalayas : Nature, Man and Culture*, Rajesh Publications, New Delhi.

Singh, S.R. and S.P. Agnihotri, (1998), *Hill Environment of India*, Chugh Publication, Allahabad.

Singh, Shiv Raj and Agnihotri Shiv Prakash (1998), *Hill Environment of India*, Chugh Publications, Allahabad, India.

Singh, Tejvir (1980), *Studies in Himalayan Ecology and Development Strategies*, The English Book Store, New Delhi.

Singh, V. (1998), Agronomy for Maximization of Forage Production in Hills, *Indian Farming*, vol. 48, No. 9, pp. 4-5.

Singh, V.R. (1970), *Landuse Pattern in Mirzapur and Environment*, Banaras Hindu University, Banaras.

Sir. E. Hillary (1984), *Ecology 2000*, Multimedia Publication, London.

Slatyer, R.D. (1973), *Plant Response to Climate Factors*, UNESCO Publication, Paris.

Smith, J.M. (1974), *Models in Ecology,* Cambridge University Press, London.

Southwick, C.H. (1976), *Ecology and Quality of Our Environment,* D. Van Nostrand Company, New York.

Southwood, T.R.E., (1978), *Ecological Methods,* The English Language Book Society, London.

Speeding, C.R.W. (1971), *Grassland Ecology,* Oxford Press, London.

Stamp, L.D. (1950), *The Land of Britain : Its Use and Misuse,* Longmans Green and Co. Ltd., London.

Stamp, L.D. (1960), *India, Pakistan, Ceylon and Burma,* Methuen and Co., London.

Sukachev, V.N. and N.V. Dylis (1964), *Fundamentals of Forests Biogeocoenology,* Nauka, Moscow.

Swaminathan, M.S. (1996), *Sustainable Agriculture Towards Food Security,* Konark Publishers Pvt. Ltd., New Delhi.

Swaroop, R. (1991), *Agricultural Economy of Himalayan Region,* Gyanodaya Prakashan, Nainital.

Syman, L. (1967), *Agricultural Geography,* G. Bell and Sons Ltd., London.

Tansley, A.G. (1926), *Practical Plant Ecology,* Allen and Unwin, London.

Tansley, A.G. (1935), The Use and Abuse of Vegetational Concepts and Terms, *Ecology,* 16, 284-307.

Tejwani, K.G. (1994), *Agroforestry in India,* Oxford and IBH Publishing Co., New Delhi.

Thakur, C. (1979), *Scientific Crop Production,* vol. 1, Foodcrops, Metropolitan Book Company Pvt. Ltd., New Delhi.

Thakur, K.S. *et al.* (1998), Scope for Increasing Rapeseed Mustard Yield in Himachal Pradesh — an assessment through frontline demonstrations, *Indian Farming,* vol. 48, No. 9, pp. 17-18.

Thomas, M.F. and Whittington, G.W. (1969), *Environment and Landuse in Africa,* Methuen and Co., London.

Tirtha, Ranjit and G. Krishan (1996), *Geography of India,* Rawat Publication, New Delhi.

Tivy, J. (1990), *Agricultural Ecology,* John and Sons, Inc., New York.

Trivedi (1988), *Ecology and Pollution of Indian Rivers,* Ashish Publishing House, New Delhi.

UNESCO, (1975), *Ecological Approaches to Landuse,* Unesco Publication, Paris.

UNESCO/UNEP, (1985), *Living in the Environment : A source book for environmental education.*

Waddell, L.A. (1899), *Among the Himalayas,* Mittal Publications, New Delhi.

Wang, J.Y. (1984), *Agriculture and Its Environment,* Kendall/Hunt Publishing Company, Iowa.

Warikoo, K. (1976), *Jammu, Kashmir and Ladakh,* Sterling Pub. Pvt. Ltd., New Delhi.

Warming, J.E.W. (1909), *Oecology of Plants,* Clarendon Press, Oxford, pp. i-xi, 1-422.

Weaver, J.E. and Clements, F.E. (1929), *Plant Ecology,* Tata McGraw-Hill Publishing Co. Ltd., New Delhi.

Webster C.C. and P.N. Wilson (1980), *Agriculture in Tropics,* Longman Group Limited, London.

Wilkinson, R.G. (1974), Reproductive Constraints in Ecologically Stabilized Societies, H.B. Parey, (eds.) *Population and Its Problems : A Plain Man's Guide,* pp. 294-299, Clerandon Press, Oxford.

Wilsie, C.P. (1974), *Crop Adaptation and Distribution,* Eurasia Publishing House Pvt. Ltd., New Delhi.

Yadaya, N.D., I.C. Gupta and S. Pal (1994), *Cropping System Research in India (1980-91),* Indian Publishers and Distributors, New Delhi.

Yawalkar, K.S. and J.P. Agarwal (1962), *Manures and Fertilizers,* Agri-Horticultural Publishing House, Nagpur.

Yugandhar, B.N. (1993), *A Perspective Plan for a Hill District with Special Reference to District Planning in India,* Heritage Publishers, New Delhi.

Journals

Abrol, I.P. (1994), Nature and Extent of Soil Degradation Problems in India, *Indian Farming,* vol. 44, No. 9, pp. 7-10.

Acharya, R.M. (1984), Research and Development in Sheep and Goat, *Indian Farming*, vol. XXXIV, No. 1, pp. 37-39.

Ahmad, A. (1968), Association Between Rainfall and Crop Landuse in the District of Bijnor, *The Geographical Observer*, vol. IV, No. 2.

Amani, K.Z. (1998), Ecology and Development, *The Geographer*, vol. XXXV, No. 1.

Asthana, A.N. (1998), Pulse Crop Research in India, *The Indian Journal of Agricultural Sciences*, Vol. 68, No. 8, pp. 447-452.

Basu, J.K.N. (1956), Landuse Planning for the Nilgirs, *Journal of Soil and Water Conservation in India*, vol. 5, No. 1.

Bhag Mal (1998), Forage Research in India – The Post-Independence Scenario, *The Indian Journal of Agricultural Sciences*, vol. 68, No. 8, pp. 439-447.

Bhan, V.M. and S. Kumar (1998), Weed Science Research in India, *The Indian Journal of Agricultural Sciences*, vol. 68, No. 8, pp. 567-582.

Bhatia, P.C. and Y.S. Ramakrishan (1998), Crop Weather Relationship – Achievements and Challenges, *Indian Farming*, vol. 48, No. 9, pp. 6-8.

Bhatia, P.C. and Y.S. Ramakrishna (1998), Crop-Weather Relationship – Achievements and Challenges, *Indian Farming*, vol. 48, No. 9, pp. 6-8.

Bhatia, R. *et. al.* (1995), Relative Abundance of Insect Visitors on Flowers of Major Subtropical Fruits in Himachal Pradesh and their Effect on fruit set, *The Indian Journal of Agricultural Sciences*, vol. 65, No. 12, pp. 907-912.

Bllas, D.J. (1985), Changing Ecology and Landuse Among the Teton Dakota Indians, *Illinois Geographical Society*, vol. 27, No. 3.

Brij Lal and S.P. Sharma, (1995), Response of Potato to Potassium Application in Mid-Hills Soils of Himachal Pradesh. *The Indian Journal of Agricultural Sciences*, vol. 65, No. 6, pp. 433-434.

Chandel R.S. and S.S. Thakur (1999), Insect Pests of Hops and their control in Lahaul Valley of Himachal Pradesh, *Indian Farming*, vol. 48, No. 3, pp. 27-28.

Chandra, N., Bisht, K.K.S. and J.V. Prasad (1999), Socio-economic and Ecological Aspects of Farming in Hills of Uttar Pradesh — A case study, *Indian Farming*, vol. 48, No. 12, p. 16.

Chawla, J.K. and S. Prakash, (1994), In Kullu – Farmer Gain Much from Krishi Vigyan Kendra, *Indian Farming*, vol. 43, No. 11, pp. 26-27.

Cole, D.N. (1989), Recreational Ecology : What we know, What Geographers can Contribute, *The Professional Geographer*, vol. 41, No. 2.

Dakshinamurthi, C. (1972), Soil Management for efficient Water Use, *Indian Farming*, vol. XXII, No. 2, pp. 508.

Dubey, Y.P. (1995), Effect of Water Regime on Growth and Water Economy in Rice Under low Hills of Himachal Pradesh, *The Indian Journal of Agricultural Sciences*, vol. 65, No. 7, pp. 676-678.

Dubey, Y.P. and B.P. Kaistha (1998), Production Potential of Lahaul Valley – Constraints and Remedies, *Indian Farming*, vol. 48, No. 9, p. 21.

Dubey, Y.P. and B.P. Kaistha, (1998), Production Potential of Lahaul Valley, *Indian Farming*, vol. 48, No. 9, p. 21.

Dubey, Y.P., B.P., Kaistha and S.K. Bhardwaj, (1998), Response of Potato to Irrigation and Nitrogen in Lahaul Valley of Himachal Pradesh. *The Indian Journal of Agricultural Sciences*, vol. 68, No. 3, pp. 174-175.

Dwivedi, S.K. and N. Singh (1999), Roots of Green Revolution in Ladakh, *Indian Farming*, vol. 49, No. 1, pp. 20-21

Gangwar, L.S. and C. Sen (1997), The Comparative Analysis of Technology Adoption and its Impact on Cropping Pattern and Productivity – Nainital Vs. U.P., *Agricultural Situation in India*, vol. LIV, No. 8.

Goswami, S.N. *et al.* (1997), Regional Imbalances in Agricultural Production and Productivity in North-East India, *Agricultural Situation in India*, vol. LIV, No. 9.

Greene, D.F. (1989), Disturbance and Vegetation Response : A Biogeographical Perspective II, *The Canadian Geographer*, vol. 33, No. 1.

Grossman, L. (1997), Man-Environment Relationship in Anthropology and Geography, *Annals of Association of American Geographers*, vol. 67, No. 1. ,

Hasan, B. and N. Ahmad (1999), Vegetable for Augmenting Green Fodder Resources during Winter of Temperate Kashmir, *Indian Farming*, vol. 48, No. 12, pp. 17-23.

Hasan, B. and R.H. Kanth (1999), Rice-wheat Cropping System in Jammu and Kashmir, *Indian Farming*, vol. 49, No. 4, pp. 7-9.

Henning, I. (1998), Climate and Vegetation, *Applied Geography and Development*, Vol. 33, No. 1.

Iqbal, Z. and S.A. Khan (1995), Regional Imbalances in Food Crop Production in Jammu and Kashmir, *Geographical Review of India*, vol. 57, No. 3, pp. 274-285.

Jaggi, R.C. and S.P. Dixit (1999), Onion Response to Sulphur in Representative Vegetable Growing Soils of Kangra Valley of Himachal Pradesh. *The Indian Journal of Agricultural Sciences*, vol. 69, No. 4, pp. 289-291.

Jalees, K. (1985), New Grasses from Kashmir, *Journal of Washington Academy of Sciences*, vol. 23, No. 3/4.

Kalloo, G. (1998), Vegetable Research in India, *The Indian Journal of Agricultural Sciences*, vol. 68, No. 8, pp. 515-526.

Kant, S. (1998), Development and the Environment in an Indian State, *Geographical Review of India*, vol. 57, No. 2, pp. 97-112.

Katherine, J. and Hasen, B. (1988), Climatic Variability and Tree response within the Forest-Alpine, Tundra Ecotone, *Annals of the Association of American Geographer*, vol. 78, No. 3.

Katiyar, R.P. (1982), Extending Arhar Cultivation in Himachal Hills, *Indian Farming*, vol. XXXII, No. 6, pp. 20-21.

Katiyar, R.P. (1984), Kulthi – A Promising Pulse crop for Himachal Hills, *Indian Farming*, vol. XXXIV, No. 9, pp. 31-36.

Krishnaiah, K. (1998), Rice Research in India. *The Indian Journal of Agricultural Sciences*, vol. 68, No. 8, pp. 385-395.

Kumar, R., Kumar, S. and B.P. Singh (1998), Eco-friendly Technologies for Sustainable Food Security Beyond 2000 A.D., *Indian Farming*, vol. 48, No. 3, pp. 4-6.

Micheal Raj, S., Singh, L.B. and S.K.Z. Ahmad (1998), A Farm Woman in Cultivation of Vegetable Crop – a success story, *Indian Farming*, vol. 48, No. 9, p. 31.

Mishra, A.S. *et al.* (1994), Social Change in Doon Valley by Lab-to-Lab Programme, *Indian Farming*, vol. 43, No. 10, pp. 13-18.

Mishra, S.K. (1998), Insect, Pest and Disease Management in Winter Season Vegetables, *Indian Farming* vol. 48, No. 9, pp. 32-36.

Motsara, M.R. and J. Singh (1981), Soil Testing for Judicious Use of Fertilizers, *Indian Farming*, vol. XXXI, No. 9, pp. 3-12.

Munshi, A.M., Sindhu, J.S. and G.H. Babu (1989), Improved Cultivation Practices for Saffron, *Indian Farming*, vol. XXXIX, No. 3, pp. 27-30.

Nagarajan, S. and G. Singh (1998), Sustaining Wheat Revolution – Retrospect and Prospect, *The Indian Journal of Agricultural Sciences*, vol. 68, No. 8, pp. 396-404.

Naidu, L.K.G. *et al.* (1998), Mapping of Agro-ecological Zones of Andhra Pradesh though Soil Resource Data. *The Indian Journal of Agricultural Sciences*, vol. 68, No. 8.

Nair, T.A.K. (1994), Changing Scenario of Wasteland Development, *Indian Farming*, vol. 44, No. 9, pp. 15-18.

Pandey, A.K. *et al.* (1998), Technologies for Baby Corn Production in Hill of North-western Himalayas, *Indian Farming*, vol. 48, No. 3.

Park Moon Myung, (1986), A Review of the History of the Japanese Agricultural Economy and its Relationship to its Present Industrial Development, *Asian Profile*, vol. 14, No. 6, pp. 509-520.

Pathania, H.S. (1957), Our Forests, *Kashmir Times Journal*, vol. 1, No. 7.

Prasad, R.R. (1971), Pollution – A Challenging Problem of Ecology, *Indian Farming*, vol. XXI, No. 8, pp. 5-8.

Raina, A.N. (1962), Our Agriculture in Facts and Figures, *Kashmir Times Journal*, vol. 6, No. 1-2.

Rana, B.S. *et al.* (1998), Sorghum Research in India, *The Indian Journal of Agricultural Sciences*, vol. 68, No. 8, pp. 405-423.

Rao, M.S.R.M. and V.N. Sharda (1994), Water Conservation Measures for Sustainable Agriculture, *Indian Farming*, vol. 44, No. 7, pp. 18-23.

Reddy, P.S. and D. Pati (1998), Indian Oilseeds : Present Status and Future Needs, *The Indian Journal of Agricultural Sciences*, vol. 68, No. 8, pp. 453-467.

Sahani, M.S. and M. Rathinasabapathy (1998), The Fast Dwindling Species of Ladakh – A Brief Note, *Indian Farming*, vol. 48, No. 4, pp. 13-14.

Saxena, A.K., Chaturvedi, O.P. and M.C. Nautiyal (1989), Forest Ecosystem of Garhwal Himalaya, *Indian Farming*, vol. XXXIX, No. 3, pp. 3-6.

Sekhon, G.S. and N.S. Randhawa (1971), Soil Fertility and Fertilizer Problems in Punjab, Haryana and Himachal Pradesh, *Indian Farming*, vol. XXI, No. 1, pp. 15-17.

Sharma, A.K. Moorti, T.V. and R.K. Sharma (1994), Kangra Irrigation network – A Great Potential for Fresh Water Aquaculture, *Indian Farming*, vol. 43, No. 11, pp. 22-25.

Sharda, N.K. and T. Raman (1997), Implications of Changing Agrarian Structure in Himachal Pradesh, *Agriculture Situation in India*, vol. LIII, No. 12.

Sharma, D.K. and K.L. Sharma (1998), Effect of Nitrogen and Potash Application on Yield and Quality of Chinahybrid tea Grown in Kangra Valley of Himachal Pradesh. *The Indian Journal of Agricultural Sciences*, vol. 68, No. 6, pp. 307-309.

Sharma, P.D. and S.P. Sharma (1994), Pastoralists of Himachal Pradesh : Need to be Rehabilitated, *Indian Farming*, vol. 44 No. 8, pp. 14-15.

Sharma, T.R. and H.K. Chaudhary (1998), Stability in Grain Yield of Buck Wheat in Dry Temperate Region of Himachal Pradesh, *The Indian Journal of Agricultural Sciences*, vol. 68, No. 6, pp. 179-180.

Sharma, T.R. *et al.* (1998), Seed Yield Stability of Indigenous and Exotic genotype of Amaranath in the North-western Himalaya, *The Indian Jounral of Agricultural Science*, Vol. 65, No. 6, pp. 328-329.

Simmong, I.G. (1966), Ecology and Landuse, *Transaction of the Institution of British Geographers*, vol. 1, No. 38.

Singh, D.V. (1998), Economics of Export Oriented Horticultural Crops in Himachal Pradesh, *Agricultural Situation in India*, vol. LIV, No. 12.

Singh, N.P. (1972), Irrigation and Fertilizer Use in Crop Production, *Indian Farming*, vol. XXII, No. 2, pp. 65-69.

Singh, P. and A.K. Vashisht (1997), Dynamics of Land Use Pattern in India, *Agricultural Situation in India*, vol. LIII, No. 11.

Singh, R.D. Bhatnagar, V.K. and J.K. Bisht (1998), Kudzu – An Alternative Source of Nutrient for Sustainability of Hill Agriculture, *Indian Farming*, vol. 48, No. 4, pp. 18-23.

Singh, S.K. (1998), Climate Change and perspective for Agriculture, *Indian Farming*, vol. 48, No. 5, pp. 49-51.

Singh, V. (1998), Agronomy for Maximization of Forage Production in Hills, *Indian Farming*, vol. 48, No. 9, pp. 4-5.

Singh, V. (1998), Agronomy, For Maximization of Foreage Production in Hills, *Indian Farming*, vol. 48, No. 9, pp. 4-5.

Siva, V. (1985), Ecological Development in India, *Development*, Rome, vol. 1, No. 3.

Solanki, K.P. (1998), Agro-forestry Research in India, *The Indian Journal of Agricultural Sciences*, vol. 68, No. 8, pp. 559-566.

Sood, K.C., Awasthi, K.R. and B.C. Sood (1999), Constraints in Rabi Pulses Production in Himachal Pradesh and Quantity Seed Production of Gram, *Indian Farming*, vol. 48, No. 8, pp. 30-31.

Sood, M.C. Sharma, R.C. and M.A. Khan (1998), Improvement of Potato Productivity through Potato-garlic system in Shimla Hills, *Indian Farn:ing*, vol. 49, No. 9, pp. 29-30.

Swaminathan, M.S. (1972), Scientific Management and Use of Water, *Indian Farming*, vol. XXII, No. 2, pp. 3-4.

Taneja, V.K. (1998), Women in Livestock Production and Economics, *Indian Farming*, vol. 48, No. 7, pp. 55-63.

Thakur, D.R. *et al.* (1997), Resource use efficiency on Sheep and Goat Farming in Himachal Pradesh – An Assessment

Through Frontline Demonstrations, *Indian Farming,* vol. 48, No. 9, pp. 19-20.

Thakur, K.S. *et al.* (1998), Scope for Increasing Rapeseed – Mustard Yield in Himachal Pradesh – An Assessment Through Frontline Demonstration, *Indian Farming,* vol. 48, No. 9, pp. 19-20.

Unpublished Ph.D. Thesis

Bhatt, J.P. (1998), *A Spatial Analysis of Mountain Environment and Agricultural Land Utilization in Bhilangana Basin – Garhwal Himalaya,* Ph.D. Thesis, Department of Geography, H.N.B. Garhwal University, Srinagar, Garhwal.

Bora, N.S. (1993), *Man-Nature Symbiosis in Development Planning Perspective : A case Study of Kota Basin,* Ph.D. Thesis, Department of Geography, D.S.B. Campus, Kumaon University, Nainital.

Joshi, K.C. (1988), *Determinants of Agricultural Productivity in Himalaya,* Ph.D. Thesis, Department of Geography, Kumaon University Campus, Nainital.

Kumar, R. (1989), *Land use of Kashmir Valley – An Agro-Ecological Analysis,* Ph.D. Thesis, Centre for the study of Regional Development School of Social Sciences, Jawaharlal Nehru University, New Delhi.

Kuniyal, J.C. (1989), *Regional Disparities in Agricultural Development in District Nainital,* Ph.D. Thesis, Department of Geography, Kumaon University, Nainital.

Mehta, C. (1993), *The Levels of Agricultural Productivity in Kumaon Himalaya – An Inter-regional Analysis,* Ph.D. Thesis, Wasteland Management Laboratory, Department of Geography, Kumaon University Campus, Nainital.

Mohindra, K. (1993), *The Land Reforms System Analysis and Environmental Resource Management in the UHL Basin (Kangra Himalaya, H.P.),* Ph.D. Thesis, Department of Geography, D.B.S. Post Graduate College, Dehradun.

Punera, A. (1993), *A Geographical Study of Environmental Changes in the Sar Valley of District Pithoragarh,* Ph.D. Thesis,

Department of Geography, Kumaon University Campus, Nainital.

Shah, S.A. (1998), *Growth and Expansion of Saffron Cultivation in Kashmir Division and Its Impact on the Economy and Society*, Department of Geography, Faculty of Natural Sciences, Jamia Millia Islamia, New Delhi.

Sharma, K.K. (1999), *Agricultural Transformation and Rural Development – A Study of Horticulture in Himachal Pradesh*, Department of Geography, Delhi, School of Economics, University of Delhi, New Delhi.

Sharma, R.K. (1994), *A Geographical Study of Environmental Hazards and Degradation in Kangra Valley, (H.P.)*, Ph.D. Thesis, Department of Geography, H.N.B. Garhwal University Campus, Pauri Garhwal.

Singh, H. (1999), *Spatial Development of Dairy Farming in Kashmir Valley and Its Socio-Economic Impact*, Ph.D. Thesis, Department of Geography and Regional Development, The University of Kashmir, Srinagar.

Singh, H. (1996), Agro-Ecological Development in Upper Kosi Catchment, Kumaon Himalaya, Department of Geography, Kumaon University, Nainital.

Upreti, T.C. (1994), *Environmental Hazards in the Kalpanigad Basin*, Ph.D. Thesis, Wasteland Management Laboratory, Department of Geography, Kumaon University Campus, Nainital.

Index